T0140154

Computer Communications and Networks

Series editor

A.J. Sammes
Centre for Forensic Computing
Cranfield University, Shrivenham Campus
Swindon, UK

The **Computer Communications and Networks** series is a range of textbooks, monographs and handbooks. It sets out to provide students, researchers, and nonspecialists alike with a sure grounding in current knowledge, together with comprehensible access to the latest developments in computer communications and networking.

Emphasis is placed on clear and explanatory styles that support a tutorial approach, so that even the most complex of topics is presented in a lucid and intelligible manner.

More information about this series at http://www.springer.com/series/4198

Thomas Herault · Yves Robert
Editors

Fault-Tolerance Techniques for High-Performance Computing

 Springer

Editors
Thomas Herault
University of Tennessee
Knoxville, TN
USA

Yves Robert
Ecole Normale Supérieure de Lyon
Lyon
France

and

University of Tennessee
Knoxville, TN
USA

ISSN 1617-7975   ISSN 2197-8433   (electronic)
Computer Communications and Networks
ISBN 978-3-319-35560-3   ISBN 978-3-319-20943-2   (eBook)
DOI 10.1007/978-3-319-20943-2

Preface

Objective

The main objective of this monograph is to provide an overview of *Fault-Tolerance Techniques for High-Performance Computing* (HPC). Resilience has already become a prominent issue on current large-scale platforms. The advent of exascale computers with millions of cores and billion-parallelism is only going to worsen the scenario. The capacity to deal with errors and faults will be a critical factor for HPC applications to be deployed efficiently.

While there are many research papers available on this hot and important topic, there was no comprehensive and easy-to-access reference available in the literature. The purpose of this monograph is to fill the gap, and to provide a detailed presentation and analysis of the various fault tolerance methods for HPC applications.

The first part of the book is made of a single survey chapter that introduces checkpoint protocols and scheduling algorithms, prediction, replication, silent error detection, and correction, together with some application-specific techniques such as Algorithm-Based Fault Tolerance (ABFT). A key feature of this survey chapter is the importance given to analytical performance models. As future extreme-scale platforms are not yet available (by definition!), a refined (and publicly available) performance model is the key to assess any resilience technique without bias nor a-priori. Various scenarios can be instantiated through selecting one's preferred model parameters, and further explored through simulations. The emphasis given to performance models explains the unusual amount of mathematical equations in the chapter, but let the reader be comforted: (i) every method is first described informally; (ii) the mathematical derivations are detailed and complemented with examples; and (iii) it is always possible to skip some proof and be back to it during a second reading.

The second part of the book is composed of four chapters, each dedicated to further investigating one topic. Chapter 2 surveys the various sources for error and faults in real large-scale systems, details their characteristics, and focuses on detection and prediction. Chapter 3 presents the spectrum of techniques that can be applied to design a fault-tolerant MPI, i.e., to enable MPI application recovery using either fully automatic or completely user-driven techniques. Chapter 4 investigates replication (coupled with checkpointing) and compares two approaches. In the first approach, entire application instances are replicated, while in the second one, each process in a single application instance is (transparently) replicated. Finally, Chap. 5 addresses the challenge of energy consumption related to fault tolerance in extreme-scale systems, and proposes a methodology to estimate the energy consumption of fault-tolerant protocols used in HPC.

The best way to read the book is to start with the overview chapter in Part I, and then to move on to the more specialized chapters of Part II. However, experienced readers may want to read a single specific chapter in Part II. To ease this approach, we have made each chapter independent of the others, at the price of some redundant information throughout the book. Cross-references between related sections of different chapters and index terms have been provided to help navigate across chapters whenever needed.

Thanks

This monograph is the follow-up of a tutorial that we gave at ICS'13. We were approached by Springer Verlag and invited to write this monograph, a task that we eventually succeeded to complete after ... some delay.

We would like to thank all chapter authors for their contribution. All of them are colleagues and Ph.D. students with whom we worked on various topics during the past few years, and all of us share many ideas on resilience for HPC. Hopefully, the reader will sense a common perspective while reading the monograph!

The tutorial that George Bosilca, Aurélien Bouteiller, and the two of us gave at SC'14 came one year after the one given at ISC'13. At that point the monograph was far from ready, and intense discussions when preparing the SC'14 tutorial have greatly influenced the overview chapter. We thank them for this.

Finally, we would like to thank Jack Dongarra and ICL for providing a unique place to collaborate and work on HPC-related research, from linear algebra to resilience and more.

Knoxville   Thomas Herault
Lyon   Yves Robert
April 2015

Contents

Contributors

Guillaume Aupy Ecole Normale Supérieure de Lyon, Lyon, France

Anne Benoit Ecole Normale Supérieure de Lyon, Lyon, France

Aurélien Bouteiller University of Tennessee, Knoxville, TN, USA

Franck Cappello Argonne National Laboratory, Lemont, USA

Henri Casanova University of Hawai'i, Manoa, USA

Mohammed El Mehdi Diouri IGA Casablanca, Casablanca, Morocco

Jack Dongarra University of Tennessee, Knoxville, TN, USA; Oak Ridge National Laboratory, Oak Ridge, USA; University of Manchester, Manchester, UK

Ana Gainaru NCSA, University of Illinois at Urbana-Champaign, Champaign, USA

Olivier Glück Université Claude Bernard de Lyon, Villeurbanne, France

Thomas Herault University of Tennessee, Knoxville, TN, USA

Laurent Lefèvre INRIA & Ecole Normale Supérieure de Lyon, Lyon, France

Yves Robert University of Tennessee, Knoxville, TN, USA; Ecole Normale Supérieure de Lyon, Lyon, France

Frédéric Vivien INRIA & Ecole Normale Supérieure de Lyon, Lyon, France

Dounia Zaidouni INRIA & Ecole Normale Supérieure de Lyon, Lyon, France

Part I
General Overview

Chapter 1
Fault Tolerance Techniques for High-Performance Computing

Jack Dongarra, Thomas Herault and Yves Robert

Abstract This chapter provides an introduction to resilience methods. The emphasis is on checkpointing, the de-facto standard technique for resilience in High Performance Computing. We present the main two protocols, namely coordinated checkpointing and hierarchical checkpointing. Then we introduce performance models and use them to assess the performance of theses protocols. We cover the Young/Daly formula for the optimal period and much more! Next we explain how the efficiency of checkpointing can be improved via fault prediction or replication. Then we move to application-specific methods, such as ABFT. We conclude the chapter by discussing techniques to cope with silent errors (or silent data corruption).

1.1 Introduction

This chapter provides an overview of fault tolerance techniques for High Performance Computing (HPC). We present scheduling algorithms to cope with faults on large-scale parallel platforms. We start with a few general considerations on resilience at scale (Sect. 1.1.1) before introducing standard failure probability distributions (Sect. 1.1.2). The main topic of study is *checkpointing*, the de-facto standard technique for resilience in HPC. We present the main protocols, *coordinated* and *hierarchical*, in Sect. 1.2. We introduce probabilistic performance models

J. Dongarra · T. Herault (✉) · Y. Robert
University of Tennessee, Knoxville, TN, USA
e-mail: herault@icl.utk.edu

J. Dongarra
Oak Ridge National Laboratory, Oak Ridge, USA
e-mail: dongarra@eecs.utk.edu

J. Dongarra
University of Manchester, Manchester, UK

Y. Robert
Ecole Normale Supérieure de Lyon, Lyon, France
e-mail: yves.robert@inria.fr

© Springer International Publishing Switzerland 2015
T. Herault and Y. Robert (eds.), *Fault-Tolerance Techniques*
for High-Performance Computing, Computer Communications and Networks,
DOI 10.1007/978-3-319-20943-2_1

to assess these protocols in Sect. 1.3. In particular, we show how to compute the optimal checkpointing period (the famous Young/Daly formula [25, 69]) and derive several extensions. Then Sect. 1.4 explains how to combine checkpointing with *fault prediction*, and discuss how the optimal period is modified when this combination is used (Sect. 1.4.1). We follow the very same approach for the combination of checkpointing with *replication* (Sect. 1.4.2).

While checkpointing (possibly coupled with fault prediction or replication) is a general-purpose method, there exist many application-specific methods. In Sect. 1.5, we present middleware adaptations to enable application-specific fault tolerance, and illustrate their use on one of the most important one, ABFT, which stands for *Algorithm based Fault Tolerance*, in Sect. 1.5.

The last technical section of this chapter (Sect. 1.6) is devoted to techniques to cope with silent errors (or silent data corruption). Section 1.7 concludes the chapter with final remarks.

1.1.1 Resilience at Scale

For HPC applications, scale is a major opportunity. Massive parallelism with 100,000+ nodes is the most viable path to achieving sustained Petascale performance. Future platforms will enroll even more computing resources to enter the Exascale era. Current plans refer to systems either with 100,000 nodes, each equipped with 10,000 cores (the *fat node* scenario), or with 1,000,000 nodes, each equipped with 1,000 cores (the *slim node* scenario) [27].

Unfortunately, scale is also a major threat, because resilience becomes a big challenge. Even if each node provides an individual MTBF (Mean Time Between Failures) of, say, one century, a machine with 100,000 such nodes will encounter a failure every 9 hours in average, which is larger than the execution time of many HPC applications. Worse, a machine with 1,000,000 nodes (also with a one-century MTBF) will encounter a failure every 53 min in average.[1] Note that a one-century MTBF per node is an optimistic figure, given that each node is composed of several hundreds or thousands of cores.

To further darken the picture, several types of errors need to be considered when computing at scale. In addition to classical fail-stop errors (such as hardware failures), silent errors (a.k.a silent data corruptions) must be taken into account. Contrary to fail-stop failures, silent errors are not detected immediately, but instead after some arbitrary *detection latency*, which complicates methods to cope with them. See Sect. 1.6 for more details.

---

[1] See Sect. 1.3.2.1 for a detailed explanation on how these values (9 h or 53 min) are computed.

1.1.2 Faults and Failures

There are many types of errors, faults, or failures. Some are transient, others are unrecoverable. Some cause a fatal interruption of the application as soon as they strike, others may corrupt the data in a silent way and will manifest only after an arbitrarily long delay. We refer to Chap. 2 for a detailed classification and analysis of error sources.

In this chapter, we mainly deal with *fail-stop failures*, which are unrecoverable failures that interrupt the execution of the application. These include all hardware faults, and some software ones. We use the terms *fault* and *failure* interchangeably. Again, silent errors are addressed at the end of the chapter, in Sect. 1.6.

Regardless of the fault type, the first question is to quantify the rate or frequency at which these faults strike. For that purpose, one uses probability distributions, and more specifically, Exponential probability distributions. The definition of $Exp(\lambda)$, the Exponential distribution law of parameter λ, goes as follows:

- The probability density function is $f(t) = \lambda e^{-\lambda t} dt$ for $t \geq 0$;
- The cumulative distribution function is $F(t) = 1 - e^{-\lambda t}$ for $t \geq 0$;
- The mean is $\mu = \frac{1}{\lambda}$.

Consider a process executing in a fault-prone environment. The time-steps at which fault strike are nondeterministic, meaning that they vary from one execution to another. To model this, we use I.I.D. (Independent and Identically Distributed) random variables $X_1, X_2, X_3, \dots$. Here X_1 is the delay until the first fault, X_2 is the delay between the first and second faults, X_3 is the delay between the second and third faults, and so on. All these random variables obey the same probability distribution $Exp(\lambda)$. We write $X_i \sim Exp(\lambda)$ to express that X_i obeys an Exponential distribution $Exp(\lambda)$.

In particular, each X_i has the same mean $\mathbb{E}(X_i) = \mu$. This amounts to say that, in average, a fault will strike every μ seconds. This is why μ is called the MTBF of the process, where MTBF stands for *Mean Time Between Faults*: one can show (see Sect. 1.3.2.1 for a proof) that the expected number of faults $N_{\text{faults}}(T)$ that will strike during T seconds is such that

$$\lim_{T \to \infty} \frac{N_{\text{faults}}(T)}{T} = \frac{1}{\mu} \tag{1.1}$$

Why are Exponential distribution laws so important? This is because of their *memoryless* property, which writes: if $X \sim Exp(\lambda)$, then $\mathbb{P}(X \geq t + s \mid X \geq s) = \mathbb{P}(X \geq t)$ for all $t, s \geq 0$. This equation means that at any instant, the delay until the next fault does not depend upon the time that has elapsed since the last fault. The memoryless property is equivalent to saying that the fault rate is constant. The fault rate at time t, RATE(t), is defined as the (instantaneous) rate of fault for the survivors to time t, during the next instant of time:

$$\text{RATE}(t) = \lim_{\Delta \to 0} \frac{F(t + \Delta) - F(t)}{\Delta} \times \frac{1}{1 - F(t)} = \frac{f(t)}{1 - F(t)} = \lambda = \frac{1}{\mu}$$

The fault rate is sometimes called a *conditional* fault rate since the denominator $1 - F(t)$ is the probability that no fault has occurred until time t, hence converts the expression into a conditional rate, given survival past time t.

We have discussed Exponential laws above, but other probability laws could be used. For instance, it may not be realistic to assume that the fault rate is constant: indeed, computers, like washing machines, suffer from a phenomenon called *infant mortality*: the probability of fault is higher in the first weeks than later on. In other words, the fault rate is not constant but instead decreasing with time. Well, this is true up to a certain point, where another phenomenon called *aging* takes over: your computer, like your car, becomes more and more subject to faults after a certain amount of time: then the fault rate increases! However, after a few weeks of service and before aging, there are a few years during which it is a good approximation to consider that the fault rate is constant, and therefore to use an Exponential law *Exp*(λ) to model the occurrence of faults. The key parameter is the MTBF $\mu = \frac{1}{\lambda}$.

Weibull distributions are a good example of probability distributions that account for infant mortality, and they are widely used to model failures on computer platforms [39, 42, 43, 54, 67]. The definition of *Weibull*(λ), the Weibull distribution law of shape parameter k and scale parameter λ, goes as follows:

- The probability density function is $f(t) = k\lambda(t\lambda)^{k-1}e^{-(\lambda t)^k} dt$ for $t \geq 0$;
- The cumulative distribution function is $F(t) = 1 - e^{-(\lambda t)^k}$;
- The mean is $\mu = \frac{1}{\lambda}\Gamma(1 + \frac{1}{k})$.

If $k = 1$, we retrieve an Exponential distribution *Exp*(λ) and the failure rate is constant. But if $k < 1$, the failure rate decreases with time, and the smaller k, the more important the decreasing. Values used in the literature are $k = 0.7$ or $k = 0.5$ [39, 54, 67].

1.2 Checkpoint and Rollback Recovery

Designing a fault-tolerant system can be done at different levels of the software stack. We call *general-purpose* the approaches that detect and correct the failures at a given level of that stack, masking them entirely to the higher levels (and ultimately to the end-user, who eventually see a correct result, despite the occurrence of failures). General-purpose approaches can target specific types of failures (e.g., message loss, or message corruption), and let other types of failures hit higher levels of the software stack. In this section, we discuss a set of well-known and recently developed protocols to provide general-purpose fault tolerance for a large set of failure types, at different levels of the software stack, but always below the application level.

These techniques are designed to work in spite of the application behavior. When developing a general-purpose fault-tolerant protocol, two adversaries must be taken into account: the occurrence of failures, that hit the system at unpredictable moments, and the behavior of the application, that is designed without taking into account the risk of failure, or the fault-tolerant protocol. All general-purpose fault tolerance technique rely on the same idea: introduce automatically computed redundant information, and use this redundancy to mask the occurrence of failures to the higher level application.

The general-purpose technique most widely used in HPC relies on checkpointing and rollback recovery: parts of the execution are lost when processes are subject to failures (either because the corresponding data is lost when the failure is a crash, or because it is corrupted due to a silent error), and the fault-tolerant protocol, when catching such errors, uses past checkpoints to restore the application in a consistent state, and recomputes the missing parts of the execution. We first discuss the techniques available to build and store process checkpoints, and then give an overview of the most common protocols using these checkpoints in a parallel application.

1.2.1 Process Checkpointing

The goal of process checkpointing is to save the current state of a process. In current HPC applications, a process consists of many user-level or system-level threads, making it a parallel application by itself. Process checkpointing techniques generally use a coarse-grain locking mechanism to interrupt momentarily the execution of all the threads of the process, giving them a global view of its current state, and reducing the problem of saving the process state to a sequential problem.

Independently of the tool used to create the checkpoint, we distinguish three parameters to characterize a process checkpoint:

- At what level of the software stack it is created;
- How it is generated;
- How it is stored.

Level of the software stack. Many process checkpointing frameworks are available: they can rely on an operating system extension [41], on dynamic libraries,[2] on compilers [50, 53, 62, 63], on a user-level API [5], or on a user-defined routine that will create an application-specific checkpoint [47]. The different approaches provide different levels of transparency and efficiency. At the lowest level, operating system extensions, like BLCR [41], provide a completely transparent checkpoint of the whole process. Such a checkpoint can be restored on the same hardware, with the same software environment (operating system, dynamic libraries, etc.). Since the entire state is saved (from CPU registers to the virtual memory map), the function call stack is also saved and restored automatically. From a programmatic point of view, the checkpoint routine returns with a different error code, to let the caller know if this calls returns from a successful checkpoint or from a successful restart.

---

[2]See https://code.google.com/p/cryopid/.

System-level checkpointing requires to save the entire memory (although an API allows to explicitly exclude pages from being saved into the checkpoint, in which case the pages are reallocated at restore time, but filled with 0), and thus the cost of checkpointing is directly proportional to the memory footprint of the process. The checkpoint routine is entirely preemptive: it can be called at any point during the execution, from any thread of the application (as long as another thread is not inside the checkpoint routine already).

At the highest level, user-defined application-specific routines are functions that a fault-tolerant protocol can call, to create a serialized view of the application, that another user-defined application-specific routine can load to restore a meaningful state of the process. Such an approach does not belong to general-purpose techniques, since it is application dependent. It is worth noting, however, that some resilient communication middleware propose this option to implement an efficient generic rollback-recovery protocol at the parallel application level. Indeed, as we will see later in the chapter, time to checkpoint is a critical parameter of the overall efficiency of a rollback-recovery technique. User-defined process checkpoints are often orders of magnitude smaller than the process memory footprint, because intermediary data, or data that is easily reconstructed from other critical data, do not need to be saved. User-defined checkpoints also benefit from a more diverse use than solely fault tolerance: they allow to do a post-mortem analysis of the application progress; they permit to restart the computation at intermediary steps, and change the behavior of the application. For these reasons, many applications provide such routines, which is the reason why fault-tolerant protocols try to also benefit from them. It is however difficult to implement a preemptive user-defined routine, capable of saving the process state at any time during the execution, which makes the use of such approach sometimes incompatible with some parallel application resilient protocols that demand to take process checkpoints at arbitrary times.

A note should be made about opened files: most existing tools to checkpoint a process do not provide an automatic way to save the content of the files opened for writing at the time of checkpoint. Files that are opened in read mode are usually reopened at the same position during the restoration routine; files that are opened in append mode can be easily truncated where the file pointer was located at the time of checkpoint during the restore; files that are opened in read/write mode, however, or files that are accessed through a memory map in read/write, must be copied at the time of checkpoint, and restored at the time of rollback. Among the frameworks that are cited above, none of them provide an automatic way of restoring the files, which remains the responsibility of the resilient protocol implementation.

How checkpoints are generated. The checkpoint routine, provided by the checkpointing framework, is usually a blocking call that terminates once the serial file representing the process checkpoint is complete. It is often beneficial, however, to be able to save the checkpoint in memory, or to allow the application to continue its progress in parallel with the I/O intensive part of the checkpoint routine. To do so, generic techniques, like process duplication at checkpoint time can be used, if enough memory is available on the node: the checkpoint can be made asynchronous by duplicating the entire process, and letting the parent process continue its execution, while the child process checkpoints and exits. This technique relies on the copy-on-write pages duplication capability of modern operating systems to ensure that if the parent process modifies a page, the child will get its own private copy, keeping the state of the process at the time of entering the checkpoint routine. Depending on the rate at which the parent process modifies its memory, and depending on the amount of available physical memory on the machine, overlapping between the checkpoint creation and the application progress can thus be achieved, or not.

How checkpoints are stored. A process checkpoint can be considered as completed once it is stored in a non-corruptible space. Depending on the type of failures considered, the available hardware, and the risk taken, this non-corruptible space can be located close to the original process, or very remote. For example, when dealing with low probability memory corruption, a reasonable risk consists of simply keeping a copy of the process checkpoint in the same physical memory; at the other extreme, the process checkpoint can be stored in a remote redundant file system, allowing any other node compatible with such a checkpoint to restart the process, even in case of machine shutdown. Current state-of-the-art libraries provide transparent multiple storage points, along a hierarchy of memory: [57], or [5], implement in-memory double-checkpointing strategies at the closest level, disk-less checkpointing, NVRAM checkpointing, and remote file system checkpointing, to feature a complete collection of storage techniques. Checkpoint transfers happen asynchronously in the background, making the checkpoints more reliable as transfers progress.

1.2.2 Coordinated Checkpointing

Distributed checkpointing protocols use process checkpointing and message passing to design rollback-recovery procedures at the parallel application level. Among them the first approach was proposed in 1984 by Chandy and Lamport, to build a possible global state of a distributed system [20]. The goal of this protocol is to build a consistent distributed snapshot of the distributed system. A distributed snapshot is a collection of process checkpoints (one per process), and a collection of in-flight messages (an ordered list of messages for each point to point channel). The protocol assumes ordered loss-less communication channel; for a given application, messages can be sent or received after or before a process took its checkpoint. A message from process p to process q that is sent by the application after the checkpoint of process p but received before process q checkpointed is said to be an *orphan* message. Orphan messages must be avoided by the protocol, because they are going to be regenerated by the application, if it were to restart in that snapshot. Similarly, a message from process p to process q that is sent by the application before the checkpoint of process p but received after the checkpoint of process q is said to be *missing*. That message must belong to the list of messages in channel p to q, or the snapshot is inconsistent. A snapshot that includes no orphan message, and for which all the saved channel

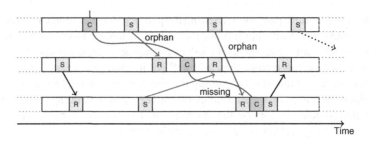

Fig. 1.1 Orphan and missing messages

messages are missing messages is consistent, since the application can be started from that state and pursue its computation correctly.

To build such snapshots, the protocol of Chandy and Lamport works as follows (see Fig. 1.1): any process may decide to trigger a checkpoint wave by taking its local process checkpoint (we say the process entered the checkpoint wave), and by notifying all other processes to participate to this wave (it sends them a notification message). Because channels are ordered, when a process receives a checkpoint wave notification, it can separate what messages belong to the previous checkpoint wave (messages received before the notification in that channel), and what belong to the new one (messages received after the notification). Messages that belong to the current checkpoint wave are appended to the process checkpoint, to complete the state of the distributed application with the content of the in-flight messages, during the checkpoint. Upon reception of a checkpoint wave notification for the first time, a process takes it local checkpoint, entering the checkpoint wave, and notifies all others that it did so. Once a notification per channel is received, the local checkpoint is complete, since no message can be left in flight, and the checkpoint wave is locally complete. Once all processes have completed their checkpoint wave, the checkpoint is consistent, and can be used to restart the application in a state that is consistent with its normal behavior.

Different approaches have been used to implement this protocol. They are discussed in detail in the case of the Message Passing Interface (MPI) in Chap. 3. The main difference is on how the content of the (virtual) communication channels is saved. A simple approach, called Blocking Coordinated Checkpointing, consists in delaying the emission of application messages after entering the checkpointing wave, and moving the process checkpointing at the end of that wave, when the process is ready to leave it (see Fig. 1.3). That way, the state of communication channels is saved within the process checkpoint itself, at the cost of delaying the execution of the application. The other approach, called Non-Blocking Coordinated Checkpointing, is a more straightforward implementation of the algorithm by Chandy and Lamport: in-flight messages are added, as they are discovered, in the process checkpoint of the receiver, and reinjected in order in the "unexpected" messages queues, when loading the checkpoint (see Fig. 1.2).

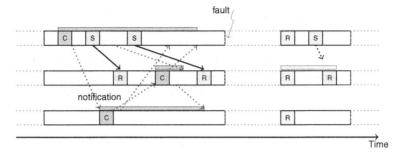

Fig. 1.2 Non-blocking coordinated rollback recovery protocol

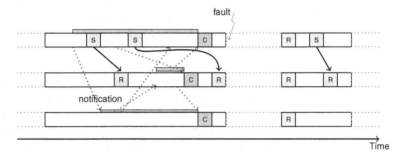

Fig. 1.3 Blocking coordinated rollback recovery protocol

At the application level, resilient application developers have often taken a very simple approach to ensure the consistency of the snapshot: since the protocol is designed knowing the application, a couple of synchronizing barriers can be used, before and after taking the process checkpoints, to guarantee that no application in-flight messages are present at the time of triggering the checkpoint wave, and thus the causal ordering of communications inside the application is used to avoid the issue entirely.

1.2.3 Uncoordinated Checkpointing

Blocking or non-blocking, the coordinated checkpointing protocols require that all processes rollback to the last valid checkpoint wave, when a failure occurs. This ensures a global consistency, at the cost of scalability: as the size of the system grows, the probability of failures increase, and the minimal cost to handle such failures also increase. Indeed, consider only the simple issue of notifying all processes that a rollback is necessary: this can hardly be achieved in constant time, independent of the number of living processes in the system. Chapter 3 will present in further details how uncoordinated checkpointing can be implemented in an MPI library (see Sect. 3.2), but we present here the general approach to compare it with the other protocols.

To reduce the inherent costs of coordinated checkpointing, uncoordinated checkpointing protocols have thus been proposed. On the failure-free part of the execution, the main idea is to remove the coordination of checkpointing, targeting a reduction of the I/O pressure when checkpoints are stored on shared space, and the reduction of delays or increased network usage when coordinating the checkpoints. Furthermore, uncoordinated protocols aim at forcing the restart of a minimal set of processes when a failure happens. Ideally, only the processes subject to a failure should be restarted. However, this requires additional steps.

Consider, for example, a naive protocol, that will let processes checkpoint their local state at any time, without coordination, and in case of failures will try to find a consistent checkpoint wave (in the sense of the Chandy-Lamport algorithm) from a set of checkpoints taken at random times. Even if we assume that all checkpoints are kept until the completion of the execution (which is unrealistic from a storage point of view), finding a consistent wave from random checkpoints might prove impossible, as illustrated by Fig. 1.4. Starting from the last checkpoint (C_1) of process p, all possible waves that include checkpoint C_2 of process q will cross the message m, thus creating another missing message. It is thus necessary to consider a previous checkpoint for p. But all waves including the checkpoint C_3 for p and the checkpoint C_2 for q will cross the message m', creating a missing message. A previous checkpoint must thus be considered for q. This effect, that will invalidate all checkpoint taken randomly, forcing the application to restart from scratch, is called the *domino* effect. To avoid it, multiple protocols have been considered, taking additional assumptions about the application into account.

1.2.3.1 Piecewise Deterministic Assumption

One such assumption is the Piecewise Deterministic Assumption (PWD). It states that a sequential process is an alternate sequence of a nondeterministic choices followed by a set of deterministic steps. As such, the PWD is not really an assumption: it

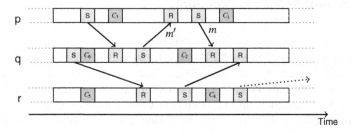

Fig. 1.4 Optimistic uncoordinated protocol: Illustration of the *domino* effect

is a way to describe a possible execution of a sequential program. The assumption resides in the fact that these nondeterministic choices can be captured and their effect replayed. Thus, under the PWD assumption, the behavior of a sequential process can be entirely guided from a given state to another deterministic state by forcing each nondeterministic choice between these two states.

Translated in the HPC world, and especially under the Message Passing Interface (MPI) paradigm, the sources of nondeterminism are rather small. Indeed, all actions that depend upon the input data (environment or user options) are not nondeterministic only in the sense of the PWD assumption: starting from the same state, the same action will follow. Pseudo-random generators fall also in this category of deterministic actions. So, in an MPI application, the only source of nondeterminism comes from time-sensitive decisions, point-to-point message reception order (and request completion order), and related actions (like probe). All these actions are captured by the MPI library (assuming the program relies only on MPI routines to measure time, if its state is time dependent), that is also capable of replaying any value that was returned by a previous call.

In most modern architectures, processes whose state depend on timing have non-deterministic actions, since with modern CPUs and network, an instruction can take a varying time, depending on the actions of other processes sharing the machine, or the operating system, and a misplaced message reception can change significantly this timing measurement. Many MPI operations have a deterministic behavior (e.g., sending a message does not change the state of the sending process; participating to a broadcast operation, seen as an atomic operation, will have a deterministic effect on the state of all processes participating to it, etc...). However, MPI allows the programmer to reorder message receptions, or to not specify an order on the messages reception (using wildcard reception tags, like MPI_ANY_TAG, or MPI_ANY_SOURCE), that enables the library to deliver the messages in an order that is the most efficient, and thus execution-dependent. These actions are then necessarily nondeterministic, since the state of the process between such two receptions depends on what reception actually happened.

Then, consider a parallel application built of sequential processes that use MPI to communicate and synchronize. In case of failure, by replaying the sequence of messages and test/probe with the same result that the process that failed obtained in the initial execution (from the last checkpoint), one can guide the execution of a process to its exact state just before the failure.

1.2.3.2 Message Logging

This leads to the concept of Message Logging (ML). The goal of message logging, in this context, is to provide a tool to capture and replay the most frequent of nondeterministic events: message receptions. To be able to reproduce a message reception, one needs to deliver it in the right order, and with the appropriate content. Message logging thus features two essential parts: a log of the event itself, and a log of the content of the message.

Events Identifiers. Events are usually identified by a few counters: based on the same idea as logical clocks of Lamport [52], these identifiers define a partial order that is sufficient to ensure the consistency of the distributed system by capturing the causality of the nondeterministic events. In most implementations, a nondeterministic message identifier consists of a 4-tuple: identifier of the message emitter, sequence number of emission in that channel, identifier of the message receiver, sequence number of delivery of that message.

The first two counters uniquely identify an outgoing message at the sender. They are used to connect that event identifier with the corresponding payload log. The second two counters make the delivery deterministic. They can only be assigned once the message is delivered by the receiver during the first execution.

A collection of event logs builds the history of a distributed application. If all event logs with the same message receiver identifier are considered, the execution of the receiver is made deterministic up to the end of the log: that process knows exactly what messages it must receive, and in which order they must be delivered.

In some applications, other nondeterministic events may be interleaved between message receptions, and the global ordering of these events on that process must be kept (as well as all information needed to replay these events). For example, in the MPI case, the evaluation of a routine like `MPI_Probe()` is nondeterministic: the routine will return `true` or `false` depending upon the internal state of the library, that depends itself upon the reception of messages. A simple event logging strategy is to remember the return value of each `MPI_Probe()`, associated with an internal event sequence number, to augment the message log with the same internal event sequence number to remember the global ordering of process-specific internal events, and to store these events in the same place as the message logs. To replay the execution, one then needs to have these routines return the same value as during the initial execution, whatever the internal state of the library, and deliver the messages in the order specified by the history. As a result, the library may have to introduce delays, reorder messages, or wait for the arrival of messages that were supposed to be delivered but are not available yet. But the process will be guided to the exact state it reached when the log was interrupted, which is the goal of message logging.

Payload Logging. To deliver messages in replay mode, the receiving process needs to have access to the message payload: its event log is not sufficient. The most widely used approach to provide this payload is to keep a copy at the sender. This is called sender-based message logging (although this is a slight abuse of language, as events can be stored at a separate place different from the sender).

The advantage of sender-based payload logging is that the local copy can be made in parallel with the network transfer, trying to minimize the impact on a failure-free execution. Its main drawback is its usage of node memory. The amount of message payload log is a function of the message throughput of the application, and memory can be exhausted quickly, so, a sender-based payload logging protocol must feature mechanisms for control flow and garbage collection.

To understand how the garbage collection mechanism works, one needs to understand first that the sender-based payload log belongs to the state of the sender process:

at any point, a receiver process may request to send back the content of previously sent messages. If the sender process was subject to a failure, and restarted somewhere in its past, it still may need to provide the payload of messages that were sent even further back in its history. Hence, when taking independent checkpoints, the message payload log must be included in the process checkpoint, as any other element of the process state.

Checkpoints, however, provide a guarantee to senders: when a receiver checkpoints, all the processes that sent it messages have the guarantee that the payload of messages delivered before that checkpoint will never be requested again. They can thus be removed from the state of the process, creating a trade-off between processes: taking a checkpoint of a process will relieve memory of processes that sent messages to it, while imposing to save all the data sent by it. In the worst case, memory can become exhausted, and remote checkpoints of sender processes must be triggered before more messages can be sent and logged by processes.

Event Logging. The last element of a Message Logging strategy has probably been the most studied: how to log the events. As described above, to replay its execution, a process needs to collect the history of all events between the restore point and the last nondeterministic event that happened during the initial execution. Since the memory of the process is lost when it is hit by a failure, this history must be saved somewhere. There are three main strategies to save the events log, called optimistic, pessimistic, and causal.

Optimistic message logging consists in sending the history to a remote event logger. That event logger must be a reliable process, either by assumption (the risk that the failure hits that specific process is inversely proportional to the number of processes in the system), or through replication. The protocol is said optimistic because while event logs are in transfer between the receiver process (that completed the event identifier when it delivered the message to the application) and the event logger, the application may send messages, and be subject to a failure.

If a failure hits the application precisely at this time, the event log might be lost. However, the message that was just sent by the application might be correctly received and delivered anyway. That message, its content, its existence, might depend on the reception whose log was lost. During a replay, the process will not find the event log, and if that reception was nondeterministic, might make a different choice, sending out a message (or doing another action), inconsistent with the rest of the application state.

The natural extension to optimistic message logging is pessimistic message logging: when a process does a nondeterministic action (like a reception), it sends the event log to the event logger, and waits for an acknowledge of logging from the event logger before it is allowed to take any action that may impact the state of the application. This removes the race condition found in optimistic message logging protocols, to the cost of introducing delays in the failure-free execution, as the latency of logging safely the event and waiting for the acknowledge must be added to every nondeterministic event.

To mitigate this issue, causal event logging protocols were designed: in a causal event logging protocol, messages carry part of the history of events that lead to their emission. When a process does a nondeterministic action, it sends the event log to the event logger, appends it to a local history slice, and without waiting for an acknowledge, continues its execution. If an acknowledge comes before any message is sent, that event log is removed from the local history slice. If the process sends a message, however, the local history slice is piggybacked to the outgoing message. That way, at least the receiving process knows of the events that may not be logged and that lead to the emission of this message.

The history slice coming with a message must be added to the history slice of a receiver process, since it is part of the history to bring the receiving process in its current state. This leads to a snowballing effect, where the local history slice of processes grows with messages, and the overhead on messages also grows with time. Multiple strategies have been devised to bound that increase, by garbage collecting events that are safely logged in the event logger from all history slices, and by detecting cycles in causality to trim redundant information from these slices.

Uncoordinated Checkpointing with Message Logging and Replay. Putting all the pieces together, all uncoordinated checkpointing with message logging and replay protocols behave similarly: processes log nondeterministic events and messages payload as they proceed along the initial execution; without strong coordination, they checkpoint their state independently; in case of failure, the failed process restarts from its last checkpoint, it collects all its log history, and enters the replay mode. Replay consists in following the log history, enforcing all nondeterministic events to produce the same effect they had during the initial execution. Message payload must be re-provided to this process for this purpose. If multiple failures happen, the multiple replaying processes may have to reproduce the messages to provide the payload for other replaying processes, but since they follow the path determined by the log history, these messages, and their contents, will be regenerated as any deterministic action. Once the history has been entirely replayed, by the piecewise deterministic assumption, the process reaches a state that is compatible with the state of the distributed application, that can continue its progress from this point on.

1.2.4 Hierarchical Checkpointing

Over modern architectures, that feature many cores on the same computing node, message logging becomes an unpractical solution. Indeed, any interaction between two threads introduces the potential for a nondeterministic event that must be logged. Shared memory also provides an efficient way to implement zero copy communication, and logging the payload of such "messages" introduces a high overhead that make this solution intractable.

In fact, if a thread fails, current operating systems will abort the entire process. If the computing node is subject to a hardware failure, all processes running on that machine fail together. Failures are then often tightly correlated, forcing all processes / threads running on that node to restart together because they crashed together. These two observations lead to the development of Hierarchical Checkpointing Protocols. Hierarchical Checkpointing tries to combine coordinated checkpoint and rollback together with uncoordinated checkpointing with message logging, keeping the best of both approaches.

The idea of Hierarchical Checkpointing is rather simple: processes are distributed in groups; processes belonging to the same group coordinate their checkpoints and rollbacks; uncoordinated checkpointing with message logging is used between groups. However, the state of a single process depends upon the interactions between groups, but also upon the interactions with other processes inside the group. Coordinated rollback guarantees that the application restarts in a consistent state; it does not guarantee that the application, if restarting from that consistent state, will reach the same state as in the initial execution, which is a condition for uncoordinated checkpointing to work. A nondeterministic group (a group of processes whose state depend upon the reception order of messages exchanged inside the group for example) cannot simply restart from the last group-coordinated checkpoint and hope that it will maintain its state globally consistent with the rest of the application.

Thus, Hierarchical Checkpointing Protocols remain uncoordinated checkpointing protocols with message logging: nondeterministic interactions between processes of the same group must be saved, but the message payload can be spared, because all processes of that group will restart and regenerate the missing message payloads, if a failure happens. Section 3.6 presents in deeper details how a specific hierarchical protocol works. In this overview, we introduce a general description of hierarchical protocols to allow for a model-based comparison of the different approaches.

Reducing the logging. There are many reasons to reduce the logging (events and payload): intragroup interactions are numerous, and treating all of them as nondeterministic introduces significant computing slowdown if using a pessimistic protocol, or memory consumption and message slowdown if using a causal protocol; intergroup interactions are less sensitive to event logging, but payload logging augments the checkpoint size, and consumes user memory.

Over the years, many works have proposed to integrate more application knowledge in the fault-tolerant middleware: few HPC applications use message ordering or timing information to take decisions; many receptions in MPI are in fact deterministic, since the source, tag, type and size, and the assumption of ordered transmission in the virtual channel make the matching of messages unique from the application level. In all these cases, logging can be avoided entirely. For other applications, although the reception is nondeterministic, the ordering of receptions will temporarily influence the state of the receiving process, but not its emissions. For example, this happens in a reduce operation written over point to point communications: if a node in the reduction receives first from its left child then from its right one, or in the other order, the state of the process after two receptions stays the same, and the message it sends up to its parent is always the same. Based on this observation, the concept of send determinism has been introduced [36], in which many events may be avoided to log.

MPI provides also a large set of collective operations. Treating these operations at the point-to-point level introduces a lot of nondeterminism, while the high-level operation itself remains deterministic. This fact is used in [14] to reduce the amount of events log.

Hierarchical Checkpointing reduces the need for coordination, allowing a load balancing policy to store the checkpoints; size of the checkpoints, however are dependent on the application message throughput and checkpointing policy (if using sender-based payload logging, as in most cases); the speed of replay, the overhead of logging the events (in message size or in latency) are other critical parameters to decide when a checkpoint must done.

In the following section, we discuss how the different checkpointing protocols can be optimized by carefully selecting the interval between checkpoints. To implement this optimization, it is first necessary to provide a model of performance for these protocols.

1.3 Probabilistic Models for Checkpointing

This section deals with probabilistic models to assess the performance of various checkpointing protocols. We start with the simplest scenario, with a single resource, in Sect. 1.3.1, and we show how to compute the optimal checkpointing *period*. Section 1.3.2 shows that dealing with a single resource and dealing with coordinated checkpointing on a parallel platform are similar problems, provided that we can compute the MTBF of the platform from that of its individual components. Section 1.3.3 deals with hierarchical checkpointing. Things get more complicated, because many parameters must be introduced in the model to account for this complex checkpointing protocol. Finally, Sect. 1.3.4 provides a model for in-memory checkpointing, a variant of coordinated checkpointing where checkpoints are kept in the memory of other processors rather than on stable storage, in order to reduce the cost of checkpointing.

1.3.1 Checkpointing with a Single Resource

We state the problem formally as follows. Let $\text{TIME}_{\text{base}}$ be the base time of the application, without any overhead (neither checkpoints nor faults). Assume that the resource is subject to faults with MTBF μ. Note that we deal with arbitrary failure distributions here, and only assume knowledge of the MTBF.

The time to take a checkpoint is C seconds (the time to upload the checkpoint file onto stable storage). We say that the checkpointing period is T seconds when a checkpoint is done each time the application has completed $T - C$ seconds of work. When a fault occurs, the time between the last checkpoint and the fault is lost. This includes useful work as well as potential fault tolerance techniques. After the fault,

Fig. 1.5 An execution

there is a *downtime* of D seconds to account for the temporary unavailability of the resource (for example rebooting, or migrating to a spare). Finally, in order to be able to resume the work, the content of the last checkpoint needs to be *recovered* which takes a time of R seconds (*e.g.*, the checkpoint file is read from stable storage). The sum of the time lost after the fault, of the downtime and of the recovery time is denoted T_{lost}. All these notations are depicted in Fig. 1.5.

To avoid introducing several conversion parameters, all model parameters are expressed in seconds. The failure inter-arrival times, the duration of a downtime, checkpoint, or recovery are all expressed in seconds. Furthermore, we assume (without loss of generality) that one work unit is executed in one second. One work-unit may correspond to any relevant application-specific quantity.

The difficulty of the problem is to trade-off between the time spent checkpointing, and the time lost in case of a fault. Let $\mathrm{TIME}_{final}(T)$ be the expectation of the total execution time of an application of size TIME_{base} with a checkpointing period of size T. The optimization problem is to find the period T minimizing $\mathrm{TIME}_{final}(T)$. However, for the sake of convenience, we rather aim at minimizing

$$\mathrm{WASTE}(T) = \frac{\mathrm{TIME}_{final}(T) - \mathrm{TIME}_{base}}{\mathrm{TIME}_{final}(T)}.$$

This objective is called the *waste* because it corresponds to the fraction of the execution time that does not contribute to the progress of the application (the time *wasted*). Of course minimizing the ratio WASTE is equivalent to minimizing the total time TIME_{final}, because we have

$$(1 - \mathrm{WASTE}(T))\, \mathrm{TIME}_{final}(T) = \mathrm{TIME}_{base},$$

but using the waste is more convenient. The waste varies between 0 and 1. When the waste is close to 0, it means that $\mathrm{TIME}_{final}(T)$ is very close to TIME_{base} (which is good), whereas, if the waste is close to 1, it means that $\mathrm{TIME}_{final}(T)$ is very large compared to TIME_{base} (which is bad). There are two sources of waste, which we analyze below.

First source of waste. Consider a *fault-free* execution of the application with periodic checkpointing. By definition, during each period of length T we take a checkpoint, which lasts for C time units, and only $T - C$ units of work are executed. Let TIME_{FF} be the execution time of the application in this setting. The fault-free execution time TIME_{FF} is equal to the time needed to execute the whole application, TIME_{base}, plus the time taken by the checkpoints:

$$\text{TIME}_{\text{FF}} = \text{TIME}_{\text{base}} + N_{\text{ckpt}} C,$$

where N_{ckpt} is the number of checkpoints taken. Additionally, we have

$$N_{\text{ckpt}} = \left\lceil \frac{\text{TIME}_{\text{base}}}{T - C} \right\rceil \approx \frac{\text{TIME}_{\text{base}}}{T - C}.$$

To discard the ceiling function, we assume that the execution time $\text{TIME}_{\text{base}}$ is large with respect to the period or, equivalently, that there are many periods during the execution. Plugging back the (approximated) value $N_{\text{ckpt}} = \frac{\text{TIME}_{\text{base}}}{T-C}$, we derive that

$$\text{TIME}_{\text{FF}} = \frac{T}{T - C} \text{TIME}_{\text{base}}. \tag{1.2}$$

Similar to the WASTE, we define WASTE_{FF}, the waste due to checkpointing in a fault-free execution, as the fraction of the fault-free execution time that does not contribute to the progress of the application:

$$\text{WASTE}_{\text{FF}} = \frac{\text{TIME}_{\text{FF}} - \text{TIME}_{\text{base}}}{\text{TIME}_{\text{FF}}} \Leftrightarrow \left(1 - \text{WASTE}_{\text{FF}}\right)\text{TIME}_{\text{FF}} = \text{TIME}_{\text{base}}. \tag{1.3}$$

Combining Eqs. (1.2) and (1.3), we get:

$$\text{WASTE}_{\text{FF}} = \frac{C}{T}. \tag{1.4}$$

This result is quite intuitive: every T seconds, we waste C for checkpointing. This calls for a very large period in a fault-free execution (even an infinite period, meaning no checkpoint at all). However, a large period also implies that a large amount of work is lost whenever a fault strikes, as we discuss now.

Second source of waste. Consider the entire execution (with faults) of the application. Let $\text{TIME}_{\text{final}}$ denote the expected execution time of the application in the presence of faults. This execution time can be divided into two parts: (i) the execution of chunks of work of size $T - C$ followed by their checkpoint; and (ii) the time lost due to the faults. This decomposition is illustrated in Fig. 1.6. The first part of the execution time is equal to TIME_{FF}. Let N_{faults} be the number of faults occurring during the execution, and let T_{lost} be the average time lost per fault. Then,

$$\text{TIME}_{\text{final}} = \text{TIME}_{\text{FF}} + N_{\text{faults}} T_{\text{lost}}. \tag{1.5}$$

In average, during a time $\text{TIME}_{\text{final}}$, $N_{\text{faults}} = \frac{\text{TIME}_{\text{final}}}{\mu}$ faults happen (recall Eq. (1.1)). We need to estimate T_{lost}. A natural estimation for the moment when the fault strikes in the period is $\frac{T}{2}$ (see Fig. 1.5). Intuitively, faults strike anywhere in the period, hence in average they strike in the middle of the period. The proof of this result for Exponential distribution laws can be found in [25]. We conclude that

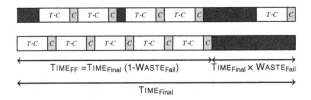

Fig. 1.6 An execution (*top*), and its reordering (*bottom*), to illustrate both sources of waste. Blackened intervals correspond to time lost due to faults: downtime, recoveries, and re-execution of work that has been lost

$T_{\text{lost}} = \frac{T}{2} + D + R$, because after each fault there is a downtime and a recovery. This leads to:

$$\text{TIME}_{\text{final}} = \text{TIME}_{\text{FF}} + \frac{\text{TIME}_{\text{final}}}{\mu}\left(D + R + \frac{T}{2}\right).$$

Let $\text{WASTE}_{\text{fault}}$ be the fraction of the total execution time that is lost because of faults:

$$\text{WASTE}_{\text{fault}} = \frac{\text{TIME}_{\text{final}} - \text{TIME}_{\text{FF}}}{\text{TIME}_{\text{final}}} \Leftrightarrow \left(1 - \text{WASTE}_{\text{fault}}\right)\text{TIME}_{\text{final}} = \text{TIME}_{\text{FF}}$$

We derive:

$$\text{WASTE}_{\text{fault}} = \frac{1}{\mu}\left(D + R + \frac{T}{2}\right). \tag{1.6}$$

Equations (1.4) and (1.6) show that each source of waste calls for a different period: a large period for WASTE_{FF}, as already discussed, but a small period for $\text{WASTE}_{\text{fault}}$, to decrease the amount of work to re-execute after each fault. Clearly, a trade-off is to be found. Here is how. By definition we have

$$\begin{aligned}
\text{WASTE} &= 1 - \frac{\text{TIME}_{\text{base}}}{\text{TIME}_{\text{final}}} \\
&= 1 - \frac{\text{TIME}_{\text{base}}}{\text{TIME}_{\text{FF}}}\frac{\text{TIME}_{\text{FF}}}{\text{TIME}_{\text{final}}} \\
&= 1 - (1 - \text{WASTE}_{\text{FF}})(1 - \text{WASTE}_{\text{fault}}).
\end{aligned}$$

Altogether, we derive the final result:

$$\text{WASTE} = \text{WASTE}_{\text{FF}} + \text{WASTE}_{\text{fault}} - \text{WASTE}_{\text{FF}}\text{WASTE}_{\text{fault}} \tag{1.7}$$

$$= \frac{C}{T} + \left(1 - \frac{C}{T}\right)\frac{1}{\mu}\left(D + R + \frac{T}{2}\right). \tag{1.8}$$

The two sources of waste do not add up, but we have:

$$(1 - \text{WASTE}) = (1 - \text{WASTE}_{\text{FF}})(1 - \text{WASTE}_{\text{fault}}),$$

just as for discount percentages in a sale: two successive 50 % rebates do not make the product free, but the final price reduction is the product of the two successive ones.

We obtain $\text{WASTE} = \frac{u}{T} + v + wT$, where $u = C\left(1 - \frac{D+R}{\mu}\right)$, $v = \frac{D+R-C/2}{\mu}$, and $w = \frac{1}{2\mu}$. It is easy to see that WASTE is minimized for $T = \sqrt{\frac{u}{w}}$. The first-order (FO) formula for the optimal period is thus:

$$T_{\text{FO}} = \sqrt{2(\mu - (D+R))C}. \tag{1.9}$$

and the optimal waste is $\text{WASTE}_{\text{FO}} = 2\sqrt{uw} + v$, therefore

$$\text{WASTE}_{\text{FO}} = \sqrt{\frac{2C}{\mu}\left(1 - \frac{D+R}{\mu}\right)} + \frac{D+R-C/2}{\mu}. \tag{1.10}$$

In 1974, Young [69] obtained a different formula, namely $T_{\text{FO}} = \sqrt{2\mu C} + C$. Thirty years later, Daly [25] refined Young's formula and obtained $T_{\text{FO}} = \sqrt{2(\mu + R)C} + C$. Equation (1.9) is yet another variant of the formula, which we have obtained through the computation of the waste. There is no mystery, though. None of the three formulas is correct! They represent different first-order approximations, which collapse into the beautiful formula $T_{\text{FO}} = \sqrt{2\mu C}$ when μ is large in front of the resilience parameters D, C and R. Below, we show that this latter condition is the key to the accuracy of the approximation.

First-order approximation of T_{FO}. It is interesting to point out why the value of T_{FO} given by Eq. (1.9) is a first-order approximation, even for large jobs. Indeed, there are several restrictions for the approach to be valid:

- We have stated that the expected number of faults during execution is $N_{\text{faults}} = \frac{\text{TIME}_{\text{final}}}{\mu}$, and that the expected time lost due to a fault is $T_{\text{lost}} = \frac{T}{2} + D + R$. Both statements are true individually, but the expectation of a product is the product of the expectations only if the random variables are independent, which is not the case here because $\text{TIME}_{\text{final}}$ depends upon the fault inter-arrival times.
- In Eq. (1.4), we have to enforce $C \leq T$ in order to have $\text{WASTE}_{\text{FF}} \leq 1$.
- In Eq. (1.6), we have to enforce $D + R \leq \mu$ in order to have $\text{WASTE}_{\text{fault}} \leq 1$. In addition, we must cap the period to enforce this latter constraint. Intuitively, we need μ to be large enough for Eq. (1.6) to make sense (see the word of caution at the end of Sect. 1.3.2.1).
- Equation (1.6) is accurate only when two or more faults do not take place within the same period. Although unlikely when μ is large in front of T, the possible occurrence of many faults during the same period cannot be eliminated.

To ensure that the condition of having at most a single fault per period is met with a high probability, we cap the length of the period: we enforce the condition $T \leq \eta\mu$, where η is some tuning parameter chosen as follows. The number of faults during a period of length T can be modeled as a Poisson process of parameter $\beta = \frac{T}{\mu}$. The probability of having $k \geq 0$ faults is $P(X = k) = \frac{\beta^k}{k!}e^{-\beta}$, where X is the random variable showing the number of faults. Hence the probability of having two or more faults is $\pi = P(X \geq 2) = 1 - (P(X = 0) + P(X = 1)) = 1 - (1 + \beta)e^{-\beta}$. To get $\pi \leq 0.03$, we can choose $\eta = 0.27$, providing a valid approximation when bounding the period range accordingly. Indeed, with such a conservative value for η, we have overlapping faults for only 3 % of the checkpointing segments in average, so that the model is quite reliable. For consistency, we also enforce the same type of bound on the checkpoint time, and on the downtime and recovery: $C \leq \eta\mu$ and $D + R \leq \eta\mu$. However, enforcing these constraints may lead to use a suboptimal period: it may well be the case that the optimal period $\sqrt{2(\mu - (D + R))C}$ of Eq. (1.9) does not belong to the admissible interval $[C, \eta\mu]$. In that case, the waste is minimized for one of the bounds of the admissible interval. This is because, as seen from Eq. (1.8), the waste is a convex function of the period.

We conclude this discussion on a positive note. While capping the period, and enforcing a lower bound on the MTBF, is mandatory for mathematical rigor, simulations in [4] show that actual job executions can always use the value from Eq. (1.9), accounting for multiple faults whenever they occur by re-executing the work until success. The first-order model turns out to be surprisingly robust!

Let us formulate our main result as a theorem:

Theorem 1.1 *The optimal checkpointing period is* $T_{\text{FO}} = \sqrt{2\mu C} + o(\sqrt{\mu})$ *and the corresponding waste is* $\text{WASTE}_{\text{FO}} = \sqrt{\frac{2C}{\mu}} + o(\sqrt{\frac{1}{\mu}})$.

Theorem 1.1 has a wide range of applications. We discuss several of them in the following sections. Before that, we explain how to compute the optimal period accurately, in the special case where failures follow an Exponential distribution law.

Optimal value of T_{FO} for Exponential distributions. There is a beautiful method to compute the optimal value of T_{FO} accurately when the failure distribution is $Exp(\lambda)$. First, we show how to compute the expected time $\mathbb{E}(\text{TIME}(T - C, C, D, R, \lambda))$ to execute a work of duration $T - C$ followed by a checkpoint of duration C, given the values of C, D, and R, and a fault distribution $Exp(\lambda)$. Recall that if a fault interrupts a given trial before success, there is a downtime of duration D followed by a recovery of length R. We assume that faults can strike during checkpoint and recovery, but not during downtime.

Proposition 1.1

$$\mathbb{E}(\text{TIME}(T - C, C, D, R, \lambda)) = e^{\lambda R}\left(\frac{1}{\lambda} + D\right)(e^{\lambda T} - 1).$$

Proof For simplification, we write TIME instead of $\text{TIME}(T - C, C, D, R, \lambda)$ in the proof below. Consider the following two cases:

(i) Either there is no fault during the execution of the period, then the time needed is exactly T;

(ii) Or there is one fault before successfully completing the period, then some additional delays are incurred. More specifically, as seen for the first order approximation, there are two sources of delays: the time spent computing by the processors before the fault (accounted for by variable T_{lost}), and the time spent for downtime and recovery (accounted for by variable T_{rec}). Once a successful recovery has been completed, there still remain $T - C$ units of work to execute.

Thus TIME obeys the following recursive equation:

$$\text{TIME} = \begin{cases} T & \text{if there is no fault} \\ T_{\text{lost}} + T_{\text{rec}} + \text{TIME} & \text{otherwise} \end{cases} \tag{1.11}$$

T_{lost} denotes the amount of time spent by the processors before the first fault, knowing that this fault occurs within the next T units of time. In other terms, it is the time that is wasted because computation and checkpoint were not successfully completed (the corresponding value in Fig. 1.5 is $T_{\text{lost}} - D - R$, because for simplification T_{lost} and T_{rec} are not distinguished in that figure).

T_{rec} represents the amount of time needed by the system to recover from the fault (the corresponding value in Fig. 1.5 is $D + R$).

The expectation of TIME can be computed from Eq. (1.11) by weighting each case by its probability to occur:

$$\mathbb{E}(\text{TIME}) = \mathbb{P}\,(\text{no fault}) \cdot T + \mathbb{P}\,(\text{a fault strikes}) \cdot \mathbb{E}\,(T_{\text{lost}} + T_{\text{rec}} + \text{TIME})$$
$$= e^{-\lambda T} T + (1 - e^{-\lambda T})\,(\mathbb{E}(T_{\text{lost}}) + \mathbb{E}(T_{\text{rec}}) + \mathbb{E}(\text{TIME}))\,,$$

which simplifies into:

$$\mathbb{E}(T) = T + (e^{\lambda T} - 1)\,(E(T_{\text{lost}}) + E(T_{\text{rec}})) \tag{1.12}$$

We have $\mathbb{E}(T_{\text{lost}}) = \int_0^\infty x \mathbb{P}(X = x | X < T) dx = \frac{1}{\mathbb{P}(X<T)} \int_0^T e^{-\lambda x} dx$, and $\mathbb{P}(X < T) = 1 - e^{-\lambda T}$. Integrating by parts, we derive that

$$\mathbb{E}(T_{\text{lost}}) = \frac{1}{\lambda} - \frac{T}{e^{\lambda T} - 1} \tag{1.13}$$

Next, the reasoning to compute $\mathbb{E}(T_{\text{rec}})$, is very similar to $\mathbb{E}(\text{TIME})$ (note that there can be no fault during D but there can be during R):

$$\mathbb{E}(T_{\text{rec}}) = e^{-\lambda R}(D + R) + (1 - e^{-\lambda R})(D + \mathbb{E}(R_{lost}) + \mathbb{E}(T_{\text{rec}}))$$

Here, R_{lost} is the amount of time lost to executing the recovery before a fault happens, knowing that this fault occurs within the next R units of time. Replacing T by R in

Eq. (1.13), we obtain $\mathbb{E}(R_{lost}) = \frac{1}{\lambda} - \frac{R}{e^{\lambda R}-1}$. The expression for $\mathbb{E}(T_{rec})$ simplifies to

$$\mathbb{E}(T_{rec}) = De^{\lambda R} + \frac{1}{\lambda}(e^{\lambda R} - 1)$$

Plugging the values of $\mathbb{E}(T_{lost})$ and $\mathbb{E}(T_{rec})$ into Eq. (1.12) leads to the desired value:

$$\mathbb{E}(\text{TIME}(T - C, C, D, R, \lambda)) = e^{\lambda R}\left(\frac{1}{\lambda} + D\right)(e^{\lambda T} - 1)$$

Proposition 1.1 is the key to proving that the optimal checkpointing strategy (with an Exponential distribution of faults) is periodic. Indeed, consider an application of duration TIME_{base}, and divide the execution into periods of different lengths T_i, each with a checkpoint at the end. The expectation of the total execution time is the sum of the expectations of the time needed for each period. Proposition 1.1 shows that the expected time for a period is a convex function of its length, hence all periods must be equal and $T_i = T$ for all i.

There remains to find the best number of periods, or equivalently, the size of each work chunk before checkpointing. With k periods of length $T = \frac{\text{TIME}_{base}}{k}$, we have to minimize a function that depends on k. Assuming k rational, one can find the optimal value k_{opt} by differentiation (and prove uniqueness using another differentiation). Unfortunately, we have to use the (implicit) Lambert function $\mathbb{L}$, defined as $\mathbb{L}(z)e^{\mathbb{L}(z)} = z$), to express the value of k_{opt}, but we can always compute this value numerically. In the end, the optimal number of periods is either $\lfloor k_{opt} \rfloor$ or $\lceil k_{opt} \rceil$, thereby determining the optimal period T_{opt}. As a sanity check, the first-order term in the Taylor expansion of T_{opt} is indeed T_{FO}, which is kind of comforting. See [12] for all details.

1.3.2 Coordinated Checkpointing

In this section we introduce a simple model for coordinated checkpointing. Consider an application executing on a parallel platform with N processors, and using coordinated checkpointing for resilience. What is the optimal checkpointing period? We show how to reduce the optimization problem with N processors to the previous problem with only one processor. Most high performance applications are *tightly-coupled* applications, where each processor is frequently sending messages to, and receiving messages from the other processors. This implies that the execution can progress only when all processors are up and running. When using coordinated checkpointing, this also implies that when a fault strikes one processor, the whole application must be restarted from the last checkpoint. Indeed, even though the other processors are still alive, they will very soon need some information from the faulty processor. But to catch up, the faulty processor must re-execute the work that it has lost, during which it had received messages from the other processors. But these

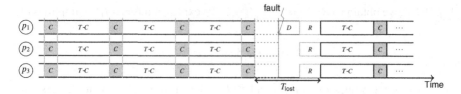

Fig. 1.7 Behavior for a tightly coupled application with coordinated checkpointing

messages are no longer available. This is why all processors have to recover from the last checkpoint and re-execute the work in parallel. On the contrary, with hierarchical checkpointing, only the group of the faulty processor must recover and re-execute (see Sect. 1.3.3 for a model of this complicated protocol).

Figure 1.7 provides an illustration of coordinated checkpointing. Each time a fault strikes somewhere on the platform, the application stops, all processors perform a downtime and a recovery, and they re-execute the work during a time T_{lost}. This is exactly the same pattern as with a single resource. We can see the whole platform as a single *super-processor*, very powerful (its speed is N times that of individual processors) but also very prone to faults: all the faults strike this super-processor! We can apply Theorem 1.1 to the super-processor and determine the optimal checkpointing period as $T_{\text{FO}} = \sqrt{2\mu C} + o(\sqrt{\mu})$, where μ now is the MTBF of the super-processor. How can we compute this MTBF? The answer is given in the next section.

1.3.2.1 Platform MTBF

With Fig. 1.8, we see that the super-processor is hit by faults N times more frequently than the individual processors. We should then conclude that its MTBF is N times smaller than that of each processor. We state this result formally:

Proposition 1.2 *Consider a platform with N identical processors, each with MTBF* μ_{ind}. *Let* μ *be the MTBF of the platform. Then*

$$\mu = \frac{\mu_{ind}}{N} \tag{1.14}$$

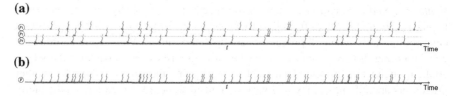

Fig. 1.8 Intuition of the proof of Proposition 1.2. **a** If three processors have around 20 faults during a time t ($\mu_{\text{ind}} = \frac{t}{20}$)... **b**...during the same time, the equivalent processor has around 60 faults ($\mu = \frac{t}{60}$)

Proof We first prove the proposition when the inter-arrival times of the faults on each individual processor are I.I.D. random variables with distribution $Exp(\lambda)$, where $\lambda = \frac{1}{\mu_{ind}}$. Recall that I.I.D. means *Independent and Identically Distributed*. In that simple case, the inter-arrival times of the faults on the super-processor are I.I.D. random variables with distribution $Exp(N\lambda)$, which proves that its MTBF is $\mu = \frac{\mu_{ind}}{N}$. To see this, the reasoning is the following:

- The arrival time of the first fault on the super-processor is a random variable $Y_1 \sim Exp(\lambda)$. This is because Y_1 is the minimum of $X_1^{(1)}$, $X_1^{(2)}$ ..., $X_1^{(N)}$, where $X_1^{(i)}$ is the arrival time of the first fault on processor P_i. But $X_1^{(i)} \sim Exp(\lambda)$ for all i, and the minimum of N random variables following an Exponential distribution $Exp(\lambda_i)$ is a random variable following an Exponential distribution $Exp(\sum_{i=1}^{N} \lambda_i)$ (see [64, p. 288]).
- The memoryless property of Exponential distributions is the key to the result for the delay between the first and second fault on the super-processor. Knowing that first fault occurred on processor P_1 at time t, what is the distribution of random variable for the occurrence of the first fault on processor P_2? The only new information if that P_2 has been alive for t seconds. The memoryless property states that the distribution of the arrival time of the first fault on P_2 is not changed at all when given this information! It is still an exponential distribution $Exp(\lambda)$. Of course this holds true not only for P_2, but for each processor. And we can use the same minimum trick as for the first fault.
- Finally, the reasoning is the same for the third fault, and so on.

This concludes the proof for exponential distributions.

We now give another proof of Proposition 1.2 that applies to any continuous probability distribution with bounded (nonzero) expectation, not just Exponential laws. Consider a single processor, say processor P_q. Let $X_i, i \geq 0$ denote the I.I.D. random variables for the fault inter-arrival times on P_q, and assume that $X_i \sim D_X$, where D_X is a continuous probability distribution with bounded (nonzero) expectation μ_{ind}. In particular, $\mathbb{E}(X_i) = \mu_{ind}$ for all i. Consider a fixed time bound F. Let $n_q(F)$ be the number of faults on P_q until time F. More precisely, the $(n_q(F) - 1)$-th fault is the last one to happen strictly before time F, and the $n_q(F)$-th fault is the first to happen at time F or after. By definition of $n_q(F)$, we have

$$\sum_{i=1}^{n_q(F)-1} X_i \leq F \leq \sum_{i=1}^{n_q(F)} X_i.$$

Using Wald's equation [64, p. 420], with $n_q(F)$ as a stopping criterion, we derive:

$$(\mathbb{E}\left(n_q(F)\right) - 1)\mu_{ind} \leq F \leq \mathbb{E}\left(n_q(F)\right)\mu_{ind},$$

and we obtain:

$$\lim_{F \to +\infty} \frac{\mathbb{E}\left(n_q(F)\right)}{F} = \frac{1}{\mu_{ind}}. \tag{1.15}$$

Now consider a platform with N identical processors, whose fault inter-arrival times are I.I.D. random variables that follow the distribution D_X. Unfortunately, if D_X is not an Exponential law, then the inter-arrival times of the faults of the whole platform, i.e., of the super-processor introduced above, are no longer I.I.D. The minimum trick used in the proof of Proposition 1.2 works only for the first fault. For the following ones, we need to remember the history of the previous faults, and things get too complicated. However, we could still define the MTBF μ of the super-processor using Eq. (1.15): this value μ must satisfy

$$\lim_{F \to +\infty} \frac{\mathbb{E}\left(n(F)\right)}{F} = \frac{1}{\mu},$$

where $n(F)$ be the number of faults on the super-processor until time F. But does the limit always exist? and if yes, what is its value?

The answer to both questions is not difficult. Let Y_i, $i \geq 1$ denote the random variables for fault inter-arrival times on the super-processor. Consider a fixed time bound F as before. Let $n(F)$ be the number of faults on the whole platform until time F, and let $m_q(F)$ be the number of these faults that strike component number q. Of course we have $n(F) = \sum_{q=1}^{N} m_q(F)$. By definition, except for the component hit by the last fault, $m_q(F) + 1$ is the number of faults on component q until time F is exceeded, hence $n_q(F) = m_q(F) + 1$ (and this number is $m_q(F) = n_q(F)$ on the component hit by the last fault). From Eq. (1.15) again, we have for each component q:

$$\lim_{F \to +\infty} \frac{\mathbb{E}\left(m_q(F)\right)}{F} = \frac{1}{\mu_{\text{ind}}}.$$

Since $n(F) = \sum_{q=1}^{N} m_q(F)$, we also have:

$$\lim_{F \to +\infty} \frac{\mathbb{E}\left(n(F)\right)}{F} = \frac{N}{\mu_{\text{ind}}}$$

which answers both questions at the same time and concludes the proof.

Note that the random variables Y_i are not I.I.D., and they do not necessarily have the same expectation, which explains why we resort to Eq. (1.15) to define the MTBF of the super-processor. Another possible asymptotic definition of the MTBF μ of the platform could be given by the equation

$$\mu = \lim_{n \to +\infty} \frac{\sum_{i=1}^{n} \mathbb{E}\left(Y_i\right)}{n}.$$

Kella and Stadje (Theorem 4, [49]) prove that this limit indeed exists and that is also equal to $\frac{\mu_{\text{ind}}}{N}$, if in addition the distribution function of the X_i is continuous (a requirement always met in practice).

Proposition 1.2 shows that scale is the enemy of fault tolerance. If we double up the number of components in the platform, we divide the MTBF by 2, and the minimum waste automatically increases by a factor $\sqrt{2} \approx 1.4$ (see Eq. (1.10)). And this assumes that the checkpoint time C remains constant. With twice as many processors, there is twice more data to write onto stable storage, hence the aggregated I/O bandwidth of the platform must be doubled to match this requirement.

We conclude this section with a word of caution: the formula $\mu = \frac{\mu_{\text{ind}}}{N}$ expresses the fact that the MTBF of a parallel platform will inexorably decrease as the number of its components increases, regardless how reliable each individual component could be. Mathematically, the expression of the waste in Eq. (1.8) is a valid approximation only if μ is large in front of the other resilience parameters. This will obviously be no longer true when the number of resources gets beyond some threshold.

1.3.2.2 Execution Time for a Parallel Application

In this section, we explain how to use Proposition 1.2 to compute the expected execution time of a parallel application using N processors. We consider the following relevant scenarios for checkpoint/recovery overheads and for parallel execution times.

Checkpoint/recovery overheads—With coordinated checkpointing, checkpoints are synchronized over all processors. We use $C(N)$ and $R(N)$ to denote the time for saving a checkpoint and for recovering from a checkpoint on N processors, respectively (we assume that the downtime D does not depend on N). Assume that the application's memory footprint is Mem, and b_{io} represents the available I/O bandwidth. bytes, with each processor holding $\frac{\text{Mem}}{N}$ bytes. We envision two scenarios:

- Proportional overhead: $C(N) = R(N) = \frac{\text{Mem}}{N b_{io}}$. This is representative of cases in which the bandwidth of the network card/link at each processor is the I/O bottleneck. In such cases, processors checkpoint their data in parallel.
- Constant overhead: $C(N) = R(N) = \frac{\text{Mem}}{b_{io}}$, which is representative of cases in which the bandwidth to/from the resilient storage system is the I/O bottleneck. In such cases, processors checkpoint their data in sequence.

Parallel work—Let $W(N)$ be the time required for a failure-free execution on N processors. We use three models:

- Embarrassingly parallel jobs: $W(N) = W/N$. Here W represents the sequential execution time of the application.
- Generic parallel jobs: $W(N) = W/N + \gamma W$. As in Amdahl's law [1], $\gamma < 1$ is the fraction of the work that is inherently sequential.
- Numerical kernels: $W(N) = W/N + \gamma W^{2/3}/\sqrt{N}$. This is representative of a matrix product (or LU/QR factorization) of size n on a 2D-processor grid, where $W = O(n^3)$. In the algorithm in [7], $N = p^2$ and each processor receives $2p$ matrix blocks of size n/p. Here γ is the communication-to-computation ratio of the platform.

We assume that the parallel job is tightly coupled, meaning that all N processors operate synchronously throughout the job execution. These processors execute the same amount of work $W(N)$ in parallel, period by period. Inter-processor messages are exchanged throughout the computation, which can only progress if all processors are available. When a failure strikes a processor, the application is missing one resource for a certain period of time of length D, the *downtime*. Then the application recovers from the last checkpoint (*recovery* time of length $R(N)$) before it re-executes the work done since that checkpoint and up to the failure. Therefore, we can compute the optimal period and the optimal waste WASTE as in Theorem 1.1 with $\mu = \frac{\mu_{\text{ind}}}{N}$ and $C = C(N)$. The (expected) parallel execution time is $Time[final] = \frac{\text{TIME}_{\text{base}}}{1 - \text{WASTE}}$, where $\text{TIME}_{\text{base}} = W(N)$.

Altogether, we have designed a variety of scenarios, some more optimistic than others, to model the performance of a parallel tightly-coupled application with coordinated checkpointing. We point out that many scientific applications are tightly-coupled, such as iterative applications with a global synchronization point at the end of each iteration. However, the fact that inter-processor information is exchanged continuously or at given synchronization steps (as in BSP-like models) is irrelevant: in steady-state mode, all processors must be available concurrently for the execution to actually progress. While the tightly-coupled assumption may seem very constraining, it captures the fact that processes in the application depend on each other and exchange messages at a rate exceeding the periodicity of checkpoints, preventing independent progress.

1.3.3 Hierarchical Checkpointing

As discussed in Sect. 1.2.4, and presented in deeper details in Sect. 3.6 later in this book, hierarchical checkpointing algorithms are capable of partial coordination of checkpoints to decrease the cost of logging, while retaining message logging capabilities to remove the need for a global restart. These hierarchical schemes partition the application processes in groups. Each group checkpoints independently, but processes belonging to the same group coordinate their checkpoints and recovery. Communications between groups continue to incur payload logging. However, because processes belonging to a same group follow a coordinated checkpointing protocol, the payload of messages exchanged between processes within the same group is not required to be logged.

The optimizations driving the choice of the size and shape of groups are varied. A simple heuristic is to checkpoint as many processes as possible, simultaneously, without exceeding the capacity of the I/O system. In this case, groups do not checkpoint in parallel. Groups can also be formed according to hardware proximity or communication patterns. In such approaches, there may be opportunity for several groups to checkpoint concurrently.

The design and analysis of a refined model for hierarchical checkpointing requires to introduce many new parameters. First, we have to account for non-blocking checkpointing, i.e., the possibility to continue execution (albeit at a reduced rate) while checkpointing. Then message logging has three consequences, two negative and one positive:

- performance degradation in a fault-free execution (negative effect)
- re-execution speed-up after a failure (positive effect)
- checkpoint size increase to store logged messages (negative effect)

The last item is the most important, because intergroup messages may rapidly increase the total size of the checkpoint as the execution progresses, thereby imposing to cap the length of the checkpointing period (see Sect. 1.2.4). The model proposed in this section captures all these additional parameters for a variety of platforms and applications, and provides formulas to compute (and compare) the waste of each checkpointing protocol and application/platform scenario. However, the curious reader must be advised that derivation of the waste becomes much more complicated than in Sects. 1.3.1 and 1.3.2.

1.3.3.1 Instantiating the Model

In this section, we detail the main parameters of the model. We consider a tightly-coupled application that executes on N processors. As before, all model parameters are expressed in seconds. However, in the previous models, one work unit was executed in one second, because we assumed that processors were always computing at full rate. However, with hierarchical checkpointing, when a processor is slowed-down by another activity related to fault tolerance (writing checkpoints to stable storage, logging messages, etc.), one work-unit takes longer than a second to complete. Also, recall that after the striking of a failure under a hierarchical scenario, the useful work resumes only when the faulty group catches up with the overall state of the application at failure time.

Blocking or non-blocking checkpoint. There are various scenarios to model the cost of checkpointing in hierarchical checkpointing protocols, so we use a flexible model, with several parameters to specify. The first question is whether checkpoints are blocking or not. On some architectures, we may have to stop executing the application before writing to the stable storage where the checkpoint data is saved; in that case checkpoint is fully blocking. On other architectures, checkpoint data can be saved on the fly into a local memory before the checkpoint is sent to the stable storage, while computation can resume progress; in that case, checkpoints can be fully overlapped with computations. To deal with all situations, we introduce a slow-down factor α: during a checkpoint of duration C, the work that is performed is αC work units, instead of C work-units if only computation takes place. In other words, $(1 - \alpha)C$ work-units are wasted due to checkpoint jitters perturbing the progress of computation. Here, $0 \leq \alpha \leq 1$ is an arbitrary parameter. The case $\alpha = 0$ corresponds to a fully blocking checkpoint, while $\alpha = 1$ corresponds to a fully overlapped checkpoint, and all intermediate situations can be represented. Note that we have resorted to fully blocking models in Sects. 1.3.1 and 1.3.2.

Periodic checkpointing strategies. Just as before, we focus on periodic scheduling strategies where checkpoints are taken at regular intervals, after some fixed amount of work-units have been performed. The execution is partitioned into periods of duration $T = W + C$, where W is the amount of time where only computations take place, while C corresponds to the amount of time where checkpoints are taken. If not slowed down for other reasons by the fault-tolerant protocol (see Sect. 1.3.3.4), the total amount of work units that are executed during a period of length T is thus $\text{WORK} = W + \alpha C$ (recall that there is a slow-down due to the overlap).

The equations that define the waste are the same as in Sect. 1.3.1. We reproduce them below for convenience:

$$\begin{aligned}
(1 - \text{WASTE}_{FF})\text{TIME}_{FF} &= \text{TIME}_{base} \\
(1 - \text{WASTE}_{fail})\text{TIME}_{final} &= \text{TIME}_{FF} \\
\text{WASTE} &= 1 - (1 - \text{WASTE}_{FF})(1 - \text{WASTE}_{fail})
\end{aligned} \tag{1.16}$$

We derive easily that

$$\text{WASTE}_{FF} = \frac{T - \text{WORK}}{T} = \frac{(1 - \alpha)C}{T} \tag{1.17}$$

As expected, if $\alpha = 1$ there is no overhead, but if $\alpha < 1$ (actual slowdown, or even blocking if $\alpha = 0$), we retrieve a fault-free overhead similar to that of coordinated checkpointing. For the time being, we do not further quantify the length of a checkpoint, which is a function of several parameters. Instead, we proceed with the abstract model. We envision several scenarios in Sect. 1.3.3.5, only after setting up the formula for the waste in a general context.

Processor groups. As mentioned above, we assume that the platform is partitioned into G groups of the same size. Each group contains q processors, hence $N = Gq$. When $G = 1$, we speak of a *coordinated* scenario, and we simply write C, D and R for the duration of a checkpoint, downtime and recovery. When $G \geq 1$, we speak of a *hierarchical* scenario. Each group of q processors checkpoints independently and sequentially in time $C(q)$. Similarly, we use $D(q)$ and $R(q)$ for the duration of the downtime and recovery. Of course, if we set $G = 1$ in the (more general) *hierarchical* scenario, we retrieve the value of the waste for the coordinated scenario. As already mentioned, we derive a general expression for the waste for both scenarios, before further specifying the values of $C(q)$, $D(q)$, and $R(q)$ as a function of q and the various architectural parameters under study.

1.3.3.2 Waste for the Coordinated Scenario ($G = 1$)

The goal of this section is to quantify the expected waste in the coordinated scenario where $G = 1$. Recall that we write C, D, and R for the checkpoint, downtime, and recovery using a single group of N processors. The platform MTBF is μ. We obtain

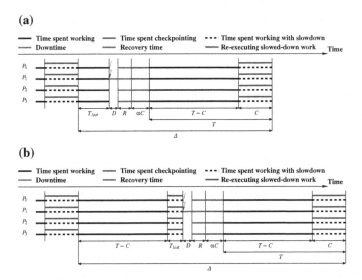

Fig. 1.9 Coordinated checkpoint: illustrating the waste when a failure occurs. **a** during the work phase; and **b** during the checkpoint phase

the following equation for the waste, which we explain briefly below and illustrate with Fig. 1.9:

$$\text{WASTE}_{\text{FF}} = \frac{(1-\alpha)C}{T} \tag{1.18}$$

$$\text{WASTE}_{\text{fail}} = \frac{1}{\mu}\left(R + D + \right.$$
$$\frac{T-C}{T}\left[\alpha C + \frac{T-C}{2}\right]$$
$$\left.+\frac{C}{T}\left[\alpha C + T - C + \frac{C}{2}\right]\right) \tag{1.19}$$

- Equation (1.18) is the portion of the execution lost in checkpointing, even during a fault-free execution, see Eq. (1.17).
- The second part of Eq. (1.19) is the overhead of the execution time due to a failure during work interval $T - C$ (see Fig. 1.9a).
- The last part of Eq. (1.19) is the overhead due to a failure during a checkpoint (see Fig. 1.9b).

After simplification of Eqs. (1.18) and (1.19), we get:

$$\text{WASTE}_{\text{fail}} = \frac{1}{\mu}\left(D + R + \frac{T}{2} + \alpha C\right) \tag{1.20}$$

Plugging this value back into Eq. (1.16) leads to:

$$\text{WASTE}_{\text{coord}} = 1 - \left(1 - \frac{(1-\alpha)C}{T}\right)\left(1 - \frac{1}{\mu}\left(D + R + \frac{T}{2} + \alpha C\right)\right) \quad (1.21)$$

The optimal checkpointing period T_{opt} that minimizes the expected waste in Eq. (1.21) is

$$T_{\text{opt}} = \sqrt{2(1-\alpha)(\mu - (D + R + \alpha C))C} \quad (1.22)$$

This value is in accordance with the first-order expression of T_{FO} in Eq. (1.9) when $\alpha = 0$ and, by construction, must be greater than C. Of course, just as before, this expression is valid only if all resilience parameters are small in front of μ.

1.3.3.3 Waste for the Hierarchical Scenario ($G \geq 1$)

In this section, we compute the expected waste for the hierarchical scenario. We have G groups of q processors, and we let $C(q)$, $D(q)$, and $R(q)$ be the duration of the checkpoint, downtime, and recovery for each group. We assume that the checkpoints of the G groups take place in sequence within a period (see Fig. 1.10a). We start by generalizing the formula obtained for the coordinated scenario before introducing several new parameters to the model.

Generalizing previous scenario with $G \geq 1$: We obtain the following intricate formula for the waste, which we illustrate with Fig. 1.10 and the discussion below:

$$\text{WASTE}_{\text{hier}} = 1 - \left(1 - \frac{T - \text{WORK}}{T}\right)\left(1 - \frac{1}{\mu}\left(D(q) + R(q) + \text{RE- EXEC}\right)\right) \quad (1.23)$$

$$\text{WORK} = T - (1-\alpha)GC(q) \quad (1.24)$$

$$\text{RE- EXEC} =$$

$$\frac{T - GC(q)}{T}\frac{1}{G}\sum_{g=1}^{G}\left[(G-g+1)\alpha C(q) + \frac{T - GC(q)}{2}\right]$$

$$+ \frac{GC(q)}{T}\frac{1}{G^2}\sum_{g=1}^{G}\Bigg[$$

$$\sum_{s=0}^{g-2}(G - g + s + 2)\alpha C(q) + T - GC(q)$$

$$+ G\alpha C(q) + T - GC(q) + \frac{C(q)}{2}$$

$$+ \sum_{s=1}^{G-g}(s + 1)\alpha C(q)\Bigg] \quad (1.25)$$

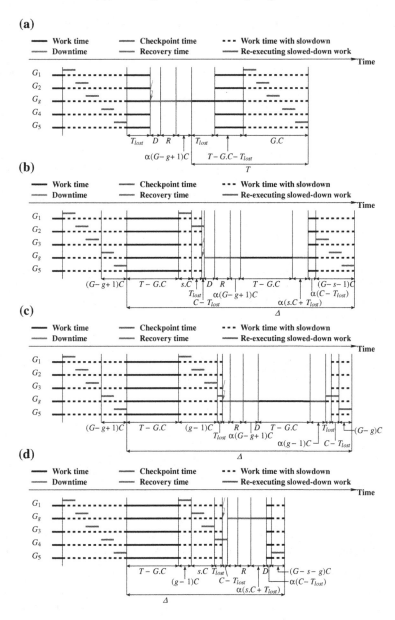

Fig. 1.10 Hierarchical checkpoint: illustrating the waste when a failure occurs. **a** during the work phase (first part of Eq. (1.25)); and during the checkpoint phase (last three parts of Eq. (1.25)), with three sub-cases: **b** before the checkpoint of the failing group (second part of Eq. (1.25)), **c** during the checkpoint of the failing group (third part of Eq. (1.25)), or **d** after the checkpoint of the failing group (last part of Eq. (1.25))

- The first term in Eq. (1.23) represents the overhead due to checkpointing during a fault-free execution (same reasoning as in Eq. (1.17)), and the second term the overhead incurred in case of failure.
- Equation (1.24) provides the amount of work units executed within a period of length T.
- The first part of Eq. (1.25) represents the time needed for re-executing the work when the failure happens in a work-only area, i.e., during the first $T - GC(q)$ seconds of the period (see Fig. 1.10a).
- The second part of Eq. (1.25) deals with the case where the fault happens during a checkpoint, i.e., during the last $GC(q)$ seconds of the period (hence the first term that represents the probability of this event).
  We distinguish three cases, depending upon what group was checkpointing at the time of the failure:

  - The third part of Eq. (1.25) is for the case when the fault happens before the checkpoint of group g (see Fig. 1.10b).
  - The fourth part of Eq. (1.25) is for the case when the fault happens during the checkpoint of group g (see Fig. 1.10c).
  - The fifth part of Eq. (1.25) is the case when the fault happens after the checkpoint of group g, during the checkpoint of group $g + s$, where $g + 1 \leq g + s \leq G$ (See Fig. 1.10d).

Of course this expression reduces to Eq. (1.21) when $G = 1$. Just as for the coordinated scenario, we enforce the constraint

$$GC(q) \leq T \tag{1.26}$$

by construction of the periodic checkpointing policy.

1.3.3.4 Refining the Model

We now introduce three new parameters to refine the model when the processors have been partitioned into several groups. These parameters are related to the impact of message logging on execution, re-execution, and checkpoint image size, respectively.

Impact of message logging on execution and re-execution. With several groups, intergroup messages need to be stored in local memory as the execution progresses, and event logs must be stored in reliable storage, so that the recovery of a given group, after a failure, can be done independently of the other groups. This induces an overhead, which we express as a slowdown of the execution rate: instead of executing one work-unit per second, the application executes only λ work-units, where $0 < \lambda < 1$. Typical values for λ are said to be $\lambda \approx 0.98$, meaning that the overhead due to payload messages is only a small percentage [14, 36].

On the contrary, message logging has a positive effect on re-execution after a failure, because intergroup messages are stored in memory and directly accessible after the recovery. Our model accounts for this by introducing a speedup factor ρ during the re-execution. Typical values for ρ lie in the interval $[1; 2]$, meaning that re-execution time can be reduced by up to half for some applications [13].

Fortunately, the introduction of λ and ρ is not difficult to account for in the expression of the expected waste: in Eq. (1.23), we replace WORK by λWORK and RE- EXEC by $\frac{\text{RE- EXEC}}{\rho}$ and obtain

$$\text{WASTE}_{\text{hier}} = 1 - \left(1 - \frac{T - \lambda\text{WORK}}{T}\right)\left(1 - \frac{1}{\mu}\left(D(q) + R(q) + \frac{\text{RE- EXEC}}{\rho}\right)\right) \quad (1.27)$$

where the values of WORK and RE- EXEC are unchanged, and given by Eqs. (1.24) and (1.25) respectively.

Impact of message logging on checkpoint size. Message logging has an impact on the execution and re-execution rates, but also on the size of the checkpoint. Because intergroup messages are logged, the size of the checkpoint increases with the amount of work per unit. Consider the hierarchical scenario with G groups of q processors. Without message logging, the checkpoint time of each group is $C_0(q)$, and to account for the increase in checkpoint size due to message logging, we write the equation

$$C(q) = C_0(q)(1 + \beta\lambda\text{WORK}) \Leftrightarrow \beta = \frac{C(q) - C_0(q)}{C_0(q)\lambda\text{WORK}} \quad (1.28)$$

As before, $\lambda\text{WORK} = \lambda(T - (1 - \alpha)GC(q))$ (see Eq. (1.24)) is the number of work units, or application iterations, completed during the period of duration T, and the parameter β quantifies the increase in the checkpoint image size per work unit, as a proportion of the application footprint. Typical values of β are given in the examples of Sect. 1.3.3.5. Combining with Eq. (1.28), we derive the value of $C(q)$ as

$$C(q) = \frac{C_0(q)(1 + \beta\lambda T)}{1 + GC_0(q)\beta\lambda(1 - \alpha)} \quad (1.29)$$

The constraint in Eq. (1.26), namely $GC(q) \leq T$, now translates into $\frac{GC_0(q)(1+\beta\lambda T)}{1+GC_0(q)\beta\lambda(1-\alpha)} \leq T$, hence

$$GC_0(q)\beta\lambda\alpha \leq 1 \text{ and } T \geq \frac{GC_0(q)}{1 - GC_0(q)\beta\lambda\alpha} \quad (1.30)$$

1.3.3.5 Case Studies

In this section, we use the previous model to evaluate different case studies. We propose three generic scenarios for the checkpoint protocols, and three application examples with different values for the parameter β.

Checkpointing algorithm scenarios.

COORD- IO —The first scenario considers a coordinated approach, where the duration of a checkpoint is the time needed for the N processors to write the memory footprint of the application onto stable storage. Let Mem denote this memory, and b_{io} represents the available I/O bandwidth. Then

$$C = C_{\text{Mem}} = \frac{\text{Mem}}{b_{io}} \tag{1.31}$$

(see the discussion on checkpoint/recovery overheads in Sect. 1.3.2.2 for a similar scenario). In most cases we have equal write and read speed access to stable storage, and we let $R = C = C_{\text{Mem}}$, but in some cases we could have different values. Recall that a constant value $D(q) = D$ is used for the downtime.

HIERARCH- IO —The second scenario uses a number of relatively large groups. Typically, these groups are composed to take advantage of the application communication pattern [32, 36]. For instance, if the application executes on a 2D-grid of processors, a natural way to create processor groups is to have one group per row (or column) of the grid. If all processors of a given row belong to the same group, horizontal communications are intragroup communications and need not to be logged. Only vertical communications are intergroup communications and need to be logged.

With large groups, there are enough processors within each group to saturate the available I/O bandwidth, and the G groups checkpoint sequentially. Hence the total checkpoint time without message logging, namely $GC_0(q)$, is equal to that of the coordinated approach. This leads to the simple equation

$$C_0(q) = \frac{C_{\text{Mem}}}{G} = \frac{\text{Mem}}{Gb_{io}} \tag{1.32}$$

where Mem denotes the memory footprint of the application, and b_{io} the available I/O bandwidth. Similarly as before, we use $R(q)$ for the recovery (either equal to $C(q)$ or not), and a constant value $D(q) = D$ for the downtime.

HIERARCH- PORT —The third scenario investigates the possibility of having a large number of very small groups, a strategy proposed to take advantage of hardware proximity and failure probability correlations [15]. However, if groups are reduced to a single processor, a single checkpointing group is not sufficient to saturate the available I/O bandwidth. In this strategy, multiple groups of q processors are allowed to checkpoint simultaneously in order to saturate the I/O bandwidth. We define q_{min} as the smallest value such that $q_{\text{min}} b_{port} \geq b_{io}$, where b_{port} is the network bandwidth of a single processor. In other words, q_{min} is the minimal size of groups so that Eq. (1.32) holds.

Small groups typically imply logging more messages (hence a larger growth factor of the checkpoint per work unit β, and possibly a larger impact on computation slowdown λ). For an application executing on a 2D-grid of processors, twice as many communications will be logged (assuming a symmetrical communication pattern along each grid direction). However, let us compare recovery times in the HIERARCH-PORT and HIERARCH- IO strategies; assume that $R_0(q) = C_0(q)$ for simplicity. In both cases Eq. (1.32) holds, but the number of groups is significantly larger for HIERARCH- PORT, thereby ensuring a much shorter recovery time.

Application examples: We study the increase in checkpoint size due to message logging by detailing three application examples that are typical scientific applications executing on 2D-or 3D-processor grids, but this exhibits a different checkpoint increase rate parameter β.

2D- STENCIL– We first consider a 2D-stencil computation: a real matrix of size $n \times n$ is partitioned across a $p \times p$ processor grid, where $p^2 = N$. At each iteration, each element is averaged with its 8 closest neighbors, requiring rows and columns that lie at the boundary of the partition to be exchanged (it is easy to generalize to larger update masks). Each processor holds a matrix block of size $b = n/p$, and sends four messages of size b (one in each grid direction). Then each element is updated, at the cost of 9 double floating-point operations. The (parallel) work for one iteration is thus WORK $= \frac{9b^2}{s_p}$, where s_p is the speed of one processor.

Here Mem $= 8n^2$ (in bytes), since there is a single (double real) matrix to store. As already mentioned, a natural (application-aware) group partition is with one group per row (or column) of the grid, which leads to $G = q = p$. Such large groups correspond to the HIERARCH- IO scenario, with $C_0(q) = \frac{C_{\text{Mem}}}{G}$. At each iteration, vertical (intergroup) communications are logged, but horizontal (intragroup) communications are not logged. The size of logged messages is thus $2pb = 2n$ for each group. If we checkpoint after each iteration, $C(q) - C_0(q) = \frac{2n}{b_{io}}$, and we derive from Eq. (1.28) that $\beta = \frac{2nps_p}{n^2 9b^2} = \frac{2s_p}{9b^3}$. We stress that the value of β is unchanged if groups checkpoint every k iterations, because both $C(q) - C_0(q)$ and WORK are multiplied by a factor k. Finally, if we use small groups of size q_{min}, we have the HIERARCH- PORT scenario. We still have $C_0(q) = \frac{C_{\text{Mem}}}{G}$, but now the value of β has doubled since we log twice as many communications.

MATRIX- PRODUCT —Consider now a typical linear-algebra kernel involving matrix products. For each matrix-product, there are three matrices involved, so Mem $= 24n^2$ (in bytes). The matrix partition is similar to previous scenario, but now each processor holds three matrix blocks of size $b = n/p$. Consider Cannon's algorithm [18] which has p steps to compute a product. At each step, each processor shifts one block vertically and one block horizontally, and WORK $= \frac{2b^3}{s_p}$. In the HIERARCH- IO scenario with one group per grid row, only vertical messages are logged: $\beta = \frac{s_p}{6b^3}$. Again, β is unchanged if groups checkpoint every k steps, or every matrix product ($k = p$). In the COORD- PORT scenario with groups of size q_{min}, the value of β is doubled.

3D- STENCIL —This application is similar to 2D- STENCIL, but with a 3D matrix of size n partitioned across a 3D-grid of size p, where $8n^3 = $ Mem and $p^3 = N$. Each processor holds a cube of size $b = n/p$. At each iteration, each pixel is averaged with its 26 closest neighbors, and WORK $= \frac{27b^3}{s_p}$. Each processor sends the six faces of its cube, one in each direction. In addition to COORD- IO, there are now three hierarchical

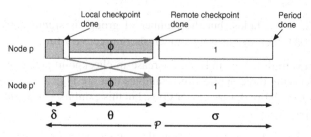

Fig. 1.11 Double checkpoint algorithm

scenarios: (A) HIERARCH- IO- PLANE where groups are horizontal planes, of size p^2. Only vertical communications are logged, which represents two faces per processor: $\beta = \frac{2s_p}{27b^3}$; (B) HIERARCH- IO- LINE where groups are lines, of size p. Twice as many communications are logged, which represents four faces per processor: $\beta = \frac{4s_p}{27b^3}$; (C) HIERARCH- PORT (groups of size q_{min}). All communications are logged, which represents six faces per processor: $\beta = \frac{6s_p}{27b^3}$. The order of magnitude of b is the cubic root of the memory per processor for 3D- STENCIL, while it was its square root for 2D- STENCIL and MATRIX- PRODUCT, so β will be larger for 3D- STENCIL.

Wrap-up. We have shown how to instantiate all the resilience parameters of the model. Now, to assess the performance of a given scenario for hierarchical checkpointing, there only remain to instantiate the platform parameters: individual MTBF μ_{ind}, number of nodes N (from which we deduce the platform MTBF μ), number of cores per node, speed of each core s_p, memory per node, fraction of that memory used for the application memory footprint Mem, I/O network and node bandwidths b_{io} and b_{port}. Then we can use the model to predict the waste when varying the number of groups and the assumptions on checkpoint time. The interested reader will find several examples in [10].

1.3.4 In-Memory Checkpointing

In this section, we briefly survey a recent protocol that has been designed to reduce the time needed to checkpoint an application. The approach to reduce checkpoint time is to avoid using any kind of stable, but slow-to-access, storage. Rather than using a remote disk system, *in-memory checkpointing* uses the main memory of the processors. This will provide faster access and greater scalability, at the price of the risk of a fatal failure in some (unlikely) scenarios.

Figure 1.11 depicts the double checkpoint algorithm of [59, 71]. Processors are arranged into pairs. Within a pair, checkpoints are replicated: each processor stores its own checkpoint and that of its *buddy* in its local memory. We use the notations of [59, 71] in Fig. 1.11, which shows the following:

• The execution is divided into periods of length $\mathscr{P}$

- At the beginning of the period, each node writes its own checkpoint in its local memory, which takes a time δ. This writing is done in blocking mode, and the execution is stopped.
- Then each node send its checkpoint to its buddy. This exchange takes a time θ. The exchange is non-blocking, and the execution can progress, albeit with a slowdown factor Φ
- During the rest of the period, for a time σ, the execution progresses at full (unit) speed

The idea of the non-blocking exchange is to use those time-steps where the application is not performing inter-processor communications to send/receive the checkpoint files, thereby reducing the overhead incurred by the application.

Let us see what happens when a failure strikes one processor, as illustrated in Fig. 1.12a. Node p is hit by a failure, and a spare node will take over. After a downtime D, the spare node starts by recovering the checkpoint file of node p, in time R. The spare receives this file from node p', the buddy of node p, most likely as fast as possible (in blocking mode) so that it can resume working. Then the spare receives the checkpoint file of node p', to ensure that the application is protected if a failure hits p' later on. As before, receiving the checkpoint file can be overlapped with the execution and takes a time Θ, but there is a trade-off to make now: as shown in Fig. 1.12b, the application is at risk until both checkpoint receptions are completed. If a failure strikes p' before that, then it is a critical failure that cannot be recovered from. Hence it might be a good idea to receive the second checkpoint (that of p') as fast as possible too, at the price of a performance degradation of the whole application: when one processor is blocked, the whole application cannot progress. A detailed

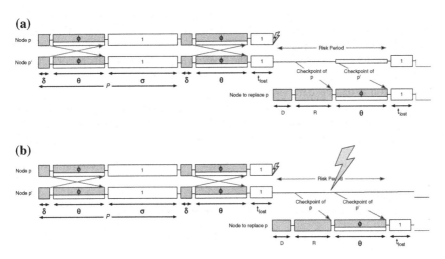

Fig. 1.12 Handling failures in the double checkpoint algorithm. **a** A failure hits node p. **b** A second failure hits node p', the buddy of node p, before the spare node had finished to receive the checkpoint file of p'. This is a fatal failure for the application

analysis is available in [28], together with extensions to a triple-checkpoint algorithm where each node has two buddies instead of one, thereby dramatically decreasing the risk of a fatal failure.

Finally, we mention that the risk of a fatal failure can be eliminated when using a *multi-level* checkpointing protocol, such as FTI. [5] or SCR. [57]. Such protocols allow to set different levels/types of checkpoints during the execution. Different checkpoint levels correspond to different recovery abilities, and also suffer from different checkpoint/recovery overheads. See [5, 57] for further details.

1.4 Probabilistic Models for Advanced Methods

In this section, we present two extensions of checkpointing performance models. Section 1.4.1 explains how to combine checkpointing with *fault prediction*, and discuss how the optimal period is modified when this combination is used. Section 1.4.2 explains how to combine checkpointing with *replication*, and discuss how the optimal period is modified when this combination is used.

1.4.1 Fault Prediction

A possible way to cope with the numerous faults and their impact on the execution time is to try and predict them. In this section we do not explain how this is done, although the interested reader will find some answers in Chap. 2 and in [35, 70, 73].

A *fault predictor* (or simply a predictor) is a mechanism that warns the user about upcoming faults on the platform. More specifically, a predictor is characterized by two key parameters, its recall r, which is the fraction of faults that are indeed predicted, and its precision p, which is the fraction of predictions that are correct (i.e., correspond to actual faults). In this section, we discuss how to combine checkpointing and prediction to decrease the platform waste.

We start with a few definitions. Let μ_P be the mean time between predicted events (both true positive and false positive), and μ_{NP} be the mean time between unpredicted faults (false negative). The relations between μ_P, μ_{NP}, μ, r and p are as follows:

- Rate of unpredicted faults: $\frac{1}{\mu_{NP}} = \frac{1-r}{\mu}$, since $1-r$ is the fraction of faults that are unpredicted;
- Rate of predicted faults: $\frac{r}{\mu} = \frac{p}{\mu_P}$, since r is the fraction of faults that are predicted, and p is the fraction of fault predictions that are correct.

To illustrate all these definitions, consider the time interval below and the different events occurring:

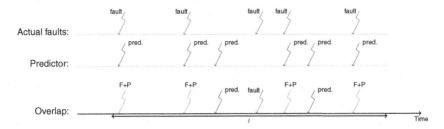

During this time interval of length t, the predictor predicts six faults, and there were five actual faults. One fault was not predicted. This gives approximately: $\mu = \frac{t}{5}$, $\mu_P = \frac{t}{6}$, and $\mu_{NP} = t$. For this predictor, the recall is $r = \frac{4}{5}$ (green arrows over red arrows), and its precision is $p = \frac{4}{6}$ (green arrows over blue arrows).

Now, given a fault predictor of parameters p and r, can we improve the waste? More specifically, how to modify the periodic checkpointing algorithm to get better results? In order to answer these questions, we introduce *proactive checkpointing*: when there is a prediction, we assume that the prediction is given early enough so that we have time for a checkpoint of size C_p (which can be different from C). We consider the following simple algorithm:

- While no fault prediction is available, checkpoints are taken periodically with period T;
- When a fault is predicted, we take a proactive checkpoint (of length C_p) as late as possible, so that it completes right at the time when the fault is predicted to strike. After this checkpoint, we complete the execution of the period (see Fig. 1.13b, c);

We compute the expected waste as before. We reproduce Eq. (1.7) below:

$$\text{WASTE} = \text{WASTE}_{\text{EFF}} + \text{WASTE}_{\text{fault}} - \text{WASTE}_{\text{EFF}}\text{WASTE}_{\text{fault}} \qquad (1.33)$$

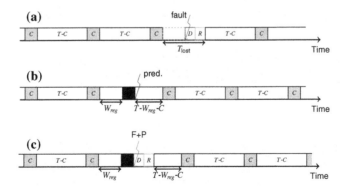

Fig. 1.13 Actions taken for the different event types. **a** Unpredicted fault, **b** Prediction taken into account—no actual fault, **c** Prediction taken into account—with actual fault

While the value of WASTE$_{FF}$ is unchanged (WASTE$_{FF} = \frac{C}{T}$), the value of WASTE$_{fault}$ is modified because of predictions. As illustrated in Fig. 1.13, there are different scenarios that contribute to WASTE$_{fault}$. We classify them as follows:

(1) Unpredicted faults: This overhead occurs each time an unpredicted fault strikes, that is, on average, once every μ_{NP} seconds. Just as in Eq. (1.6), the corresponding waste is $\frac{1}{\mu_{NP}} \left[\frac{T}{2} + D + R \right]$.

(2) Predictions: We now compute the overhead due to a prediction. If the prediction is an actual fault (with probability p), we lose $C_p + D + R$ seconds, but if it is not (with probability $1 - p$), we lose the unnecessary extra checkpoint time C_p. Hence

$$T_{lost} = p(C_p + D + R) + (1 - p)C_p = C_p + p(D + R)$$

We derive the final value of WASTE$_{fault}$:

$$
\begin{aligned}
\text{WASTE}_{fault} &= \frac{1}{\mu_{NP}} \left(\frac{T}{2} + D + R \right) + \frac{1}{\mu_P} \left(C_p + p(D + R) \right) \\
&= \frac{1 - r}{\mu} \left(\frac{T}{2} + D + R \right) + \frac{r}{p\mu} \left(C_p + p(D + R) \right) \\
&= \frac{1}{\mu} \left((1 - r)\frac{T}{2} + D + R + \frac{rC_p}{p} \right)
\end{aligned}
$$

We can now plug this expression back into Eq. (1.33):

$$
\begin{aligned}
\text{WASTE} &= \text{WASTE}_{FF} + \text{WASTE}_{fault} - \text{WASTE}_{FF}\text{WASTE}_{fault} \\
&= \frac{C}{T} + \left(1 - \frac{C}{T} \right) \frac{1}{\mu} \left(D + R + \frac{rC_p}{p} + \frac{(1 - r)T}{2} \right).
\end{aligned}
$$

To compute the value of T_{FO}^p, the period that minimizes the total waste, we use the same reasoning as in Sect. 1.3.1 and obtain:

$$T_{FO}^p = \sqrt{ \frac{ 2 \left(\mu - \left(D + R + \frac{rC_p}{p} \right) \right) C }{ 1 - r } }.$$

We observe the similarity of this result with the value of T_{FO} from Eq. (1.9). If μ is large in front of the resilience parameters, we derive that $T_{FO}^p = \sqrt{\frac{2\mu C}{1 - r}}$. This tells us that the recall is more important than the precision. If the predictor is capable of predicting, say, 84 % of the faults, then $r = 0.84$ and $\sqrt{1 - r} = 0.4$. The optimal period is increased by 40 %, and the waste is decreased by the same factor. Prediction can help!

Going further. The discussion above has been kept overly simple. For instance when a fault is predicted, sometimes there is not enough time to take proactive actions, because we are already checkpointing. In this case, there is no other choice than ignoring the prediction.

Furthermore, a better strategy should take into account at what point in the period does the prediction occur. After all, there is no reason to always trust the predictor, in particular if it has a bad precision. Intuitively, the later the prediction takes place in the period, the more likely we are inclined to trust the predictor and take proactive actions. This is because the amount of work that we could lose gets larger as we progress within the period. On the contrary, if the prediction happens in the beginning of the period, we have to trade-off the possibility that the proactive checkpoint may be useless (if we indeed take a proactive action) with the small amount of work that may be lost in the case where a fault would actually happen. The optimal approach is to never trust the predictor in the beginning of a period, and to always trust it in the end; the crossover point $\frac{C_p}{p}$ depends on the time to take a proactive checkpoint and on the precision of the predictor. See [4] for details.

Finally, it is more realistic to assume that the predictor cannot give the exact moment where the fault is going to strike, but rather will provide an interval of time for that event, a.k.a. a prediction window. More information can be found in [2].

1.4.2 Replication

Another possible way to cope with the numerous faults and their impact on the execution time is to use replication. Replication consists in duplicating all computations. Processors are grouped by pairs, such as each processor has a *replica* (another processor performing exactly the same computations, receiving the same messages, etc.). See Fig. 1.14 for an illustration. We say that the two processes in a given pair

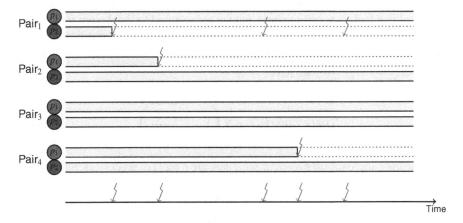

Fig. 1.14 Processor pairs for replication: each *blue* processor is paired with a *red* processor. In each pair, both processors do the same work

are *replicas*. When a processor is hit by a fault, its replica is not impacted. The execution of the application can still progress, until the replica itself is hit by a fault later on. This sounds quite expensive: by definition, half of the resources are wasted (and this does not include the overhead of maintaining a consistent state between the two processors of each pair). At first sight, the idea of using replication on a large parallel platform is puzzling: who is ready to waste half of these expensive supercomputers?

In this section, we explain how replication can be used in conjunction with checkpointing and under which conditions it becomes profitable. In order to do this, we compare the checkpointing technique introduced earlier to the replication technique.

A *perfectly parallel application* is an application such that in a failure-free, checkpoint-free environment, the time to execute the application ($\text{TIME}_{\text{Base}}$) decreases linearly with the number of processors. More precisely:

$$\text{TIME}_{\text{base}}(N) = \frac{\text{TIME}_{\text{base}}(1)}{N}.$$

Consider the execution of a perfectly parallel application on a platform with $N = 2n$ processors, each with individual MTBF μ_{ind}. As in the previous sections, the optimization problem is to find the strategy minimizing $\text{TIME}_{\text{final}}$. Because we compare two approaches using a different number of processors, we introduce the THROUGHPUT, which is defined as the total number of useful flops per second:

$$\text{THROUGHPUT} = \frac{\text{TIME}_{\text{base}}(1)}{\text{TIME}_{\text{final}}}$$

Note that for an application executing on N processors,

$$\text{THROUGHPUT} = N\left(1 - \text{WASTE}\right)$$

The *standard* approach, as seen before, is to use all $2n$ processors so the execution of the application benefits from the maximal parallelism of the platform. This would be optimal in a fault-free environment, but we are required to checkpoint frequently because faults repeatedly strike the N processors. According to Proposition 1.2, the platform MTBF is $\mu = \frac{\mu_{\text{ind}}}{N}$. According to Theorem 1.1, the waste is (approximately) $\text{WASTE} = \sqrt{\frac{2C}{\mu}} = \sqrt{\frac{2CN}{\mu_{\text{ind}}}}$. We have:

$$\text{THROUGHPUT}_{\text{Std}} = N\left(1 - \sqrt{\frac{2CN}{\mu_{\text{ind}}}}\right) \tag{1.34}$$

The second approach uses *replication*. There are n pairs of processors, all computations are executed twice, hence only half the processors produce useful flops. One way to see the replication technique is as if there were half the processors using only the checkpoint technique, with a different (potentially higher) mean time between faults, μ_{rep}. Hence, the throughput $\text{THROUGHPUT}_{\text{Rep}}$ of this approach writes:

$$\text{THROUGHPUT}_{\text{Rep}} = \frac{N}{2}\left(1 - \sqrt{\frac{2C}{\mu_{\text{rep}}}}\right) \tag{1.35}$$

In fact, rather than MTBF, we should say MTTI, for *Mean Time To Interruption*. As already mentioned, a single fault on the platform does not interrupt the application, because the replica of the faulty processor is still alive. What is the value of *MNFTI*, the *Mean Number of Faults To Interruption*, i.e., the mean number of faults that should strike the platform until there is a replica pair whose processors have both been hit? If we find how to compute *MNFTI*, we are done, because we know that

$$\mu_{\text{rep}} = MNFTI \times \mu = MNFTI \times \frac{\mu_{\text{ind}}}{N}$$

We make an analogy with a balls-into-bins problem to compute *MNFTI*. The classical problem is the following: what is the expected number of balls that you will need, if you throw these balls randomly into n bins, until one bins gets two balls? The answer to this question is given by Ramanujans Q-Function [34], and is equal to $\lceil q(n) \rceil$ where $q(n) = \frac{2}{3} + \sqrt{\frac{\pi n}{2}} + \sqrt{\frac{\pi}{288n}} - \frac{4}{135n} + \dots$. When $n = 365$, this is the birthday problem where balls are persons and bins are calendar dates; in the best case, one needs two persons; in the worst case, one needs $n + 1 = 366$ persons; on average, one needs $\lceil q(n) \rceil = 25$ persons.[3]

In the replication problem, the bins are the processor pairs, and the balls are the faults. However, the analogy stops here. The problem is more complicated, see Fig. 1.15 to see why. Each processor pair is composed of a blue processor and of a red processor. Faults are (randomly) colored blue or red too. When a fault strikes a processor pair, we need to know which processor inside that pair: we decide that it is the one of the same color as the fault. Blue faults strike blue processors, and red faults strike red processors. We now understand that we may need more than two faults hitting the same pair to interrupt the application: we need one fault of each color. The balls-and-bins problem to compute *MNFTI* is now clear: what is the expected number of red and blue balls that you will need, if you throw these balls randomly into n bins, until one bins gets one red ball and one blue ball? To the best of our knowledge, there is no closed-form solution to answer this question, but a recursive computation does the job:

Proposition 1.3 *MNFTI* $= \mathbb{E}(NFTI|0)$ *where*

$$\mathbb{E}(NFTI|n_f) = \begin{cases} 2 & \text{if } n_f = N, \\ \frac{2N}{2N-n_f} + \frac{2N-2n_f}{2N-n_f}\mathbb{E}\left(NFTI|n_f + 1\right) & \text{otherwise.} \end{cases}$$

Proof Let $\mathbb{E}(NFTI|n_f)$ be the expectation of the number of faults needed to interrupt the application, knowing that the application is still running and that faults have

---

[3]As a side note, one needs only 23 persons for the probability of a common birthday to reach 0.5 (a question often asked in geek evenings).

Pair$_1$ Pair$_2$ Pair$_3$ Pair$_4$

Fig. 1.15 Modeling the state of the platform of Fig. 1.14 as a balls-into-bins problem. We put a *red* ball in bin Pair$_i$ when there is a fault on its *red* processor p_1, and a *blue* ball when there is a fault on its *blue* processor p_2. As long as no bin has received a ball of each *color*, the game is on

already hit n_f different processor pairs. Because each pair initially has 2 replicas, this means that n_f different pairs are no longer replicated, and that $N - n_f$ are still replicated. Overall, there are $n_f + 2(N - n_f) = 2N - n_f$ processors still running.

The case $n_f = N$ is simple. In this case, all pairs have already been hit, and all pairs have only one of their two initial replicas still running. A new fault will hit such a pair. Two cases are then possible:

1. The fault hits the running processor. This leads to an application interruption, and in this case $\mathbb{E}(NFTI|N) = 1$.
2. The fault hits the processor that has already been hit. Then the fault has no impact on the application. The *MNFTI* of this case is then: $\mathbb{E}(NFTI|N) = 1 + \mathbb{E}(NFTI|N)$.

The probability of fault is uniformly distributed between the two replicas, and thus between these two cases. Weighting the values by their probabilities of occurrence yields:

$$\mathbb{E}(NFTI|N) = \frac{1}{2} \times 1 + \frac{1}{2} \times (1 + \mathbb{E}(NFTI|N)),$$

hence $\mathbb{E}(NFTI|N) = 2$.

For the general case $0 \leq n_f \leq N - 1$, either the next fault hits a new pair, i.e., a pair whose 2 processors are still running, or it hits a pair that has already been hit, hence with a single processor running. The latter case leads to the same sub-cases as the $n_f = N$ case studied above. The fault probability is uniformly distributed among the $2N$ processors, including the ones already hit. Hence the probability that the next fault hits a new pair is $\frac{2N - 2n_f}{2N}$. In this case, the expected number of faults needed to interrupt the application fail is one (the considered fault) plus $\mathbb{E}(NFTI|n_f + 1)$. Altogether we have:

$$\mathbb{E}(NFTI|n_f) = \frac{2N - 2n_f}{2N} \times (1 + \mathbb{E}(NFTI|n_f + 1))$$
$$+ \frac{2n_f}{2N} \times \left(\frac{1}{2} \times 1 + \frac{1}{2} \left(1 + \mathbb{E}(NFTI|n_f) \right) \right).$$

Therefore,

$$\mathbb{E}\left(NFTI|n_f\right) = \frac{2N}{2N - n_f} + \frac{2N - 2n_f}{2N - n_f}\mathbb{E}\left(NFTI|n_f + 1\right).$$

Let us compare the throughput of each approach with an example. From Eqs. (1.34) and (1.35), we have

$$\text{THROUGHPUT}_{\text{Rep}} \geq \text{THROUGHPUT}_{\text{Std}} \Leftrightarrow (1 - \sqrt{\frac{2CN}{MNFTI\,\mu_{\text{ind}}}}) \geq 2(1 - \sqrt{\frac{2CN}{\mu_{\text{ind}}}})$$

which we rewrite into

$$C \geq \frac{\mu_{\text{ind}}}{2N} \frac{1}{(2 - \frac{1}{\sqrt{MNFTI}})^2} \tag{1.36}$$

Take a parallel machine with $N = 2^{20}$ processors. This is a little more than one million processors, but this corresponds to the size of the largest platforms today. Using Proposition 1.3, we compute $MNFTI = 1284.4$ Assume that the individual MTBF is 10 years, or in seconds $\mu_{\text{ind}} = 10 \times 365 \times 24 \times 3600$. After some painful computations, we derive that replication is more efficient if the checkpoint time is greater than 293 seconds (around 6 minutes). This sets a target both for architects and checkpoint protocol designers.

Maybe you would say say that $\mu_{\text{ind}} = 10$ years is pessimistic, because we rather observe that $\mu_{\text{ind}} = 100$ years in current supercomputers. Since $\mu_{\text{ind}} = 100$ years allows us to use a checkpointing period of one hour, you might then decide that replication is not worth it. On the contrary, maybe you would say that $\mu_{\text{ind}} = 10$ years is optimistic for processors equipped with thousands of cores and rather take $\mu_{\text{ind}} = 1$ year. In that case, unless you checkpoint in less than 30 s, better be prepared for replication. The beauty of performance models is that you can decide which approach is better *without bias nor a priori*, simply by plugging your own parameters into Eq. (1.36).

Going further. There are two natural options "counting" faults. The option chosen above is to allow new faults to hit processors that have already been hit. This is the option chosen in [33], who introduced the problem. Another option is to count only faults that hit *running processors*, and thus effectively kill replica pairs and interrupt the application. This second option may seem more natural as the running processors are the only ones that are important for executing the application. It turns out that both options are almost equivalent, the values of their *MNFTI* only differ by one [19].

We refer the interested reader to Chap. 4 for a full analysis of replication. For convenience, we provide a few bibliographical notes in the following lines. Replication has long been used as a fault tolerance mechanism in distributed systems [38], and in the context of volunteer computing [51]. Replication has recently received attention in the context of HPC (High Performance Computing) applications [31, 33, 66, 72]. While replicating all processors is very expensive, replicating only critical processes, or only a fraction of all processes, is a direction being currently explored under the name *partial replication*.

Speaking of critical processes, we make a final digression. The de-facto standard to enforce fault tolerance in critical or embedded systems is *Triple Modular Redundancy* and voting, or TMR [56]. Computations are triplicated on three different processors, and if their results differ, a voting mechanism is called. TMR is not used to protect from fail-stop faults, but rather to detect and correct errors in the execution of the application. While we all like, say, safe planes protected by TMR, the cost is tremendous: by definition, two thirds of the resources are wasted (and this does not include the overhead of voting when an error is identified).

1.5 Application-Specific Fault Tolerance Techniques

All the techniques presented and evaluated so far are *general* techniques: the assumptions they make on the behavior of the application are as little constraining as possible, and the protocol to tolerate failures considered two adversaries: the occurrence of failures, which can happen at the worst possible time, and also the application itself, which can take the worst possible action at the worst possible moment.

We now examine the case of application-specific fault tolerance techniques in HPC: when the application itself may use redundant information inherent of its coding of the problem, to tolerate misbehavior of the supporting platform. As one can expect, the efficiency of such approaches can be orders of magnitude better than the efficiency of general techniques; their programming, however, becomes a much harder challenge for the final user.

First, the application must be programmed over a middleware that not only tolerates failures for its internal operation, but also exposes them in a manageable way to the application; then, the application must maintain redundant information exploitable in case of failures during its execution. We will present a couple of cases of such applicative scenarios. Finally, we will discuss the portability of such approaches, and present a technique that allows the utilization of application-specific fault tolerance technique inside a more general application, preserving the fault tolerance property while exhibiting performance close to the one expected from application-specific techniques.

1.5.1 Fault-Tolerant Middleware

The first issue to address, to consider application-specific fault tolerance, is how to allow failures to be presented to the application. Even in the case of fail-stop errors, that can be detected easily under the assumption of pseudo-synchronous systems usually made in HPC, the most popular programming middleware, MPI, does not allow to expose failures in a portable way.

The MPI-3 specification has little to say about failures and their exposition to the user:

It is the job of the implementor of the MPI subsystem to insulate the user from this unreliability, or to reflect unrecoverable errors as failures. Whenever possible, such failures will be reflected as errors in the relevant communication call. Similarly, MPI itself provides no mechanisms for handling processor failures.

MPI Standard, v3.0, p. 20, l. 36:39

This fist paragraph would allow implementations to expose the failures, limiting their propagation to the calls that relate to operations that cannot complete because of the occurrence of failures. However, later in the same standard:

This document does not specify the state of a computation after an erroneous MPI call has occurred.

MPI Standard v3.0, p. 21, l. 24:25

Unfortunately, most Open Source MPI implementations, and the numerous vendor-specific MPI implementations that derive from them, chose, by lack of demand from their users, and by lack of consensus, to interpret these paragraphs in a way that limits the opportunities for the user to tolerate failures: in the worst case, even if all communicators hit by failures are marked to return in case of error, the application is simply shutdown by the runtime system, as a cleanup procedure; in the best case, the control is given back to the user program, but no MPI call that involves a remote peer is guaranteed to perform any meaningful action for the user, leaving the processes of the application as separate entities that have to rely on external communication systems to tolerate failures.

The Fault Tolerance Working Group of the MPI Forum has been constituted to address this issue. With the dawn of extreme scale computing, at levels where failures become expected occurrences in the life of an application, MPI has set a cap to evolve towards more scalability. Capacity for the MPI implementation to continue its service in case of failures, and capacity for the MPI language to present these failures to the application, or to the software components that wish to handle these failures directly, are key among the milestones to remove technological locks towards scalability. Chapter 3 details the User-Level Failures Mitigation (ULFM) proposal of the FTWG of the MPI Forum in its Sect. 3.8. We present here its main features, as an introduction.

There are two main issues to address to allow applications written in MPI to tolerate failures:

- Detect and report failures
- Provide service after the occurrence of failures

ULFM exposes failures to the application through MPI *exceptions*. It introduces a couple of error classes that are returned by pertaining MPI calls if a failure strikes, and prevents their completion (be it because the failure happened before or during the call). As per traditional MPI specification, exceptions are raised only if the user defined a specific error handler for the corresponding communicator, or if it specified to use the predefined error handler that makes exceptions return an error code.

In those cases, the ULFM proposal states that no MPI call should block indefinitely because of the occurrence of failures. Collective calls must return either a success code if they did complete despite the failure, or an error code if their completion was compromised; point to point operations must also return. This raises two issues:

- the same collective call may return with success or fail, depending on the rank. For example, a broadcast operation is often implemented using a broadcast tree to provide logarithmic overheads. If a node low in the broadcast tree is subject to a failure, the root of the tree may not notice the failure and succeed completing all its local operations, while trees under the failed node will not receive the information. In all cases, all processes must enter the broadcast operation, as the meaning of collective is not changed, and all processes must leave the operation, as none could stall forever because of a failure. Nodes under the failed process may raise an exception, while nodes above it may not notice the failure during this call.
- in the case of point to point operations, it may become hard for the implementation to decide whether an operation will complete or not. Take the example of a receive from any source operation: any process in the communicator may be the sender that would, in a failure-free execution, send the message that would match this reception. As a consequence, if a single process failed, the MPI implementation cannot safely decide (unless it finds incoming messages to match the reception) if the reception is going to complete or not. Since the specification does not allow for a process to stall forever because of the occurrence of failures, the implementation should raise an exception. However, the reception operation cannot be marked as failed, since it is possible that the matching send comes later from a different process. The specification thus allows the implementation to delay the notification for as long as seems fit, but for a bounded time, after which the reception must return with a special exception that marks the communication as undecided, thus giving back the control to the application to decide if that message is going to come or not.

To take such decisions, the application has access to a few additional routines. The application can acknowledge the presence of failures in a communicator (using MPI_Comm_failure_ack, and resume its operation over the same communicator that holds failed processes. Over such a communicator, any operation that involves a failed process will fail. Thus, collective operations that involve all processes in the communicator will necessarily fail. Point to point communications, on the other hand, may succeed if they are not a specific emission to a failed process or reception from a failed process. Receptions from any source will succeed and wait for a matching message, as the user already acknowledged the presence of some failures. If the user wanted to cancel such a reception, she can decide by requesting the MPI implementation to provide the list of failed processes after an acknowledgment (via MPI_Comm_get_acked). If more processes fail after the acknowledgment, more exceptions will be raised and can be acknowledged. Point to point communications will thus continue to work after a failure, as long as they do not directly involve an acknowledged failed process.

The application may also need to fix the communicator, in order to allow for collective operations to succeed. In order to clearly separate communications that happened before or after a set of failures, ULFM does not provide a way to fix the communicator. Instead, it provides a routine that exclude the failed processes from a communicator and creates a new one, suitable for the whole range of MPI routines (the routine `MPI_Comm_shrink`). This communicator creation routine is specified to work despite the occurrence of failures. The communicator that it creates must exclude failures that were acknowledged before entering the routine, but since failures may happen at any time, the newly created communicator may itself include failed processes, for example if a failure happened just after its creation.

The last routine provided by the ULFM proposal is a routine to allow resolution of conflicts after a failure. `MPI_Comm_agree` provides a consensus routine over the surviving ranks of a communicator. It is critical to determine an agreement in the presence of failures, since collective operations have no guarantee of consistent return values if a failure happens during their execution. Its usage is documented more closely in Chap. 3, as it interacts with `MPI_Comm_failure_ack` to enable the user to construct a low cost group membership service, that provides a global view of processes that survived a set of failures.

The leading idea of ULFM was to complement the MPI specification with a small set of routines, and extended specification for the existing routines, in case of process failures, enabling the user application or library to notice failures, react and continue the execution of the application despite the occurrence of these failures. The specification targets a lean set of changes, not promoting any specific model to tolerate failures, but providing the minimal building blocks to implement, through composition of libraries or directly in the application, a large spectrum of application-specific fault tolerance approaches. In the following, we discuss a few typical cases that were implemented over this ULFM proposal.

1.5.2 ABFT for Dense Matrix Factorization

Algorithm-Based Fault Tolerance (ABFT) was introduced by Abraham and Huang in 1984 [45] to tolerate possible memory corruptions during the computation of a dense matrix factorization. It is a good example of application-specific fault tolerance technique that is not simplistic, but provides an extreme boost in performance when used (compared to a general technique, like rollback-recovery). ABFT and disk-less checkpointing have been combined to apply to basic matrix operations like matrix-matrix multiplication [8, 22, 23] and have been implemented on algorithms similar to those of ScaLAPACK [24], which is widely used for dense matrix operations on parallel distributed memory systems, or the High Performance Linpack (HPL) [26] and to the Cholesky factorization [40].

An ABFT scheme for dense matrix factorization was introduced in [16, 29], and we explain it here, because it combines many application-level techniques, including replication, user-level partial checkpointing, and ABFT itself. We illustrate this technique with the LU factorization algorithm, which is the most complex due to its pivoting, but the approach applies to other direct methods of factorization.

Section 3.7.3 of Chap. 3 presents a similar algorithm for the QR factorization in the context of a less supportive communication middleware.

To support fail-stop errors, an ABFT scheme must be built on top of a fault-aware middleware. We assume a failure, defined in this section as a process that *completely* and *definitely* stops responding, triggering the loss of a critical part of the global application state, could occur at any moment and can affect any part of the application's data.

Algorithm Based Fault Tolerance. The general idea of ABFT is to introduce information redundancy in the data, and maintain this redundancy during the computation. Linear algebra operations over matrices are well suited to apply such a scheme: the matrix (original data of the user) can be extended by a number of columns, in which checksums over the rows are stored. The operation applied over the initial matrix can then be extended to apply at the same time over the initial matrix and its extended columns, maintaining the checksum relation between data in a row and the corresponding checksum column(s). Usually, it is sufficient to extend the scope of the operation to the checksum rows, although in some cases the operation must be redefined.

If a failure hits processes during the computation, the data host by these processes is lost. However, in theory, the checksum relation being preserved, if enough information survived the failure between the initial data held by the surviving processes and the checksum columns, a simple inversion of the checksum function is sufficient to reconstruct the missing data and pursue the operation.

No periodical checkpoint is necessary, and more importantly the recovery procedure brings back the missing data at the point of failure, without introducing a period of re-execution as the general techniques seen above impose, and a computational cost that is usually linear with the size of the data. Thus, the overheads due to ABFT are expected to be significantly lower than those due to rollback-recovery.

LU Factorization: The goal of a factorization operation is usually to transform a matrix that represents a set of equations into a form suitable to solve the problem $Ax = b$, where A and b represent the equations, A being a matrix and b a vector of same height. Different transformations are considered depending on the properties of the matrix, and the LU factorization transforms $A = LU$ where L is a lower triangular matrix, and U an upper triangular matrix. This transformation is done by blocks of fixed size inside the matrix to improve the efficiency of the computational kernels. Figure 1.16 represents the basic operations applied to a matrix during a block LU factorization. The GETF2 operation is a panel factorization, applied on a block column. This panel operation factorizes the upper square, and scales the lower rectangle accordingly. The output of that operation is then used to the right of the factored block to scale it accordingly using a triangular solve (TRSM), and the trailing matrix is updated accordingly using a matrix-matrix multiplication (GEMM). The block column and the block row are in their final LU form, and that trailing matrix must be transformed using the same algorithm, until the last block of the matrix is in the LU form. Technically, each of these basic steps is usually performed by applying a parallel Basic Linear Algebra Subroutine (PBLAS).

Fig. 1.16 Operations applied on a matrix, during the LU factorization. A' is the trailing matrix, that needs to be factorized using the same method until the entire initial matrix is in the form LU

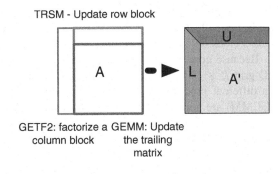

TRSM - Update row block

GETF2: factorize a column block   GEMM: Update the trailing matrix

Fig. 1.17 Block cyclic distribution of a $8m_b \times 8n_b$ matrix over a 2×3 process grid

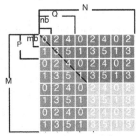

Data Distribution. For a parallel execution, the data of the matrix must be distributed among the different processors. For dense matrix factorization, the data is distributed following a 2D block cyclic distribution: processes are arranged over a 2D cyclic processor grid of size $P \times Q$, the matrix is split in blocks of size $m_b \times n_b$, and the blocks are distributed among the processes cyclically. Figure 1.17 shows how the blocks are distributed in a case of a square matrix of size $8m_b \times 8n_b$, and a process grid of size 2×3.

Reverse Neighboring Scheme: If one of the processes is subject of failure, many blocks are lost. As explained previously, the matrix is extended with checksum columns to introduce information redundancy. Figure 1.18 presents how the matrix is extended with checksum columns following a *reverse neighboring scheme*. The reverse neighboring scheme is a peculiar arrangement of data that simplifies significantly the design of the ABFT part of the algorithm.

The data matrix has 8×8 blocks and therefore the size of checksum is 8×3 blocks with an extra 8×3 blocks copy. Checksum blocks are stored on the right of

Fig. 1.18 Reverse neighboring scheme of checksum storage

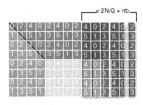

the data matrix. In the example, the first 3 block columns produce the checksum in the last two block columns (hence making 2 duplicate copies of the checksum); the next 3 block columns then produce the next 2 rightmost checksum columns, etc.

Because copies are stored in consecutive columns of the process grid, for any 2D grid $P \times Q$ with $Q > 1$, the checksum duplicates are guaranteed to be stored on different processors. The triangular solve (TRSM) and trailing matrix update (GEMM) are applied to the whole checksum area until the first three columns are factored. In the following factorization steps, the two last block columns of checksum are excluded from the TRSM and GEMM scope. Since TRSM and GEMM claim most of the computation in the LU factorization, shrinking the update scope greatly reduces the overhead of the ABFT mechanism by diminishing the amount of (useless) extra computations; meanwhile, the efficiency of the update operation itself remains optimal as, thanks to the reverse storage scheme, the update still operates on a contiguous memory region and can be performed by a single PBLAS call.

Checksum blocks are duplicated for a reason: since they are stored on the same processes as the matrix and following the same block cyclic scheme, when a process is subject to a failure, blocks of initial data are lost, but also blocks of checksums. Because of the cyclic feature of the data distribution, all checksum blocks must remain available to recover the missing data. Duplicating them guarantees that if a single failure happens, one of the copies will survive. In the example, checksum blocks occupy almost as much memory as the initial matrix once duplicated. However, the number of checksum block column necessary is $2N/(Q \times n_b)$, thus decreases linearly with the width of the process grid.

To simplify the figures, in the following we will represent the checksum blocks over a different process grid, abstracting the duplication of these blocks as if they were hosted by virtual processes that are not subject to failures. We consider here an algorithm that can tolerate only one simultaneous failure (on the same process row), hence at least one of the two checksum blocks will remain available.

Q-**Panel**: The idea of the ABFT factorization is that by extending the scope of the operation to the checksum blocks, the checksum property is maintained between the matrix and the checksum blocks: a block still represents the sum of the blocks of the initial matrix. This is true for the compute-intensive update operations, like GEMM and TRSM. Unfortunately, this is not true for the GETF2 operation that cannot be extended to span over the corresponding checksum blocks.

To deal with this, a simplistic approach would consist in changing the computational kernel to go update the checksum blocks during the GETF2 operation. We avoid doing this because this would introduce more synchronization, having more processes participate to this operation (as the processes spanning over the corresponding checksum blocks are not necessarily involved in a given GETF2 operation). The GETF2 operation is already a memory-bound operation, that require little computation compared to the update operations. It also sits in the critical path of the execution, and is a major blocker to performance, so introducing more synchronization and more delay is clearly detrimental to the performance.

That is the reason why we introduced the concept of Q-panel update. Instead of maintaining the checksum property at all time for all blocks, we will let some of

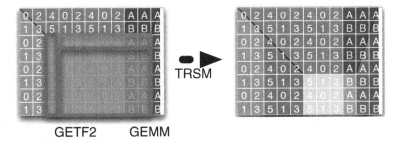

Fig. 1.19 Q-panel update of the ABFT LU factorization

the checksum blocks drift away, for a bounded time, and tolerate the risk for these
Q-panel blocks with another approach. Then, when the cost of checksum update can
be safely absorbed with maximal parallelism, we will let the algorithm update the
checksums of the drifted away blocks, and pursue the computation.

ABFT LU Factorization: We now present the steps of the ABFT LU factorization
using Q-panel update :

1. At a beginning of a Q-panel, when process $(0, 0)$ hosts the first block on which
   GETF2 is going to be applied, processes take a partial checkpoint of the matrix:
   the first Q-block columns of the trailing matrix are copied, as well as the block
   column of corresponding checksums.
2. Then, the usual operations of LU are applied, using the first block column of the
   trailing matrix as a block panel (see Fig. 1.19): GETF2 is applied on that block
   column, then TRSM extended to the corresponding checksums, and GEMM, also
   extended on the corresponding checksums, producing a smaller trailing matrix.
   The checksums that correspond to the previously factored part of the matrix are
   left untouched, as the corresponding data in the matrix, so the checksum property
   is preserved for them. The checksums that were just updated with TRSM and
   GEMM also preserve the checksum property, as the update operations preserve
   the checksum property.
   The part of the checksum represented in red in the figure, however, violates the
   checksum property: the block column on which GETF2 was just applied hold
   values that are not represented in the corresponding block column in the reserve
   neighboring storing scheme.
3. The algorithm iterates, using the second block column of the Q-panel as a panel,
   until Q panels have been applied. In that case, the checksum property is preserved everywhere, except between the blocks that belong to the Q-panel, and
   the corresponding checksum block column. A checksum update operation is then
   executed, to recompute this checksum, the checkpoint saved at the beginning of
   this Q-panel loop can be discarded, and the next Q-panel loop can start.

Failure Handling. When a failure occurs, it is detected by the communication middleware, and the normal execution of the algorithm is interrupted. The ABFT fac-

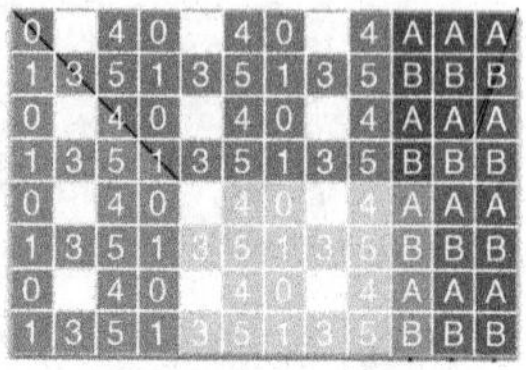

Fig. 1.20 Single failure
during a Q-panel update of
the ABFT LU factorization

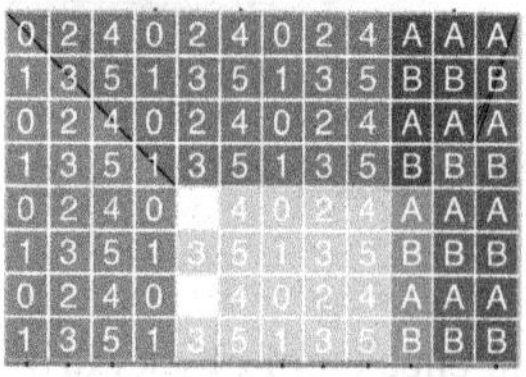

Fig. 1.21 Data restored
using valid checksums

torization enters its recovery routine. Failures can occur at any point during the
execution. The first step of the recovery routine is to gather the status of all surviving
processes, and determine when the failure happened. Spare processes can then be
reclaimed to replace the failed ones, or dynamic process management capabilities of
the communication middleware are used to start new processes that will replace the
missing ones.

In the general case, the failure happened while some blocks have been updated,
and others not, during one of the Q-panels (see Fig. 1.20). Since the checksum blocks
are replicated on adjacent processes, one copy survived the failure, so they are not
missing. For all blocks where the checksum property holds, the checksum blocks are
used to reconstruct the missing data.

The checkpoint of the Q-panel at the beginning of the last Q-panel step also lost
blocks, since a simple local copy is kept. But because the processes also copied the
checksum blocks corresponding to this Q-panel, they can rebuild the missing data
for the checkpoint (Fig. 1.21).

The matrix is then overwritten with the restored checkpoint; the corresponding
checksum blocks are also restored to their checkpoint. Then, the processes re-execute
part of the update and factorization operations, but limiting their scope to the Q-panel
section, until they reach the step when the Q-panel factorization was interrupted. At
this point, all data has been restored to the time of failure, and the processes continue
their execution, and are in a state to tolerate another failure.

If a second failure happens before the restoration is complete (or if multiple failures happen), the application may enter a state where recovery is impossible. This can
be mitigated by increasing the number of checksum block columns, and by replacing checksum copies with linearly independent checksum functions. Then, when
multiple failures occur, the restoration process consists of solving a small system of
equations for each block, to determine the missing values. More importantly, this
exhibits one of the features of application-specific fault tolerance: the overheads are
a function of the risk the developer or user is ready to take.

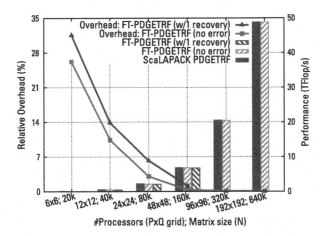

Fig. 1.22 Single failure during a Q-panel update of the ABFT LU factorization

Performance of ABFT LU. Fig. 1.22 (from [16]) shows a weak scalability study of the ABFT scheme that we presented above. On the left axis, the lines show the relative overhead of the ABFT-LU implementation in a failure-free and 1-failure/1-recovery scenario, compared to the non fault-tolerant implementation. On the right axis, the bar graphs show the raw performance of each scenario. This is a weak-scaling experiment, and the matrix size progresses with the process grid size, so that in each case, each processor is responsible for the same amount of data. We denote by $Q \times Q$; N in the x-axis the process grid size ($P \times Q$) and the matrix size (N).

That experiment was conducted on the NSF Kraken supercomputer, hosted at the National Institute for Computational Science (NICS). At the time of the experiment, this machine featured 112,896 2.6 GHz AMD Opteron cores, 12 cores per node, with the Seastar interconnect. At the software level, to serve as a comparison base, we used the non fault-tolerant ScaLAPACK LU in double precision with block size $m_b = n_b = 100$.

The recovery procedure adds a small overhead that also decreases when scaled to large problem size and process grid. For largest setups, only 2–3 percent of the execution time is spent recovering from a failure. Due to the introduction of checksum, operations counts and communication have been increased, as update operation span on a larger matrix comprised of the original trailing matrix and the checksums. During checkpointing and recovery, extra workload is performed and this all together leads to higher computing complexity than the original implementation in ScaLAPACK.

For simplicity of description, we consider square data matrices of size $N \times N$ distributed on a square grid $Q \times Q$. The operation count ration for LU factorization without and with checksum is:

$$R = \frac{\frac{2}{3}N^3 - \frac{1}{2}N^2 + \frac{5}{6}N}{\frac{2}{3}(N + \frac{N}{Q})^3 - \frac{1}{2}(N + \frac{N}{Q})^2 + \frac{5}{6}(N + \frac{N}{Q})}$$

$$= \frac{\frac{2}{3} - \frac{1}{2N} + \frac{5}{6N^2}}{\frac{2}{3}(1 + \frac{1}{Q})^3 - \frac{1}{2N}(1 + \frac{1}{Q})^2 + \frac{5}{6N^2}(1 + \frac{1}{Q})} \quad (1.37)$$

Clearly $\lim_{Q \to +\infty} R = 1$. Hence for systems with high number of processes, the extra flops for updating checksum columns is negligible with respect to the normal flops realized to compute the result.

In addition, checksums must be generated, once at the start of the algorithm, the second time at the completion of a Q-wide panel scope. Both these activities account for $O(N^2)$ extra computations, but can be computed at maximal parallelism, since there is no data dependency.

1.5.3 Composite Approach: ABFT and Checkpointing

ABFT is a useful technique for production systems, offering protection to important infrastructure software. As we have seen, ABFT protection and recovery activities are not only inexpensive, but also have a negligible asymptotic overhead when increasing node count, which makes them extremely scalable. This is in sharp contrast with checkpointing, which suffers from increasing overhead with system size. Many HPC applications do spend quite a significant part of their total execution time inside a numerical library, and in many cases, these numerical library calls can be effectively protected by ABFT.

However, typical HPC applications do spend some time where they perform computations and data management that are incompatible with ABFT protection. The ABFT technique, as the name indicates, allows for tolerating failures only during the execution of the algorithm that features the ABFT properties. Moreover, it then protects only the part of the user dataset that is managed by the ABFT algorithm. In case of a failure outside the ABFT-protected operation, all data is lost; in case of a failure during the ABFT-protected operation, only the data covered by the ABFT scheme is restored. Unfortunately, these ABFT-incompatible phases force users to resort to general-purpose (presumably checkpoint based) approaches as their sole protection scheme.

A composition scheme proposed in [9, 11], protects the application partly with general fault tolerance techniques, and partly with application-specific fault tolerance techniques, harnessing the best of each approach. Performance is close to ABFT, as the ABFT-capable routines dominate the execution, but the approach is generic enough to be applied to any application that uses for at least a part of its execution ABFT-capable routines, so generality is not abandoned, and the user is not forced to rely only on generic rollback-recovery. We present this scheme below, because the underlying approach is key to the adoption of application-specific fault tolerance

```
          while( !converged() ) {
          /* Extract data from the simulator, create the LA problem */
GENERAL   sim2mat();

          /* Factorize the matrix, and solve the problem */
LIBRARY   dgetrf();
          dsolve();

GENERAL   /* Update simulation with result vector */
          vec2sim();
          }
```

Fig. 1.23 Pseudo-code of a typical application using Linear Algebra routines

methods in libraries: without a generic composition scheme, simply linking with different libraries that provide internal resilience capabilities to protect their data from a process crash will not make an application capable of resisting such crashes: process failure breaks the separation introduced by library composition in the software stack, and non protected data, as well as the call stack itself, must be protected by another mean.

As an illustration, consider an application that works as the pseudo-code given in Fig. 1.23. The application has two data: a matrix, on which linear algebra operations are performed, and a simulated state. It uses two libraries: a simulation library that changes the simulated state, and formulates a problem as an equation problem, and a linear algebra library that solves the problem presented by the simulator. The first library is not fault-tolerant, while there is an ABFT scheme to tolerate failures in the linear algebra library.

To abstract the reasoning, we distinguish two phases during the execution: during GENERAL phases, we have no information about the application behavior, and an algorithm-agnostic fault tolerance technique, namely checkpoint and rollback recovery, must be used. On the contrary, during LIBRARY phases, we know much more about the behavior of the library, and we can apply ABFT to ensure resiliency.

ABFT&PERIODICCKPT Algorithm. During a GENERAL phase, the application can access the whole memory; during a LIBRARY phase, only the LIBRARY dataset (a subset of the application memory, which is passed as a parameter to the library call) is accessed. The REMAINDER dataset is the part of the application memory that does not belong to the LIBRARY dataset.

The ABFT&PERIODICCKPT composite approach (see Fig. 1.24) consists of alternating between periodic checkpointing and rollback recovery on one side, and ABFT on the other side, at different phases of the execution. Every time the application enters a LIBRARY phase (that can thus be protected by ABFT), a partial checkpoint is taken to protect the REMAINDER dataset. The LIBRARY dataset, accessed by the ABFT algorithm, need not be saved in that partial checkpoint, since it will be reconstructed by the ABFT algorithm inside the library call.

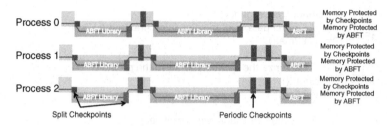

Fig. 1.24 ABFT&PERIODICCKPT composite approach

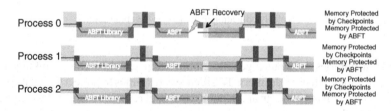

Fig. 1.25 Fault handling during a LIBRARY phase

When the call returns, a partial checkpoint covering the modified LIBRARY dataset is added to the partial checkpoint taken at the beginning of the call, to complete it and to allow restarting from the end of the terminating library call. In other words, the combination of the partial entry and exit checkpoints forms a split, but complete, coordinated checkpoint covering the entire dataset of the application.

If a failure is detected while processes are inside the library call (Fig. 1.25), the crashed process is recovered using a combination of rollback recovery and ABFT. ABFT recovery is used to restore the LIBRARY dataset before all processes can resume the library call, as would happen with a traditional ABFT algorithm. The partial checkpoint is used to recover the REMAINDER dataset (everything except the data covered by the current ABFT library call) at the time of the call, and the process stack, thus restoring it before quitting the library routine. The idea of this strategy is that ABFT recovery will spare some of the time spent redoing work, while periodic checkpointing can be completely de-activated during the library calls.

During GENERAL phases, regular periodic coordinated checkpointing is employed to protect against failures (Fig. 1.26). In case of failure, coordinated rollback recovery brings all processes back to the last checkpoint (at most back to the split checkpoint capturing the end of the previous library call).

ABFT&PERIODICCKPT Algorithm Optimization. Recall from Sect. 1.3.2 that a critical component to the efficiency of periodic checkpointing algorithms is the duration of the checkpointing interval. A short interval increases the algorithm overheads, by introducing many coordinated checkpoints, during which the application experiences slowdown, but also reduces the amount of time lost when there is a failure: the last checkpoint is never long ago, and little time is spent re-executing part of the application. Conversely, a large interval reduces overhead, but increases the

Fig. 1.26 Fault handling during a GENERAL phase

time lost in case of failure. In the ABFT&PERIODICCKPT algorithm, we interleave periodic checkpointing protected phases with ABFT protected phases, during which periodic checkpointing is de-activated. Thus, different cases have to be considered:

- When the time spent in a GENERAL phase is larger than the optimal checkpoint interval, periodic checkpointing is used during these phases in the case of ABFT-&PERIODICCKPT;
- When the time spent in a GENERAL phase is smaller than the optimal checkpoint interval, the ABFT&PERIODICCKPT algorithm already creates a complete valid checkpoint for this phase (formed by combining the entry and exit partial checkpoints), so the algorithm will not introduce additional checkpoints.

Moreover, the ABFT&PERIODICCKPT algorithm forces (partial) checkpoints at the entry and exit of library calls; thus if the time spent in a library call is very small, this approach will introduce more checkpoints than a traditional periodic checkpointing approach. The time complexity of library algorithms usually depends on a few input parameters related to problem size and resource number, and ABFT techniques have deterministic, well known time overhead complexity. Thus, when possible, the ABFT&PERIODICCKPT algorithm features a safeguard mechanism: if the projected duration of a library call with ABFT protection (computed at runtime thanks to the call parameters and the algorithm complexity) is smaller than the optimal periodic checkpointing interval, then ABFT is not activated, and the corresponding LIBRARY phase is protected using the periodic checkpointing technique only.

1.5.3.1 Performance Model of ABFT&PERIODICCKPT

The execution of the application is partitioned into epochs of total duration T_0. Within an epoch, there are two phases: the first phase is spent outside the library (it is a GENERAL phase, of duration T_G), and only periodic checkpointing can be employed to protect from failures during that phase. Then the second phase (a LIBRARY phase of duration T_L) is devoted to a library routine that has the potential to be protected by ABFT. Let α be the fraction of time spent in a LIBRARY phase: then we have $T_L = \alpha \times T_0$ and $T_G = (1 - \alpha) \times T_0$.

As mentioned earlier, another important parameter is the amount of memory that is accessed during the LIBRARY phase (the LIBRARY dataset). This parameter is important because the cost of checkpointing in each phase is directly related to the amount of memory that needs to be protected. The total memory footprint is M, and the associated checkpointing cost is C (we assume a finite checkpointing bandwidth, so $C > 0$). We write $M = M_L + M_{\overline{L}}$, where M_L is the size of the LIBRARY dataset, and $M_{\overline{L}}$ is the size of the REMAINDER dataset. Similarly, we write $C = C_L + C_{\overline{L}}$, where C_L is the cost of checkpointing M_L, and $C_{\overline{L}}$ the cost of checkpointing $M_{\overline{L}}$. We can define the parameter ρ that defines the relative fraction of memory accessed during the LIBRARY phase by $M_L = \rho M$, or, equivalently, by $C_L = \rho C$.

Fault-free execution. During the GENERAL phase, we separate two cases. First, if the duration T_G of this phase is short, *i.e.* smaller than $P_G - C_{\overline{L}}$, which is the amount of work during one period of length P_G (and where P_G is determined below), then we simply take a partial checkpoint at the end of this phase, before entering the ABFT-protected mode. This checkpoint is of duration $C_{\overline{L}}$, because we need to save only the REMAINDER dataset in this case. Otherwise, if T_G is larger than $P_G - C_{\overline{L}}$, we rely on periodic checkpointing during the GENERAL phase: more specifically, the regular execution is divided into periods of duration $P_G = W + C$. Here W is the amount of work done per period, and the duration of each periodic checkpoint is $C = C_{\overline{L}} + C_L$, because the whole application footprint must be saved during a GENERAL phase. The last period is different: we execute the remainder of the work, and take a final checkpoint of duration $C_{\overline{L}}$ before switching to ABFT-protected mode. The optimal (approximated) value of P_G will be computed below.

Altogether, the length T_G^{ff} of a fault-free execution of the GENERAL phase is the following:

- If $T_G \leq P_G - C_{\overline{L}}$, then $T_G^{\text{ff}} = T_G + C_{\overline{L}}$
- Otherwise, we have $\lfloor \frac{T_G}{\text{WORK}} \rfloor$ periods of length P_G, plus possibly a shorter last period if T_G is not evenly divisible by W. In addition, we need to remember that the last checkpoint taken is of length $C_{\overline{L}}$ instead of C.

This leads to

$$
T_G^{\text{ff}} = \begin{cases} T_G + C_{\overline{L}} & \text{if } T_G \leq P_G - C_{\overline{L}} \\ \lfloor \frac{T_G}{P_G - C} \times P_G \rfloor + (T_G \bmod W) + C_{\overline{L}} & \text{if } T_G > P_G - C_{\overline{L}} \text{ and } T_G \bmod W \neq 0 \\ \frac{T_G}{P_G - C} \times P_G - C_L & \text{if } T_G > P_G - C_{\overline{L}} \text{ and } T_G \bmod W = 0 \end{cases}
$$
$$(1.38)$$

Now consider the LIBRARY phase: we use the ABFT-protection algorithm, whose cost is modeled as an affine function of the time spent: if the computation time of the library routine is t, its execution with the ABFT-protection algorithm becomes $\phi \times t$. Here, $\phi > 1$ accounts for the overhead paid per time-unit in ABFT-protected mode. This linear model for the ABFT overhead fits the existing algorithms for linear algebra, but other models could be considered. In addition, we pay a checkpoint C_L when exiting the library call (to save the final result of the ABFT phase). Therefore, the fault-tree execution time is

$$T_L^{\text{ff}} = \phi \times T_L + C_L \tag{1.39}$$

Finally, the fault-free execution time of the whole epoch is

$$T^{\text{ff}} = T_G^{\text{ff}} + T_L^{\text{ff}} \tag{1.40}$$

where T_G^{ff} and T_L^{ff} are computed according to the Eqs. (1.38) and (1.39).

Cost of failures. Next we have to account for failures. For each phase, we have a similar equation: the final execution time is the fault-free execution time, plus the number of failures multiplied by the (average) time lost per failure:

$$T_G^{\text{final}} = T_G^{\text{ff}} + \frac{T_G^{\text{final}}}{\mu} \times t_G^{\text{lost}} \tag{1.41}$$

$$T_L^{\text{final}} = T_L^{\text{ff}} + \frac{T_L^{\text{final}}}{\mu} \times t_L^{\text{lost}} \tag{1.42}$$

Equations (1.41) and (1.42) correspond to Eq. (1.5) in Sect. (1.3.1). Equation (1.41) reads as follows: T_G^{ff} is the failure-free execution time, to which we add the time lost due to failures; the expected number of failures is $\frac{T_G^{\text{final}}}{\mu}$, and t_G^{lost} is the average time lost per failure. We have a similar reasoning for Eq. (1.42). Then, t_G^{lost} and t_L^{lost} remain to be computed. For t_G^{lost} (GENERAL phase), we discuss both cases:

- If $T_G \leq P_G - C_{\overline{L}}$: since we have no checkpoint until the end of the GENERAL phase, we have to redo the execution from the beginning of the phase. On average, the failure strikes at the middle of the phase, hence the expectation of loss is $\frac{T_G^{\text{ff}}}{2}$ time units. We then add the downtime D (time to reboot the resource or set up a spare) and the recovery R. Here R is the time needed for a complete reload from the checkpoint (and $R = C$ if read/write operations from/to the stable storage have the same speed). We derive that:

$$t_G^{\text{lost}} = D + R + \frac{T_G^{\text{ff}}}{2} \tag{1.43}$$

- If $T_G > P_G - C_{\overline{L}}$: in this case, we have periodic checkpoints, and the amount of execution which needs to be redone after a failure corresponds to half a checkpoint period on average, so that:

$$t_G^{\text{lost}} = D + R + \frac{P_G}{2} \tag{1.44}$$

For t_L^{lost} (LIBRARY phase), we derive that

$$t_L^{\text{lost}} = D + R_{\overline{L}} + \text{Recons}_{\text{ABFT}}$$

Here, $R_{\overline{L}}$ is the time for reloading the checkpoint of the REMAINDER dataset (and in many cases $R_{\overline{L}} = C_{\overline{L}}$). As for the LIBRARY dataset, there is no checkpoint to retrieve, but instead it must be reconstructed from the ABFT checksums, which takes time $\text{Recons}_{\text{ABFT}}$.

Optimization: finding the optimal checkpoint interval in GENERAL phase.

We verify from Eqs. (1.39) and (1.42) that T_L^{final} is always a constant. Indeed, we derive that:

$$T_L^{\text{final}} = \frac{1}{1 - \dfrac{D + R_{\overline{L}} + \text{Recons}_{\text{ABFT}}}{\mu}} \times (\phi \times T_L + C_L) \qquad (1.45)$$

As for T_G^{final}, it depends on the value of T_G: it is constant when T_G is small. In that case, we derive that:

$$T_G^{\text{final}} = \frac{1}{1 - \dfrac{D + R + \frac{T_G + C_{\overline{L}}}{2}}{\mu}} \times \left(T_G + C_{\overline{L}}\right) \qquad (1.46)$$

The interesting case is when T_G is large: in that case, we have to determine the optimal value of the checkpointing period P_G which minimizes T_G^{final}. We use an approximation here: we assume that we have an integer number of periods, and the last periodic checkpoint is of size C. Note that the larger T_G, the more accurate the approximation. From Eqs. (1.38), (1.41) and (1.44), we derive the following simplified expression:

$$T_G^{\text{final}} = \frac{T_G}{X} \text{ where } X = \left(1 - \frac{C}{P_G}\right)\left(1 - \frac{D + R + \frac{P_G}{2}}{\mu}\right) \qquad (1.47)$$

We rewrite:

$$X = \left(1 - \frac{C}{2\mu}\right) - \frac{P_G}{2\mu} - \frac{C(\mu - D - R)}{\mu P_G}$$

The maximum of X gives the optimal period P_G^{opt} . Differentiating X as a function of P_G, we find that it is obtained for:

$$P_G^{\text{opt}} = \sqrt{2C(\mu - D - R)} \qquad (1.48)$$

We retrieve Eq. 1.9 of Sect. 1.3.1 (as expected). Plugging the value of P_G^{opt} back into Eq. (1.47) provides the optimal value of T_G^{final} when T_G is large. We conclude this with reminding the word of caution given at the end of Sect. 1.3.2.1: the optimal value of the waste is only a first-order approximation, not an exact value. Just as in [25, 69], the formula only holds when μ, the value of the MTBF, is large with respect to the other resilience parameters. Owing to this hypothesis, we can neglect the probability of several failures occurring during the same checkpointing period.

Comparison of the scalability of approaches. The ABFT&PERIODICCKPT approach is expected to provide better performance when a significant time is spent in the LIBRARY phase, *and* when the failure rate implies a small optimal checkpointing period. If the checkpointing period is large (because failures are rare), or if the duration of the LIBRARY phase is small, then the optimal checkpointing interval becomes larger than the duration of the LIBRARY phase, and the algorithm automatically resorts to the periodic checkpointing protocol. This can also be the case when the epoch itself is smaller than (or of the same order of magnitude as) the optimal checkpointing interval (i.e., when the application does a fast switching between LIBRARY and GENERAL phases).

However, consider such an application that frequently switches between (relatively short) LIBRARY and GENERAL phases. When porting that application to a future larger scale machine, the number of nodes that are involved in the execution will increase, and at the same time, the amount of memory on which the ABFT operation is applied will grow (following Gustafson's law [37]). This has a double impact: the time spent in the ABFT routine increases, while at the same time, the MTBF of the machine decreases. As an illustration, we evaluate quantitatively how this scaling factor impacts the relative performance of the ABFT&PERIODICCKPT and a traditional periodic checkpointing approach.

First, we consider the case of an application where the LIBRARY and GENERAL phases scale at the same rate. We take the example of linear algebra kernels operating on $2D$-arrays (matrices), that scale in $O(n^3)$ of the array order n (in both phases). Following a weak scaling approach, the application uses a fixed amount of memory M_{ind} per node, and when increasing the number x of nodes, the total amount of memory increases linearly as $M = x M_{ind}$. Thus $O(n^2) = O(x)$, and the parallel completion time of the $O(n^3)$ operations, assuming perfect parallelism, scales in $O(\sqrt{x})$.

To instantiate this case, we take an application that would last a thousand minutes at 100,000 nodes (the scaling factor corresponding to an operation in $O(n^3)$ is then applied when varying the number of nodes), and consisting for 80 % of a LIBRARY phase, and 20 % of a GENERAL phase. We set the duration of the complete checkpoint and rollback (C and R, respectively) to 1 minute when 100,000 nodes are involved, and we scale this value linearly with the total amount of memory, when varying the number of nodes. The MTBF at 100,000 nodes is set to 1 failure every day, and this also scales linearly with the number of components. The ABFT overheads, and the downtime, are set to the same values as in the previous section, and 80 % of the application memory (M_L) is touched by the LIBRARY phase.

Given these parameters, Fig. 1.27 shows (i) the relative waste of periodic checkpointing and ABFT&PERIODICCKPT, as a function of the number of nodes, and (ii) the average number of faults that each execution will have to deal with to complete. The expected number of faults is the ratio of the application duration by the platform MTBF (which decreases when the number of nodes increases, generating more fail-

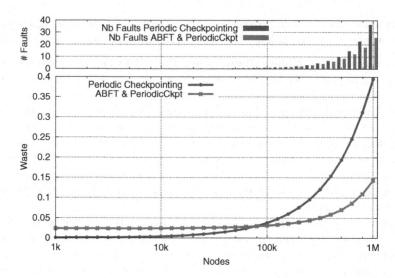

Fig. 1.27 Total waste for periodic checkpointing and ABFT&PERIODICCKPT, when considering the weak scaling of an application with a fixed ratio of 80 % spent in a LIBRARY routine

ures). The fault-free execution time increases with the number of nodes (as noted above), and the fault-tolerant execution time is also increased by the waste due to the protocol. Thus, the total execution time of periodic checkpointing is larger at 1 million nodes than the total execution time of ABFT&PERIODICCKPT at the same scale, which explains why more failures happen for these protocols.

Up to approximately 100,000 nodes, the fault-free overhead of ABFT negatively impacts the waste of the ABFT&PERIODICCKPT approach, compared to periodic checkpointing. Because the MTBF on the platform is very large compared to the application execution time (and hence to the duration of each LIBRARY phase), the periodic checkpointing approach has a very large checkpointing interval, introducing very few checkpoints, thus a small failure-free overhead. Because failures are rare, the cost due to time lost at rollbacks does not overcome the benefits of a small failure-free overhead, while the ABFT technique must pay the linear overhead of maintaining the redundancy information during the whole computation of the LIBRARY phase.

Once the number of nodes reaches 100,000, however, two things happen: failures become more frequent, and the time lost due to failures starts to impact rollback recovery approaches. Thus, the optimal checkpointing interval of periodic check-pointing becomes smaller, introducing more checkpointing overheads. During 80 % of the execution, however, the ABFT&PERIODICCKPT approach can avoid these over-heads, and when they reach the level of linear overheads due to the ABFT technique, ABFT&PERIODICCKPT starts to scale better than both periodic checkpointing approaches.

All protocols have to resort to checkpointing during the GENERAL phase of the application. Thus, if failures hit during this phase (which happens 20 % of the time in

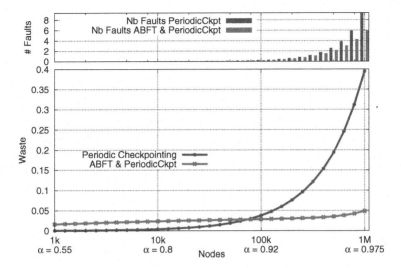

Fig. 1.28 Total waste for ABFT&PERIODICCKPT and periodic checkpointing when considering the weak scaling of an application with variable ratio of time spent in a LIBRARY routine

this example), they will all have to resort to rollbacking and lose some computation time. Hence, when the number of nodes increases and the MTBF decreases, eventually, the time spent in rollbacking and recomputing, which is linear in the number of faults, will increase the waste of all algorithms. However, one can see that this part is better controlled by the ABFT&PERIODICCKPT algorithm.

Next, we consider the case of an unbalanced GENERAL phase: consider an application where the LIBRARY phase has a cost $O(n^3)$ (where n is the problem size), as above, but where the GENERAL phase consists of $O(n^2)$ operations. This kind of behavior is reflected in many applications where matrix data is updated or modified between consecutive calls to computation kernels. Then, the time spent in the LIBRARY phase will increase faster with the number of nodes than the time spent in the GENERAL phase, varying α. This is what is represented in Fig. 1.28. We took the same scenario as above for Fig. 1.27, but α is a function of the number of nodes chosen such that at 100,000 nodes, $\alpha = T_L^{\text{final}}/T^{\text{final}} = 0.8$, and everywhere, $T_L^{\text{final}} = O(n^3) = O(\sqrt{x})$, and $T_{PC}^{\text{final}} = O(n^2) = O(1)$. We give the value of α under the number of nodes, to show how the fraction of time spent in LIBRARY phases increases with the number of nodes.

The periodic checkpointing protocol is not impacted by this change, and behaves exactly as in Fig. 1.27. Note, however, that $T^{\text{final}} = T_L^{\text{final}} + T_{PC}^{\text{final}}$ progresses at a lower rate in this scenario than in the previous scenario, because T_{PC}^{final} does not increase with the number of nodes. Thus, the average number of faults observed for all protocols is much smaller in this scenario.

The efficiency on ABFT&PERIODICCKPT, however, is more significant. The latter protocol benefits from the increased α ratio in both cases: since more time is

Fig. 1.29 Error and
detection latency

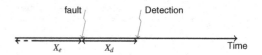

spent in the LIBRARY phase, periodic checkpointing is de-activated for relatively longer periods. Moreover, this increases the probability that a failure will happen during the LIBRARY phase, where the recovery cost is greatly reduced using ABFT techniques. Thus, ABFT&PERIODICCKPT is capable of mitigating failures at a much smaller overhead than simple periodic checkpointing, and more importantly with better scalability.

1.6 Silent Errors

This section deals with techniques to cope with silent errors. We focus on a general-purpose approach that combines checkpointing and (abstract) verification mechanisms. Section 1.6.1 provides some background, while Sect. 1.6.2 briefly surveys different approaches form the literature. Then Sect. 1.6.3 details the performance model for the checkpoint/verification approach and explains how to determine the optimal pattern minimizing the waste.

1.6.1 Motivation

Checkpoint and rollback recovery techniques assume reliable error detection, and therefore apply to fail-stop failures, such as for instance the crash of a resource. In this section, we revisit checkpoint protocols in the context of *silent* errors, also called silent data corruption. Such errors must be accounted for when executing HPC applications [58, 61, 74–76]. The cause for silent errors may be for instance soft efforts in L1 cache, or bit flips due to cosmic radiation. The problem is that the detection of a silent error is not immediate, but will only manifest later as a failure, once the corrupted data has impacted the result (see Fig. 1.29). If the error stroke before the last checkpoint, and is detected after that checkpoint, then the checkpoint is corrupted, and cannot be used to restore the application. In the case of fail-stop failures, a checkpoint cannot contain a corrupted state, because a process subject to failure will not create a checkpoint or participate to the application: failures are naturally contained to failed processes; in the case of silent errors, however, faults can propagate to other processes and checkpoints, because processes continue to participate and follow the protocol during the interval that separates the error and its detection.

To alleviate this issue, one may envision to keep several checkpoints in memory, and to restore the application from the last *valid* checkpoint, thereby rolling back to the last *correct* state of the application [55]. This multiple-checkpoint approach has three major drawbacks. First, it is very demanding in terms of stable storage: each checkpoint typically represents a copy of the entire memory footprint of the application, which may well correspond to several terabytes. The second drawback is the possibility of fatal failures. Indeed, if we keep k checkpoints in memory, the approach assumes that the error that is currently detected did not strike before all the checkpoints still kept in memory, which would be fatal: in that latter case, all live checkpoints are corrupted, and one would have to re-execute the entire application from scratch. The probability of a fatal failure is evaluated in [3] for various error distribution laws and values of k. The third drawback of the approach is the most serious, and applies even without memory constraints, i.e., if we could store an infinite number of checkpoints in storage. The critical question is to determine which checkpoint is the last valid one. We need this information to safely recover from that point on. However, because of the detection latency (which is unknown), we do not know when the silent error has indeed occurred, hence we cannot identify the last valid checkpoint, unless some verification system is enforced.

This section introduces algorithms coupling verification and checkpointing, and shows how to analytically determine the best balance of verifications between checkpoints so as to minimize platform waste. In this (realistic) model, silent errors are detected only when some verification mechanism is executed. This approach is agnostic of the nature of this verification mechanism (checksum, error correcting code, coherence tests, etc.). This approach is also fully general-purpose, although application-specific information, if available, can always be used to decrease the cost of verification.

The simplest protocol (see Fig. 1.30) would be to perform a verification just before taking each checkpoint. If the verification succeeds, then one can safely store the checkpoint and mark it as *valid*. If the verification fails, then an error has struck since the last checkpoint, which was duly verified, and one can safely recover from that checkpoint to resume the execution of the application. This protocol with verifications eliminates fatal errors that would corrupt all live checkpoints and cause to restart execution from scratch. However, we still need to assume that both checkpoints and verifications are executed in a reliable mode.

There is room for optimization. Consider the second pattern illustrated in Fig. 1.31 with three verifications per checkpoint. There are three chunks of size w, each followed by a verification. Every third verification is followed by a checkpoint. We assume that $w = W/3$ to ensure that both patterns correspond to the same amount

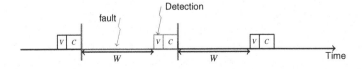

Fig. 1.30 The first pattern with one verification before each checkpoint

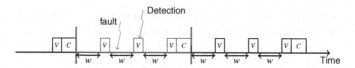

Fig. 1.31 The second pattern with three verifications per checkpoint

of work, W. Just as for the first pattern, a single checkpoint needs to be kept in memory, owing to the verifications. Also, as before, each error leads to re-executing the work since the last checkpoint. But detection occurs much more rapidly in the second pattern, owing to the intermediate verifications. If the error strikes in the first of the three chunks, it is detected by the first verification, and only the first chunk is re-executed. Similarly, if the error strikes in the second chunk (as illustrated in the figure), it is detected by the second verification, and the first two chunks are re-executed. The entire pattern of work needs to be re-executed only if the error strikes during the third chunk. On average, the amount of work to re-execute is $(1 + 2 + 3)w/3 = 2w = 2W/3$. On the contrary, in the first pattern of Fig. 1.30, the amount of work to re-execute always is W, because the error is never detected before the end of the pattern. Hence the second pattern leads to a 33 % gain in re-execution time. However, this comes at the price of three times as many verifications. This overhead is paid in every failure-free execution, and may be an overkill if the verification mechanism is too costly.

This little example shows that the optimization problem looks difficult. It can be stated as follows: given the cost of checkpointing C, recovery R, and verification V, what is the optimal strategy to minimize the (expectation of the) waste? A strategy is a periodic pattern of checkpoints and verifications, interleaved with work segments, that repeats over time. The length of the work segments also depends upon the platform MTBF μ. For example, with a single checkpoint and no verification (which corresponds to the classical approach for fail-stop failures), recall from Theorem 1.1 that the optimal length of the work segment can be approximated as $\sqrt{2\mu C}$. Given a periodic pattern with checkpoints and verifications, can we extend this formula and compute similar approximations?

We conclude this introduction by providing a practical example of the checkpoint and verification mechanisms under study. A nice instance of this approach is given by Chen [21], who deals with sparse iterative solvers. Chen considers a simple method such as the PCG, the Preconditioned Conjugate Gradient method, and aims at protecting the execution from arithmetic errors in the ALU. Chen's approach performs a periodic verification every d iterations, and a periodic checkpoint every $d \times c$ iterations, which is a particular case of the pattern with $p = 1$ and $q = c$. For PCG, the verification amounts to checking the orthogonality of two vectors and to recomputing and checking the residual, while the cost of checkpointing is that of storing three vectors. The cost of a checkpoint is smaller than the cost of the verification, which itself is smaller than the cost of an iteration, especially when the preconditioner requires much more flops than a sparse matrix-vector product. In this context, Chen [21] shows how to numerically estimate the best values of the parameters d and c. The results given in Sect. 1.6.3 show using equidistant verifications, as suggested in [21], is asymptotically optimal when using a pattern with a single checkpoint ($p = 1$), and enable to determine the best pattern with p checkpoints and q verifications as a function of C, R, and V, and the MTBF μ.

1.6.2 Other Approaches

In this section, we briefly survey other approaches to detect and/or correct silent errors. Considerable efforts have been directed at error-checking to reveal silent errors. Error detection is usually very costly. Hardware mechanisms, such as ECC memory, can detect and even correct a fraction of errors, but in practice they are complemented with software techniques. General-purpose techniques are based on replication, which we have already met in Sect. 1.4.2: using replication [31, 33, 66, 72], one can compare the results of both replicas and detect a silent error. Using TMR [56] would allow to correct the error (by voting) after detection. Note that another approach based on checkpointing and replication is proposed in [60], in order to detect and enable fast recovery of applications from both silent errors and hard errors.

Coming back to verification mechanisms, application-specific information can be very useful to enable ad-hoc solutions, that dramatically decrease the cost of detection. Many techniques have been advocated. They include memory scrubbing [48], but also ABFT techniques [8, 46, 68], such as coding for the sparse-matrix vector multiplication kernel [68], and coupling a higher-order with a lower-order scheme for Ordinary Differential Equations [6]. These methods can only detect an error but do not correct it. Self-stabilizing corrections after error detection in the conjugate gradient method are investigated by Sao and Vuduc [65]. Also, Heroux and Hoemmen [44] design a fault-tolerant GMRES capable of converging despite silent errors, and Bronevetsky and de Supinski [17] provide a comparative study of detection costs for iterative methods. Elliot et al. [30] combine partial redundancy and checkpointing, and confirm the benefit of dual and triple redundancy. The drawback is that twice the number of processing resources is required (for dual redundancy).

As already mentioned, the combined checkpoint/verification approach is agnostic of the underlying error-detection technique and takes the cost of verification as an input parameter to the model.

1.6.3 Optimal Pattern

In this section, we detail the performance model to assess the efficiency of any checkpoint/verification pattern. Then we show how to determine the best pattern.

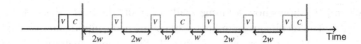

Fig. 1.32 The BALANCEDALGORITHM with five verifications for two checkpoints

1.6.3.1 Model for Patterns

Consider a periodic pattern with p checkpoints and q verifications, and whose total length is $S = pC + qV + W$. Here, W is the work that is executed during the whole pattern, and it is divided into several chunks that are each followed by a verification, or a checkpoint, or both. Checkpoints and verifications are at arbitrary location within the pattern. The only constraint is that the pattern always ends by a verification immediately followed by a checkpoint: this is to enforce that the last checkpoint is always valid, thereby ruling out the risk of a fatal failure. In the example of Fig. 1.31, we have three chunks of same size w, hence $W = 3w$ and $S = C + 3V + 3w$. The example of Fig. 1.32 represents a more complicated pattern, with two checkpoints and five verifications. The two checkpoints are equidistant in the pattern, and so are the five verifications, hence the six chunks of size either w or $2w$, for a total work $W = 10w$, and $S = 2C + 5V + 10w$. The rationale for using such chunk sizes in Fig. 1.32 is given in Sect. 1.6.3.2.

We compute the waste incurred by the use of a pattern similarly to what we did for fail-stop failures in Sect. 1.3.1. We consider a periodic pattern with p checkpoints, q verifications, work W, and total length $S = pC + qV + W$. We assume a a *selective reliability* model where checkpoint, recovery and verification are error-free operations. The input parameters are the following:

- the cost V of the verification mechanism;
- the cost C of a checkpoint;
- the cost R of a recovery;
- the platform MTBF μ.

We aim at deriving the counterpart of Eq. (1.33) for silent errors. We easily derive that the waste in a fault-free execution is $\text{WASTE}_{\text{ff}} = \frac{pC+qV}{S}$, and that the waste due to silent errors striking during execution. is which is the waste due to checkpointing is $\text{WASTE}_{\text{fail}} = \frac{T_{\text{lost}}}{\mu}$, where T_{lost} is the expected time lost due to each error. The value of T_{lost} is more complicated to compute than for fail-stop errors, because it depends upon which pattern is used. Before computing T_{lost} for arbitrary values of p and q in Sect. 1.6.3.2, we give two examples.

The first example is for the simple pattern of Fig. 1.30. We have $p = q = 1$, a single chunk of size $w = W$, and a pattern of size $S = C + V + W$. Computing T_{lost} for this pattern goes as follows: whenever an error strikes, it is detected at the end of the work, during the verification. We first recover from the last checkpoint, then re-execute the entire work, and finally redo the verification. This leads to $T_{\text{lost}} = R + W + V = R + S - C$. From Eq. (1.33), we obtain that

$$\text{WASTE} = 1 - \left(1 - \frac{R + S - C}{\mu}\right)\left(1 - \frac{C + V}{S}\right) = aS + \frac{b}{S} + c, \qquad (1.49)$$

where $a = \frac{1}{\mu}$, $b = (C + V)(1 + \frac{C-R}{\mu})$ and $c = \frac{R-V-2C}{\mu}$. The value that minimizes the waste is $S = S_{\text{opt}}$, and the optimal waste is $\text{WASTE}_{\text{opt}}$, where

$$S_{\text{opt}} = \sqrt{\frac{b}{a}} = \sqrt{(C + V)(\mu + C - R)} \quad \text{and} \quad \text{WASTE}_{\text{opt}} = 2\sqrt{ab} + c. \quad (1.50)$$

Just as for fail-stop failures, we point out that this approach leads to a first-order approximation of the optimal pattern, not to an optimal value. As always, the approach is valid when μ is large in front of S, and of all parameters R, C and V. When this is the case, we derive that $S_{\text{opt}} \approx \sqrt{(C + V)\mu}$ and $\text{WASTE}_{\text{opt}} \approx 2\sqrt{\frac{C+V}{\mu}}$. It is very interesting to make a comparison with the optimal checkpointing period T_{FO} (see Eq. (1.9)) when dealing with fatal failures: we had $T_{\text{FO}} \approx \sqrt{2C\mu}$. In essence, the factor 2 comes from the fact that we re-execute only half the period on average with a fatal failure, because the detection is instantaneous. In our case, we always have to re-execute the entire pattern. And of course, we have to replace C by $C + V$, to account for the cost of the verification mechanism.

The second example is for the BALANCEDALGORITHM illustrated in Fig. 1.32. We have $p = 2$, $q = 5$, six chunks of size w or $2w$, $W = 10w$, and a pattern of size $S = 2C + 5V + W$. Note that it may now be the case that we store an invalid checkpoint, if the error strikes during the third chunk (of size w, just before the non-verified checkpoint), and therefore we must keep two checkpoints in memory to avoid the risk of fatal failures. When the verification is done at the end of the fourth chunk, if it is correct, then we can mark the preceding checkpoint as valid and keep only this checkpoint in memory. Because $q > p$, there are never two consecutive checkpoints without a verification between them, and at most two checkpoints need to be kept in memory. The time lost due to an error depends upon where it strikes:

- With probability $2w/W$, the error strikes in the first chunk. It is detected by the first verification, and the time lost is $R + 2w + V$, since we recover, and re-execute the work and the verification.
- With probability $2w/W$, the error strikes in the second chunk. It is detected by the second verification, and the time lost is $R + 4w + 2V$, since we recover, re-execute the work and both verifications.
- With probability w/W, the error strikes in the third chunk. It is detected by the third verification, and we roll back to the last checkpoint, recover and verify it. We find it invalid, because the error struck before taking it. We roll back to the beginning of the pattern and recover from that checkpoint. The time lost is $2R + 6w + C + 4V$, since we recover twice, re-execute the work up to the third verification, redo the checkpoint and the three verifications, and add the verification of the invalid checkpoint.
- With probability w/W, the error strikes in the fourth chunk. It is detected by the third verification. We roll back to the previous checkpoint, recover and verify it.

In this case, it is valid, since the error struck after the checkpoint. The time lost is
$R + w + 2V$.

- With probability $2w/W$, the error strikes in the fifth chunk. Because there was a valid verification after the checkpoint, we do not need to verify it again, and the time lost is $R + 3w + 2V$.
- With probability $2w/W$, the error strikes in the sixth and last chunk. A similar reasoning shows that the time lost is $R + 5w + 3V$.

Averaging over all cases, we derive that $T_{lost} = \frac{11R}{10} + \frac{35w}{10} + \frac{C}{10} + \frac{22V}{10}$. We then proceed as with the first example to derive the optimal size S of the pattern. We obtain $S_{opt} = \sqrt{\frac{b}{a}}$ and $WASTE_{opt} = 2\sqrt{ab} + c$ (see Eq.(1.50)), where $a = \frac{7\mu}{20}$, $b = (2C + 5V)(1 - \frac{1}{20\mu}(22R - 12C + 9V))$ and $c = \frac{1}{20\mu}(22R - 26C - 17V)$. When μ is large, we have $S_{opt} \approx \sqrt{\frac{20}{7}(2C + 5V)\mu}$ and $WASTE_{opt} \approx 2\sqrt{\frac{7(2C+5V)}{20\mu}}$.

1.6.3.2 Optimal Pattern

In this section, we generalize from the examples and provide a generic expression for the waste when the platform MTBF μ is large in front of all resilience parameters R, C and V. Consider a general pattern of size $S = pC + qV + W$, with $p \leq q$. We have $WASTE_{ff} = \frac{o_{ff}}{S}$, where $o_{ff} = pC + qV$ is the fault-free overhead due to inserting p checkpoints and q verifications within the pattern. We also have $WASTE_{fail} = \frac{T_{lost}}{\mu}$, where T_{lost} is the time lost each time an error strikes and includes two components: re-executing a fraction of the total work W of the pattern, and computing additional verifications, checkpoints and recoveries (see the previous examples). The general form of T_{lost} is thus $T_{lost} = f_{re}W + \alpha$ where f_{re} stands for *fraction* of work that is *re-executed* due to failures; α is a constant that is a linear combination of C, V and R. For the first example (Fig. 1.30), we have $f_{re} = 1$. For the second example (Fig. 1.32), we have $f_{re} = \frac{7}{20}$ (recall that $w = W/10$). For convenience, we use an equivalent form

$$T_{lost} = f_{re}S + \beta, \tag{1.51}$$

where $\beta = \alpha - f_{re}(pC + qV)$ is another constant. When the platform MTBF μ is large in front of all resilience parameters R, C and V, we can identify the dominant term in the optimal waste $WASTE_{opt}$. Indeed, in that case, the constant β becomes negligible in front of μ, and we derive that

$$S_{opt} = \sqrt{\frac{o_{ff}}{f_{re}}} \times \sqrt{\mu} + o(\sqrt{\mu}), \tag{1.52}$$

and that the optimal waste is

$$WASTE_{opt} = 2\sqrt{o_{ff}f_{re}}\sqrt{\frac{1}{\mu}} + o(\sqrt{\frac{1}{\mu}}). \tag{1.53}$$

This equation shows that the optimal pattern when μ is large is obtained when the product $o_{\text{ff}} f_{\text{re}}$ is minimal. This calls for a trade-off, as a smaller value of o_{ff} with few checkpoints and verifications leads to a larger re-execution time, hence to a larger value of f_{re}. For instance, coming back to the examples of Figs. 1.30 and 1.32, we readily see that the second pattern is better than the first one for large values of μ whenever $V > 2C/5$, which corresponds to the condition $\frac{7}{20} \times (5V + 2C) > 1 \times (V + C)$.

For a general pattern of size $S = pC + qV + W$, with $p \leq q$, we always have $o_{\text{ff}} = o_{\text{ff}}(p, q) = pC + qV$ and we aim at (asymptotically) minimizing $f_{\text{re}} = f_{\text{re}}(p, q)$, the expected fraction of the work that is re-executed, by determining the optimal size of each work segment. It turns out that $f_{\text{re}}(p, q)$ is minimized when the pattern has pq same-size intervals and when the checkpoints and verifications are equally spaced among these intervals as in the BALANCEDALGORITHM, in which case $f_{\text{re}}(p, q) = \frac{p+q}{2pq}$. We first prove this result for $p = 1$ before moving to the general case. Finally, we explain how to choose the optimal pattern given values of C and V.

Theorem 1.2 *The minimal value of $f_{\text{re}}(1, q)$ is obtained for same-size chunks and it is $f_{\text{re}}(1, q) = \frac{q+1}{2q}$.*

Proof For $q = 1$, we already know from the study of the first example that $f_{\text{re}}(1, 1) = 1$. Consider a pattern with $q \geq 2$ verifications, executing a total work W. Let $\alpha_i W$ be the size of the i-th chunk, where $\sum_{i=1}^{q} \alpha_i = 1$ (see Fig. 1.33). We compute the expected fraction of work that is re-executed when a failure strikes the pattern as follows. With probability α_i, the failure strikes in the i-th chunk. The error is detected by the i-th verification, we roll back to the beginning of the pattern, so we re-execute the first i chunks. Altogether, the amount of work that is re-executed is $\sum_{i=1}^{q} \left(\alpha_i \sum_{j=1}^{i} \alpha_j W \right)$, hence

$$f_{\text{re}}(1, q) = \sum_{i=1}^{q} \left(\alpha_i \sum_{j=1}^{i} \alpha_j \right). \tag{1.54}$$

What is the minimal value of $f_{\text{re}}(1, q)$ in Eq. (1.54) under the constraint $\sum_{i=1}^{q} \alpha_i = 1$? We rewrite

$$f_{\text{re}}(1, q) = \frac{1}{2} \left(\sum_{i=1}^{q} \alpha_i \right)^2 + \frac{1}{2} \sum_{i=1}^{q} \alpha_i^2 = \frac{1}{2} \left(1 + \sum_{i=1}^{q} \alpha_i^2 \right),$$

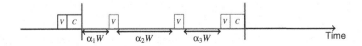

Fig. 1.33 A pattern with different-size chunks, for $p = 1$ and $q = 3$

and by convexity, we see that f_{re} is minimal when all the α_i's have the same value $1/q$. In that case, we derive that $f_{re}(1, q) = \frac{1}{2}(1 + \sum_{i=1}^{q} \frac{1}{q^2}) = \frac{q+1}{2q}$, which concludes the proof.

When $p = 1$, BALANCEDALGORITHM uses q same-size chunks. Theorem 1.2 shows that this is optimal.

Theorem 1.3 *For a pattern with $p \geq 1$, the minimal value of $f_{re}(p, q)$ is $f_{re}(p, q) = \frac{p+q}{2pq}$, and it is obtained with the BALANCEDALGORITHM.*

Proof Consider an arbitrary pattern with p checkpoints, $q \geq p$ verifications and total work W. The distribution of the checkpoints and verifications is unknown, and different-size chunks can be used. The only assumption is that the pattern ends by a verification followed by a checkpoint.

The main idea of the proof is to compare the gain in re-execution time due to the $p - 1$ intermediate checkpoints. Let $f_{re}^{(p)}$ be the fraction of work that is re-executed for the pattern, and let $f_{re}^{(1)}$ be the fraction of work that is re-executed for the same pattern, but where the $p - 1$ first checkpoints have been suppressed. Clearly, $f_{re}^{(p)}$ is smaller than $f_{re}^{(1)}$, because the additional checkpoints save some roll-backs, and we aim at maximizing their difference.

In the original pattern, let $\alpha_i W$ be the amount of work before the i-th checkpoint, for $1 \leq i \leq p$ (and with $\sum_{i=1}^{p} \alpha_i = 1$). Figure 1.34 presents an example with $p = 3$. What is the gain due to the presence of the $p - 1$ intermediate checkpoints? If an error strikes before the first checkpoint, which happens with probability α_1, there is no gain, because we always rollback from the beginning of the pattern. This is true regardless of the number and distribution of the q verifications in the pattern. If an error strikes after the first checkpoint and before the second one, which happens with probability α_2, we do have a gain: instead of rolling back to the beginning of the pattern, we rollback only to the first checkpoint, which saves $\alpha_1 W$ units of re-executed work. Again, this is true regardless of the number and distribution of the q verifications in the pattern. For the general case, if an error strikes after the $(i - 1)$-th checkpoint and before the i-th one, which happens with probability α_i, the gain is $\sum_{j=1}^{i-1} \alpha_j W$. We derive that

$$f_{re}^{(1)} - f_{re}^{(p)} = \sum_{i=1}^{p} \left(\alpha_i \sum_{j=1}^{i-1} \alpha_j \right).$$

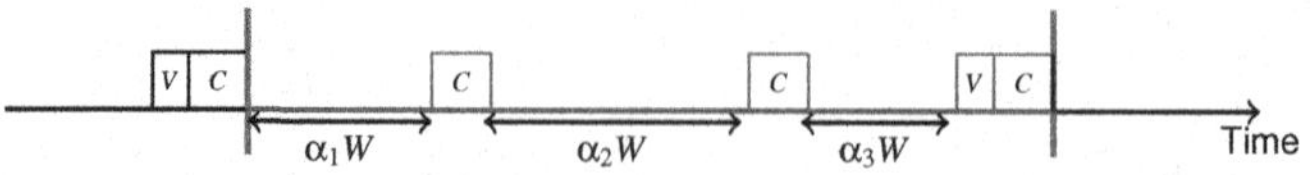

Fig. 1.34 A pattern with different-size chunks, with 3 checkpoints (we do not show where intermediate verifications are located)

Similarly to the proof of Theorem 1.2, we have

$$\sum_{i=1}^{p}\left(\alpha_i \sum_{j=1}^{i-1}\alpha_j\right) = \frac{1}{2}\left(\left(\sum_{i=1}^{p}\alpha_i\right)^2 - \sum_{i=1}^{p}\alpha_i^2\right) = \frac{1}{2}\left(1 - \sum_{i=1}^{p}\alpha_i^2\right)$$

and by convexity, the difference $f_{re}^{(1)} - f_{re}^{(p)}$ is maximal when $\alpha_i = 1/p$ for all i. In that latter case, $f_{re}^{(1)} - f_{re}^{(p)} = \sum_{i=1}^{p}(i-1)/p^2 = (p-1)/p^2$. This result shows that the checkpoints should be equipartitioned in the pattern, regardless of the location of the verifications.

To conclude the proof, we now use Theorem 1.2: to minimize the value of $f_{re}^{(1)}$, we should equipartition the verifications too. In that case, we have $f_{re}^{(1)} = \frac{q+1}{2q}$ and $f_{re}^{(p)} = \frac{q+1}{2q} - \frac{p-1}{2p} = \frac{q+p}{2pq}$, which concludes the proof.

Theorem 1.3 shows that BALANCEDALGORITHM is the optimal pattern with p checkpoints and q verifications when μ is large. An important consequence of this result is that we never need to keep more than two checkpoints in memory when $p \leq q$, because it is optimal to regularly interleave checkpoints and verifications.

To conclude this study, we outline a simple procedure to determine the best pattern. We start with the following result:

Theorem 1.4 *Assume that μ is large in front of C, R and V, and that $\sqrt{\frac{V}{C}}$ is a rational number $\frac{u}{v}$, where u and v are relatively prime. Then the optimal pattern S_{opt} is obtained with the BALANCEDALGORITHM, using $p = u$ checkpoints, $q = v$ verifications, and pq equal-size chunks of total length $\sqrt{\frac{2pq(pC+qV)\mu}{p+q}}$.*

We prove this theorem before discussing the case where $\sqrt{\frac{V}{C}}$ is not a rational number.

Proof Assume that $V = \gamma C$, where $\gamma = \frac{u^2}{v^2}$, with u and v relatively prime integers. Then, the product $o_{ff}f_{re}$ can be expressed as

$$o_{ff}f_{re} = \frac{p+q}{2pq}(pC + qV) = C \times \frac{p+q}{2}\left(\frac{1}{q} + \frac{\gamma}{p}\right).$$

Therefore, given a value of C and a value of V, i.e., given γ, the goal is to minimize the function $\frac{p+q}{2}\left(\frac{1}{q} + \frac{\gamma}{p}\right)$ with $1 \leq p \leq q$, and p, q taking integer values.

Let $p = \lambda \times q$. Then we aim at minimizing

$$\frac{1+\lambda}{2}\left(1 + \frac{\gamma}{\lambda}\right) = \frac{\lambda}{2} + \frac{\gamma}{2\lambda} + \frac{1+\gamma}{2},$$

and we obtain $\lambda_{opt} = \sqrt{\gamma} = \sqrt{\frac{V}{C}} = \frac{u}{v}$. Hence the best pattern is that returned by the BALANCEDALGORITHM with $p = u$ checkpoints and $q = v$ verifications. This pattern uses pq equal-size chunks whose total length is given by Eq. (1.52), hence the result.

For instance, for $V = 4$ and $C = 9$, we obtain $\lambda_{opt} = \sqrt{\frac{V}{C}} = \frac{2}{3}$, and a balanced pattern with $p = 2$ and $q = 3$ is optimal. This pattern will have 6 equal-size chunks whose total length is $\sqrt{\frac{12(2C+3V)\mu}{5}} = 6\sqrt{2\mu}$. However, if $V = C = 9$, then $\lambda_{opt} = 1$ and the best solution is the base algorithm with $p = q = 1$ and a single chunk of size $\sqrt{(C + V)\mu} = \sqrt{13\mu}$.

In some cases, $\lambda_{opt} = \sqrt{\frac{V}{C}}$ may not be a rational number, and we need to find good approximations of p and q in order to minimize the asymptotic waste. A solution is to try all reasonable values of q, say from 1 to 50, and to compute the asymptotic waste achieved with $p_1 = \lfloor \lambda_{opt} \times q \rfloor$ and $p_2 = \lceil \lambda_{opt} \times q \rceil$, hence testing at most 100 configurations (p, q). Altogether, we can compute the best pattern with $q \leq 50$ in constant time.

1.7 Conclusion

This chapter presented an overview of the fault tolerance techniques most frequently used in HPC. Large-scale machines consist of components that are robust but not perfectly reliable. They combine a number of components that grows exponentially and will suffer from failures at a rate inversely proportional to that number. Thus, to cope with such failures, we presented two sets of approaches:

• On the one hand, middleware, hardware, and libraries can implement general techniques to conceal failures from higher levels of the software stack, enabling the execution of genuine applications not designed for fault tolerance. Behind such approaches, periodic checkpointing with rollback-recovery is the most popular technique used in HPC, because of its multiple uses (fault tolerance, but also post-mortem analysis of behavior, and validation), and potential better usage of resources. We presented many variations on the protocols of these methods, and discussed practical issues, like checkpoint creation and storage.

At the heart of periodic checkpointing with rollback recovery lays an optimization problem: rollback happens when failures occurs; it induces re-execution, hence resource consumption to tolerate the failure that occurred; frequent checkpointing reduces that resource consumption. However, checkpointing also consumes resources, even when failures do not occur; thus checkpointing too often becomes a source of inefficiency. We presented probabilistic performance models that allow to deduce the optimal trade-off between frequent checkpoints and high failure-free overheads for the large collection of protocols that we presented before.

The costs of checkpointing and coordinated rollback are the major source of overheads in these protocols: future large scale systems can hope to rely on rollback recovery only if the time spent in checkpointing and rolling-back can be kept orders of magnitude under the time between failures. The last protocol we studied, that uses the memory of the peers to store checkpoints, aims precisely at this. But since the checkpoints become stored in memory, that storage becomes unreliable, and mitigating the risk of a non-recoverable failure reenters the trade-off. Here again, probabilistic models allow to quantify this risk, to guide the decision of resource usage optimization.

- On the other hand, by letting the hardware and middleware expose the failures to the higher-level libraries and the application (while tolerating failures at their level to continue providing their service), we showed how a much better efficiency can be expected. We presented briefly the current efforts pursued in the MPI standardization body to allow such behavior in high-performance libraries and application. Then, we illustrated over complex realistic examples how some applications can take advantage of failures awareness to provide high efficiency and fault tolerance. Because these techniques are application-specific, many applications may not be capable of using them. To address this issue, we presented a composition technique that enables libraries to mask failures that are exposed to them from a non fault-tolerant application. That composition relies on the general rollback-recovery technique, but allows to disable periodic checkpointing during long phases where the library controls the execution, featuring the high-efficiency of application-specific techniques together with the generality of rollback-recovery.

To conclude, we considered the case of silent errors: silent errors, by definition, do not manifest as a failure at the moment they strike; the application may slowly diverge from a correct behavior, and the data be corrupted before the error is detected. Because of this, they pose a new challenge to fault tolerance techniques. We presented how multiple rollback points may become necessary, and how harder it becomes to decide when to rollback. We also presented how application-specific techniques can mitigate these issue by providing data consistency checkers (validators), allowing to detect the occurrence of a silent error not necessarily when it happens, but before critical steps.

Designing a fault-tolerant system is a complex task that introduces new programming and optimization challenges. However, the combination of the whole spectrum of techniques, from application-specific to general tools, at different levels of the software stack, allows to tolerate a large range of failures with the high efficiency expected in HPC. In the rest of this book, experts in fault tolerance and HPC have contributed with technical chapters, in which they dig deeper into some of the topics that were overviewed in this chapter. Their contributions present the most recent advances in an intellectually buoyant research field. We hope they will inspire innovative solutions and the adoption of sound approaches to tolerate failures at large scale.

Acknowledgments Yves Robert is with the Institut Universitaire de France. The research presented in this chapter was supported in part by the French ANR (*Rescue* project) and by contracts with the DOE through the SUPER-SCIDAC project, and the CREST project of the Japan Science and Technology Agency (JST). This chapter has borrowed material from publications co-authored with many colleagues and PhD students, and the authors would like to thank Guillaume Aupy, Anne Benoit, George Bosilca, Aurélien Bouteiller, Aurélien Cavelan, Franck Cappello, Henri Casanova, Amina Guermouche, Saurabh K. Raina, Hongyang Sun, Frédéric Vivien, and Dounia Zaidouni.

References

1. Amdahl G (1967) The validity of the single processor approach to achieving large scale computing capabilities. In: AFIPS conference proceedings, vol 30. AFIPS Press, pp 483–485
2. Aupy G, Robert Y, Vivien F, Zaidouni D (2013) Checkpointing strategies with prediction windows. In: IEEE 19th Pacific Rim international symposium on dependable computing (PRDC), 2013. IEEE, pp 1–10
3. Aupy G, Benoit A, Herault T, Robert Y, Vivien F, Zaidouni D (2013) On the combination of silent error detection and checkpointing. In: PRDC 2013, the 19th IEEE Pacific Rim international symposium on dependable computing. IEEE Computer Society Press
4. Aupy G, Robert Y, Vivien F, Zaidouni D (2014) Checkpointing algorithms and fault prediction. J Parallel Distrib Comput 74(2):2048–2064
5. Bautista-Gomez L, Tsuboi S, Komatitsch D, Cappello F, Maruyama N, Matsuoka S (2011) FTI: high performance fault tolerance interface for hybrid systems. In: International conference high performance computing, networking, storage and analysis SC'11
6. Benson AR, Schmit S, Schreiber R (2013) Silent error detection in numerical time-stepping schemes. CoRR, abs/1312.2674
7. Blackford LS, Choi J, Cleary A, D'Azevedo E, Demmel J, Dhillon I, Dongarra J, Hammarling S, Henry G, Petitet A, Stanley K, Walker D, Whaley RC (1997) ScaLAPACK users' guide. SIAM
8. Bosilca G, Delmas R, Dongarra J, Langou J (2009) Algorithm-based fault tolerance applied to high performance computing. J Parallel Distrib Comput 69(4):410–416
9. Bosilca G, Bouteiller A, Herault T, Robert Y, Dongarra JJ (2014) Assessing the impact of ABFT and checkpoint composite strategies. In: 2014 IEEE international parallel and distributed processing symposium workshops, Phoenix, AZ, USA, May 19–23 2014, pp 679–688
10. Bosilca G, Bouteiller A, Brunet E, Cappello F, Dongarra J, Guermouche A, Herault T, Robert Y, Vivien F, Zaidouni D (2014) Unified model for assessing checkpointing protocols at extreme-scale. Concurr Comput Pract Exp 26(17):925–957
11. Bosilca G, Bouteiller A, Herault T, Robert Y, Dongarra JJ (2015) Composing resilience techniques: ABFT, periodic and incremental checkpointing. IJNC 5(1):2–25
12. Bougeret M, Casanova H, Rabie M, Robert Y, Vivien F (2011) Checkpointing strategies for parallel jobs. In: Proceedings of SC'11
13. Bouteiller A, Herault T, Krawezik G, Lemarinier P, Cappello F (2006) MPICH-V: a multiprotocol fault tolerant MPI. IJHPCA 20(3):319–333
14. Bouteiller A, Bosilca G, Dongarra J (2010) Redesigning the message logging model for high performance. Concurr Comput Pract Exp 22(16):2196–2211
15. Bouteiller A, Herault T, Bosilca G, Dongarra JJ (2011) Correlated set coordination in fault tolerant message logging protocols. In: Proceedings of Euro-Par'11 (II). LNCS, vol 6853. Springer, pp 51–64
16. Bouteiller A, Herault T, Bosilca G, Du P, Dongarra J (2015) Algorithm-based fault tolerance for dense matrix factorizations, multiple failures and accuracy. ACM Trans Parallel Comput 1(2):10:1–10:28
17. Bronevetsky G, de Supinski B (2008) Soft error vulnerability of iterative linear algebra methods. In: Proceedings of 22nd international conference on supercomputing, ICS '08. ACM, pp 155–164
18. Cannon LE (1969) A cellular computer to implement the Kalman filter algorithm. Ph.D. thesis, Montana State University
19. Casanova H, Robert Y, Vivien F, Zaidouni D (2012) Combining process replication and checkpointing for resilience on exascale systems. Research report RR-7951, INRIA
20. Chandy KM, Lamport L (1985) Distributed snapshots: determining global states of distributed systems. ACM Trans Comput Syst 3(1):63–75
21. Chen Z (2013) Online-ABFT: an online algorithm based fault tolerance scheme for soft error detection in iterative methods. In: Proceedings of 18th ACM SIGPLAN symposium on principles and practice of parallel programming, PPoPP '13. ACM, pp 167–176

22. Chen Z, Dongarra J (2006) Algorithm-based checkpoint-free fault tolerance for parallel matrix computations on volatile resources. In: Proceedings of the 20th international conference on parallel and distributed processing, IPDPS'06, Washington, DC, USA. IEEE Computer Society, pp 97–97

23. Chen Z, Dongarra J (2008) Algorithm-based fault tolerance for fail-stop failures. IEEE TPDS 19(12):1628–1641

24. Choi J, Demmel J, Dhillon I, Dongarra J, Ostrouchov S, Petitet A, Stanley K, Walker D, Whaley R (1996) ScaLAPACK: a portable linear algebra library for distributed memory computers-design issues and performance. Comput Phys Commun 97(1–2):1–15

25. Daly JT (2004) A higher order estimate of the optimum checkpoint interval for restart dumps. FGCS 22(3):303–312

26. Davies T, Karlsson C, Liu H, Ding C, Chen Z (2011) High performance linpack benchmark: a fault tolerant implementation without checkpointing. In: Proceedings of the international conference on supercomputing, ICS '11. ACM, New York, pp 162–171

27. Dongarra J, Beckman P, Aerts P, Cappello F, Lippert T, Matsuoka S, Messina P, Moore T, Stevens R, Trefethen A, Valero M (2009) The international exascale software project: a call to cooperative action by the global high-performance community. Int J High Perform Comput Appl 23(4):309–322

28. Dongarra J, Herault T, Robert Y (2014) Performance and reliability trade-offs for the double checkpointing algorithm. Int J Netw Comput 4(1):23–41

29. Du P, Bouteiller A, Bosilca G, Herault T, Dongarra J (2012) Algorithm-based fault tolerance for dense matrix factorizations. In: Proceedings of the 17th ACM SIGPLAN symposium on principles and practice of parallel programming, PPOPP 2012, New Orleans, LA, USA, 25–29 February 2012, pp 225–234

30. Elliott J, Kharbas K, Fiala D, Mueller F, Ferreira K, Engelmann C (2012) Combining partial redundancy and checkpointing for HPC. In: Proceedings of ICDCS '12. IEEE Computer Society

31. Engelmann C, Ong HH, Scorr SL (2009) The case for modular redundancy in large-scale highh performance computing systems. In: Proceedings of the 8th IASTED international conference on parallel and distributed computing and networks (PDCN), pp 189–194

32. Esteban Meneses CLM, Kalé LV (2010) Team-based message logging: preliminary results. In: Workshop resilience in clusters, clouds, and grids (CCGRID 2010)

33. Ferreira K, Stearley J, Laros JHI, Oldfield R, Pedretti, K, Brightwell R, Riesen R, Bridges, PG, Arnold D (2011) Evaluating the viability of process replication reliability for exascale systems. In: Proceedings of the ACM/IEEE on supercomputing

34. Flajolet P, Grabner PJ, Kirschenhofer P, Prodinger H (1995) On Ramanujan's Q-function. J Comput Appl Math 58:103–116

35. Gainaru A, Cappello F, Kramer W (2012) Taming of the shrew: modeling the normal and faulty behavior of large-scale HPC systems. In: Proceedings of IPDPS'12

36. Guermouche A, Ropars T, Snir M, Cappello F (to appear) HydEE: failure containment without event logging for large scale send-deterministic MPI applications. In: Proceedings of IEEE IPDPS 2012

37. Gustafson JL (1988) Reevaluating Amdahl's law. IBM Syst J 31(5):532–533

38. Gärtner F (1999) Fundamentals of fault-tolerant distributed computing in asynchronous environments. ACM Comput Surv 31(1):1–26

39. Hacker TJ, Romero F, Carothers CD (2009) An analysis of clustered failures on large supercomputing systems. J Parallel Distrib Comput 69(7):652–665

40. Hakkarinen D, Chen Z (2010) Algorithmic cholesky factorization fault recovery. In: 2010 IEEE International symposium on parallel distributed processing (IPDPS). IEEE, Atlanta, pp 1–10

41. Hargrove PH, Duell JC (2006) Berkeley lab checkpoint/restart (BLCR) for linux clusters. In: Proceedings of SciDAC 2006

42. Heath T, Martin RP, Nguyen TD (2002) Improving cluster availability using workstation validation. SIGMETRICS Perform Eval Rev 30(1):217–227

43. Heien R, Kondo D, Gainaru A, LaPine D, Kramer B, Cappello F (2011) Modeling and tolerating heterogeneous failures on large parallel system. In: Proceedings of the IEEE/ACM conference on supercomputing (SC)
44. Heroux M, Hoemmen M (2011) Fault-tolerant iterative methods via selective reliability. Research report SAND2011-3915 C, Sandia National Laboratories
45. Huang K-H, Abraham J (1984) Algorithm-based fault tolerance for matrix operations. IEEE Trans Comput C-33(6):518–528
46. Huang K-H, Abraham JA (1984) Algorithm-based fault tolerance for matrix operations. IEEE Trans Comput 33(6):518–528
47. Hursey J, Squyres J, Mattox T, Lumsdaine A (2007) The design and implementation of check-point/restart process fault tolerance for open MPI. In: IEEE international parallel and distributed processing symposium, 2007. IPDPS. pp 1–8
48. Hwang AA, Stefanovici IA, Schroeder B (2012) Cosmic rays don't strike twice: understanding the nature of dram errors and the implications for system design. SIGARCH Comput Archit News 40(1):111–122
49. Kella O, Stadje W (2006) Superposition of renewal processes and an application to multi-server queues. Stat Probab Lett 76(17):1914–1924
50. Kingsley G, Beck M, Plank JS (1995) Compiler-assisted checkpoint optimization using SUIF. In: First SUIF compiler workshop
51. Kondo D, Chien A, Casanova H (2007) Scheduling task parallel applications for rapid application turnaround on enterprise desktop grids. J Grid Comput 5(4):379–405
52. Lamport L (1978) Time, clocks, and the ordering of events in a distributed system. Commun ACM 21(7):558–565
53. Li C-C, Fuchs W (1990) Catch-compiler-assisted techniques for checkpointing. In: 20th international symposium fault-tolerant computing, 1990. FTCS-20. Digest of papers, pp 74–81
54. Liu Y, Nassar R, Leangsuksun C, Naksinehaboon N, Paun M, Scott S (2008) An optimal checkpoint/restart model for a large scale high performance computing system. In: IPDPS'08. IEEE
55. Lu G, Zheng Z, Chien AA (2013) When is multi-version checkpointing needed. In: 3rd Workshop for fault-tolerance at extreme scale (FTXS). ACM Press. https://sites.google.com/site/uchicagolssg/lssg/research/gvr
56. Lyons RE, Vanderkulk W (1962) The use of triple-modular redundancy to improve computer reliability. IBM J Res Dev 6(2):200–209
57. Moody A, Bronevetsky G, Mohror K, Supinski B (2010) Design, modeling, and evaluation of a scalable multi-level checkpointing system. In: International conference high performance computing, networking, storage and analysis SC'10
58. Moody A, Bronevetsky G, Mohror K, Supinski BR de (2010) Design, modeling, and evaluation of a scalable multi-level checkpointing system. In: Proceedings of the ACM/IEEE conference SC, pp 1–11
59. Ni X, Meneses E, Kalé LV (2012) Hiding checkpoint overhead in HPC applications with a semi-blocking algorithm. In: Proceedings of IEEE international conference on cluster computing. IEEE Computer Society
60. Ni X, Meneses E, Jain N, Kalé LV (2013) ACR: automatic checkpoint/restart for soft and hard error protection. In: Proceedings of international conference high performance computing, networking, storage and analysis, SC '13. ACM
61. O'Gorman T (1994) The effect of cosmic rays on the soft error rate of a DRAM at ground level. IEEE Trans Electron Devices 41(4):553–557
62. Plank JS, Beck M, Kingsley G (1995) Compiler-assisted memory exclusion for fast check-pointing. IEEE Tech Comm Oper Syst Appl Environ 7:10–14
63. Rodríguez G, Martín MJ, González P, Touriño J, Doallo R (2010) CPPC: a compiler-assisted tool for portable checkpointing of message-passing applications. Concurr Comput Pract Exp 22(6):749–766
64. Ross SM (2009) Introduction to probability models, 8th edn. Academic Press, San Diego
65. Sao P, Vuduc R (2013) Self-stabilizing iterative solvers. In: Proceedings of ScalA '13. ACM

66. Schroeder B, Gibson G (2007) Understanding failures in petascale computers. J Phys Conf Ser 78(1):188–198
67. Schroeder B, Gibson GA (2006) A large-scale study of failures in high-performance computing systems. In: Proceedings of DSN, pp 249–258
68. Shantharam M, Srinivasmurthy S, Raghavan P (2012) Fault tolerant preconditioned conjugate gradient for sparse linear system solution. In: Proceedings of ICS '12. ACM
69. Young JW (1974) A first order approximation to the optimum checkpoint interval. Commun ACM 17(9):530–531
70. Yu L, Zheng Z, Lan Z, Coghlan S (2011) Practical online failure prediction for blue gene/p: period-based vs event-driven. In: Dependable systems and networks workshops (DSN-W), pp 259–264
71. Zheng G, Shi L, Kalé LV (2004) FTC-Charm++: an in-memory checkpoint-based fault tolerant runtime for Charm++ and MPI. In: Proceedings of IEEE international conference on cluster computing. IEEE Computer Society
72. Zheng Z, Lan Z (2009) Reliability-aware scalability models for high performance computing. In: Proceedings of the IEEE conference on cluster computing
73. Zheng Z, Lan Z, Gupta R, Coghlan S, Beckman P (2010) A practical failure prediction with location and lead time for blue gene/p. In: Dependable systems and networks workshops (DSN-W), pp 15–22
74. Ziegler J, Muhlfeld H, Montrose C, Curtis H, O'Gorman T, Ross J (1996) Accelerated testing for cosmic soft-error rate. IBM J Res Dev 40(1):51–72
75. Ziegler J, Nelson M, Shell J, Peterson R, Gelderloos C, Muhlfeld H, Montrose C (1998) Cosmic ray soft error rates of 16-Mb DRAM memory chips. IEEE J Solid-State Circuits 33(2):246–252
76. Ziegler JF, Curtis HW, Muhlfeld HP, Montrose CJ, Chin B (1996) IBM experiments in soft fails in computer electronics. IBM J Res Dev 40(1):3–18

Part II
Technical Contributions

Chapter 2
Errors and Faults

Ana Gainaru and Franck Cappello

Abstract Understanding the behavior of failures in large-scale systems is important in order to design techniques to tolerate them. Reliability knowledge of resources can be used in numerous ways by scientist of systems administrators: (1) it can be used to improve the quality of service of the machine; (2) to reduce performance loss due to unexpected failures either by reliability-aware scheduling or by reliability-aware checkpointing; (3) to design more resilient applications, programming models or machines in the future. This chapter focuses on offering an overview of failures observed in real large-scale systems and their characteristics, with an emphasis on modeling, detection, and prediction.

2.1 Introduction

As large-scale systems evolve toward post-Petascale computing to accommodate applications' increasing demands for computational capabilities, many new challenges need to be faced, among which fault tolerance is a crucial one.

The number of system components in current and future supercomputers increases faster than component reliability. In the future, projections show larger systems with even more failure-prone hardware and more complex system and application codes. Near threshold logic is considered as a candidate technology for future system. It has advantages in power consumption but it increases error rates. Even in classic CMOS technology, soft errors can cause one or multiple bits to spontaneously flip to the opposite state, due to multiple reasons such as alpha particles from package decay or cosmic rays. Although techniques such as error correcting codes (ECCs) have been implemented in memory, in reality, some bit flips still manage to pass undetected.

A. Gainaru (✉)
NCSA, University of Illinois at Urbana-Champaign, Champaign, USA
e-mail: againaru@ncsa.illinois.edu

F. Cappello
Argonne National Laboratory, Lemont, USA
e-mail: cappello@mcs.anl.gov

© Springer International Publishing Switzerland (outside the USA) 2015   89
T. Herault and Y. Robert (eds.), *Fault-Tolerance Techniques*
for High-Performance Computing, Computer Communications and Networks,
DOI 10.1007/978-3-319-20943-2_2

Moreover, processor caches are not protected by ECC in general. In addition, the constant need to reduce component size and voltage, limits the use of soft-error mitigation techniques. The overall consequence is a decreasing Mean Time Between Failures (MTBF) for future extreme-scale systems.

Failures in supercomputers are assumed to be uniformly distributed in time. However, recent studies show that failures in high-performance computing systems are partially correlated in time, generating periods of higher failure density. Understanding the inter-arrival patterns of failures is crucial in optimizing current fault tolerance approaches and decreasing the impact of failures on execution time to a minimum. This chapter provides characterization of failures and their pattern in extreme-scale computers with a focus on modeling, detection, and prediction. We present currently used detection mechanisms for several national laboratories in the US as well as state-of-the-art research for more sophisticated mechanism in Sect. 2.3. Failure detection is valuable for system management, replication, load balancing, and other maintenance services. The inter-arrival distribution of failures has been the study of many research programs. Failure modeling is an important research direction used in guiding reliability-aware resource allocation and optimizing fault-tolerant protocols in order to minimize the performance loss due to failures. We present the most recent findings and their impact on resiliency protocols in Sect. 2.5.

There are important lessons to be learned form the statistical information of failures and events generated by current large-scale systems (Sect. 2.4). These lessons will guide the design of future extreme-scale platforms and can be used to predict the direction of improvements in technological solutions for future system and application software development. Finally failure prediction is a field that complements current resiliency methods and has the potential of improving the performance of fault-tolerant protocols. A survey of prediction methods (starting from methods that assess the future reliability of a system to methods that pinpoint the exact time and space occurrence of the next failure) and their impact on checkpointing is presented in Sect. 2.6.

2.2 Definitions

The absence of consistent definitions and metrics for supercomputer reliability, availability and serviceability has hindered meaningful collaborations in the community [72]. In order to avoid this problem, the workshop organized by the Institute for Computing Sciences on August 2012 proposed a taxonomy of terms to be used as standard. The definitions that were proposed are based almost entirely on [5]. We will use the terms as defined in the workshop's report [71]. This section enumerates the most important ones.

Resilience is defined as the ability of a system to keep applications running and maintain an acceptable level of service in the face of transient, intermittent, and permanent faults. The term "fault tolerance" refers to the ability of a system to continue performing its intended function properly in the face of faults.

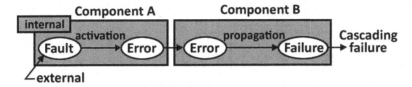

Fig. 2.1 Error propagation and cascading failures

Figure 2.1 shows the propagation chain from faults to failures in a system. A fault represents the cause of an error, like a bit flip due to an alpha particle. An error is the part of total state that might lead to a failure and the failure is a transition to incorrect service. Faults can be active or inactive, depending on whether they cause errors or not. For example, a software bug that is never exercised is called inactive while a bit flip in the processor cache that leads to an application crash is called an active fault. In general, a fault is local to one component, either software or hardware, while errors and failures may propagate from one component to another. In case of failures, this propagation is called cascading failures.

More generally, a failure can be defined as the event that occurs when the service delivered deviates from the correct service operation or when at least one external state of the system deviates from the correct service state. Faults may be caused by complex combinations of internal states and external conditions that occur rarely and are difficult to reproduce.

By error identification, we mean the process of discovering the presence of an error but without necessarily identifying which part of the system state is incorrect, and what fault caused this error. By definition, every fault causes an error. Almost always, the fault is detected by detecting the error the fault caused. Therefore, fault detection or error detection often refers to the same thing. Latent or silent errors are errors that are not detected.

There are several means of dealing with faults divided in four separate classes:

- Tolerance is used to avoid failures in the presence of faults
- Removal is used to reduce the fault number and severity
- Forecasting is used to estimate the present number, future incidence and likely consequences of faults
- Prevention is used to prevent fault occurrences

The term "Time to Failure" or TTF represents the interval between the end of the last failure and the beginning of the consecutive failure. Time between Failures (TBF) represents the interval between the beginnings of two consecutive failures and the Time to Repair (TTR) is synonymous with the Downtime and it defines the interval between the beginning of a failure and its end. Formally:

MTBF (Mean time between failures) = $\frac{TotalTime}{NumFailures}$

MTTF (Mean time to failure) = $\frac{Uptime}{NumFailures}$

MTTR (Mean time to repair) = $\frac{UnscheduledDowntime}{NumFailures}$

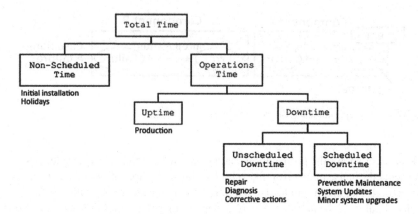

Fig. 2.2 Diagram of possible states in a supercomputer's life

An incredibly dense state diagram would be required to characterize all possible states for a supercomputer lifespan and its workload. For the purpose of resiliency, we will consider the diagram in Fig. 2.2. The diagram is a simplified version of the one proposed by [72].

The nonscheduled time represents the interval when the system is not scheduled to be utilized by production or system engineering users. Otherwise, the system is considered in its operation times. This interval includes production time or unscheduled and scheduled downtime. The unscheduled downtime occurs when a system is not available due to unplanned events, like failures, while the scheduled downtime occurs when the system is not available due to planned events, for example system testing in order to verify that a component is functioning properly.

Prediction is defined as the activity of estimating the presence of a failure. A prediction where the predicted failures occurred in the given time interval and on the given location (or set of locations) is called a true positive. A false positive happens when the predicted failure either does not occur in the given time frame or the predicting location is wrong. Failures that occur without being predicted are false negatives. These three values define the three metrics that we use to measure the performance of a failure predictor: precision, recall and F-measure. Precision defines the quality of all the predictions made by the method and it is equal to the ratio of true positives to the total of number of predictions made. Recall represents the coverage of a predictor and defines the ratio of failures predicted to the total failures in the system. A measure that combines precision and recall is the harmonic mean of the two, called F-measure.

2.3 Detection

In order for management systems to be able to analyze the characteristics of failures, to try to reduce the fault number and severity or to implement forecasting and prevention mechanisms, the system needs to be able to detect failures. In general, HPC system vendors integrate management systems that aim at ensuring job completion and providing system administrator with a way to check the sanity of the machine. Examples of such management systems are Cray's Node Health Checker or IBM's Cluster Systems Management.

Cray's node health checker is automatically invoked by the scheduler upon the termination of an application. The scheduler gives the monitoring system a list of compute nodes associated with the terminated application on which the node health checker performs specified tests to determine if the given compute nodes are healthy enough to support running subsequent applications. If a node fails one of the tests, it gets removed from the available resource pool. In addition, subsequent monitoring systems are deployed for other components as well. For example, the Gemini inter-connect technology used by Cray, has hardware and software support that allows the system to handle certain types of failures without requiring a system reboot. Another example is the CLFS Lustre Monitor tool that implements detection and fault-tolerant methods in order to keep the file systems available in the event of a Meta Data Server (MDS) and/or Object Storage Server (OSS) failures.

Similarly, IBM's Cluster System Management (CSM) provides automatic error detection through heartbeats that is implemented in conjunction with problem avoid-ance, resolution, and recovery. Disk-heartbeat networks work by exchanging heart-beat messages on a reserved portion of a shared disk. Moreover, current HPC systems implement various optimization methods that provide a way of reducing the time it takes for a node failure to be sent throughout the cluster. For example, IBM's CSM uses disk-heartbeat so that the node that fails puts a departing message on a shared disk so its neighbors will be immediately aware of the node failure (without waiting for missed heartbeats).

There are some failure situations in which heartbeat monitoring cannot determine what exactly failed. For example, lets imagine a cluster node failure due to a problem in a critical hardware component such as a processor. The whole machine can go down without giving the cluster resource service on that node an opportunity to notify other cluster nodes of the failure. The other nodes can only see a failure in the heartbeat monitoring. They are unable to know if it was due to a node failure or a failure in some part of the communication path (for example a router or an adapter).

More advanced node failure detection methods are provided outside of the vendors management systems. These solutions are implemented by system administrators and researchers and are used to reduce the number of failure scenarios which result in system partitions. In general, failures that are not detected at the system level but that propagate and crash an application, are analyzed post-mortem by system administrators. Rules and alerts are afterwards created in order to capture future occurrences of the same problem. This process is, in general, manual and requires a considerable effort.

There are several general methods used in the literature in order to detect failures in an automatic way, from modeling the hardware system and finding outliers in this model, to implementing fault detection at the application and programming model levels. We will briefly present, in this section, a few examples from each method. Most studies are not integrated in any production machine, but some are used as external tools by several national laboratories.

Firstly, there are external approaches for detecting a node failure by using a similar method offered by HPC vendors, namely based on the use of heartbeats, as a way of constantly monitoring a system. These are, in general, hardware health monitoring methods (e.g., IPMI an open standard hardware management specification that defines a set of common interfaces to hardware and firmware). For example, some network failures partition the filesystem into two or more groups of nodes that can only see the nodes in their group. These types of failures can be easily detected through a hardware heartbeat protocol. Software health monitoring [61] systems are also implemented on several large-scale systems by using timeouts to detect node problems.

One method used extensively in the past was to measure each node's behavior and compare it to all other nodes executing similar workloads. An event is categorized as a failure in case of a significant deviation [73, 88]. The method can be applied on performance metrics as well as log entries. After recording performance metrics at a fixed time interval from various components in the system, this information can then be aggregated into a single large matrix. Similarly, for system logs, homogeneous nodes correctly executing similar workloads will tend to generate almost identical logs. The same matrix as for performance metrics can be created by indexing the logs and using nonzero values in the matrix to indicate how many times word i occurs during hour j. By normalizing and performing PCA (principal component analysis), the methods are able to determine anomalies in these matrices. Such a method has been used by researchers at the Sandia National Laboratory in order to complement the available software integrated on their Spirit cluster. The results are in general good and show that the method is able to detect several known fault conditions. On Sandia's 512-node "Spirit" Linux cluster, the detection algorithm was able to localize 50 % of faults with 75 % precision.

Another similar type of method models the components and their interactions and then monitors the model. Most examples are using pattern recognition algorithms [65, 83] to model the system. Others include context free grammars [13] and mathematical equations [2]. Some methods are better suited for analyzing performance metrics (like regression models), while other can be applied on both log files and performance metrics. Some analyze the entire system, while others focus on a specific component. For example, Markov models can be used to implement network failure symptom detection and event correlation discovery. The failure detection's results, when applying the method on the entire system, are decent but less impressive than using the previous method. However, when focusing on one component, specifically networks, path analysis contributes to the discovery of the system structure and the detection of subtle behavioral differences across system versions. Path analysis is an extension of regression methods that estimates the magnitude and significance of causal connections between sets of variables. The method complements existing failure detection methods offered by vendors. Similarly, several studies use, as their core abstraction, runtime paths followed by requests as they move through the system. Based on this, they characterize component interactions. Automated statistical analysis of multiple paths allows to detect and put a diagnosis on complex application level failures.

In the Clouds field, online failure detection is done by using entropies [80]. The algorithm works in three phases: metric collection, entropy time series construction, and entropy time series processing like spike detection, signal processing, or subspace methods, in order to find anomaly patterns. The method uses performance metrics by collecting values from all components on each physical level of the system's hierarchy. A leaf component collects raw metric data from its local sensors. A nonleaf component collects not only its local metric data but also entropy time series data from its child nodes. The method is not integrated in production at this point, but the preliminary results on smaller systems are extremely good showing a 90 % recall with an 80 % precision.

More specific methods focus on the software stack of an HPC system. One example is Rani et al. [61], where the authors propose a fault-tolerant approach that provides the ability to detect and self-recover the parallel runtime environment in cases of compute node failure. Their solution consists of a lightweight heartbeat protocol (BHB) that addresses the scalability issues in system monitoring and failure detection. Their focus is common fault tolerance issues in large scale systems, especially due to permanent component failure.

Application level failure detection is in general used for large servers or for large web application. For example, in [42], the authors present a generic framework for using statistical learning techniques to detect and localize likely application-level failures in component-based Internet services. In the HPC field, Kharbas et al. [41] propose a study on generic fault detection capabilities at the MPI level. The authors implement multiple detectors at various layers of the software stack: at the MPI communication layer and a separate one as stand-alone processes across nodes.

2.4 Observations

The design of extreme-scale platforms that are expected to reach Exascale in the next several years will represent an improvement in technological solutions and will push the boundary of algorithm and application software development. The precise details of these new designs are not yet known. However, there are numerous papers that look at the configurations and properties of existing systems and make predictions regarding the future of HPC systems. There are important lessons to be learned form the statistical information of failures and events by analyzing generated log files and performance/environmental metrics from current systems.

There are several Exascale/Petascale reports that focus on presenting directions for resiliency and programming models for future Exascale systems. The DARPA white paper on system resilience at extreme scale from 2008 [21] points out that current high-end systems waste in average 20 % of their computing capacity on failure and recovery. The paper outlines possible research in order to bring this number down to 2 %, but current methods are not yet at this point. The DOD/DOE report issued in 2009 [16] identifies resilience as a major emerging issue for HPC. It proposes research in five thrust areas: theoretical foundations, enabling infrastructure, fault prediction and detection, monitoring and control and end-to-end data integrity. The paper published in the International Journal of High Performance Computing Applications in 2009 [11] describes the challenges resiliency faces in the Exascale era and possible directions in order to address these needs. The DOD/DOE report [81] issued in 2012 identifies six high priorities: fault characterization, detection, fault-tolerant algorithms, fault-tolerant programming models, fault-tolerant system services and tools. The DOE workshop from 2012 [78] describes the required HPC resilience for critical DOE mission needs and details what HPC resilience research is already being done at the DOE national labs and what is expected to be done by industry and other groups. Also, the workshop focused on determining what fault management research is a priority for DOEs Office of Science and NNSA over the next five years. The Exascale report from March 2013 [71] gathered the main points discussed at the workshop organized by the Institute for Computing Sciences on August 2012. The report analyzes the state of resiliency for HPC and proposes three designs for approaches in this field: (1) business as usual where the global checkpoint/restart is used; (2) system-level resilience where vendors do not provide sufficiently low SDC rates at an acceptable acquisition and operation cost and a combination of hardware and software technologies is needed to hide the increased failure rates from the application; (3) application-level resilience for which there is an assumption that application codes will need to be modified in order to handle the increased failure rate. The paper makes a couple of recommendations for each design in order for them to become solutions for future systems.

The International Exascale Software Project (IESP) Workshop [19], held in Kobe, Japan on April 12–13, 2012 discussed what will be the major obstacles that the climate community will face at Exascale and proposed and evaluated possible ways to overcome these obstacles. The focus of the workshop was on node-level performance, scalability and resilience. The European Exascale Software Initiative EESI2 [62] is a collaborative project that aims to build and consolidate a vision and roadmap at the European level including applications both from academia and industry to address the challenge of performing scientific computing on the new generation super-computers. In September 2013, the project released the report on the first technical workshop where experts in the areas of software development, performance analysis, applications knowledge, funding models and governance aspects in High Performance Computing provided recommendations and roadmaps for the future of HPC in Europe.

Continuous availability of HPC systems has become a primary concern with the continuous increase of system size to thousands of components. Understanding the behavior of failures in current systems is increasingly important in order to design more reliable systems. To this extent, failure data analysis of current HPC systems can serve three purposes. Firstly, it can highlight dependability bottlenecks and might serve as a guideline for designing more reliable systems in the future. Moreover, real data can be used to develop performance models and simulations, which are an essential part of reliability engineering. As we will see in the following sections, these models can be used to predict node availability, which is useful for resource characterization and scheduling. Reliability knowledge of resources can reduce performance loss due to unexpected failures, and can improve QoS (Quality of Service) either by reliability-aware scheduling, where the systems allocate a priority job to get maximum reliability or by reliability-aware checkpointing where the optimal checkpointing interval can be computed based on the reliability of a set of nodes (see Chap. 1, and Sect. 1.3 in particular).

There are several papers that study the statistics of the data, including the root cause of failures, the mean time between failures, and the mean time to repair. Work on characterizing failures in computer systems differs in the type of data used; the type and number of systems under study; the time of data collection; and the number of failure or error records in the data set. Most of these statistics are based on reliability, availability and serviceability (RAS) data mainly provided by major HPC laboratories and centers in the USA: Los Alamos National Laboratory (LANL), National Energy Research Scientific Computing Center (NERSC), Pacific Northwest National Laboratory, Sandia National Laboratory (SNL), Laurence Livermore National Laboratory (LLNL), and the National Center for Supercomputing Applications (NCSA).

Most studies divide failures into two major categories: software and hardware, each having separately different subcategories. Reliability monitoring and analysis considers failures that affect a single node and also failures affecting a group of nodes. The studies also look at failures that may affect applications or important services. Table 2.1 presents an overview of the categories used in literature when looking at the broad overall image of a system.

Table 2.2 presents a summary of the current studies presenting failure statistics for different machines. Depending on the study and on the analyzed system, the table shows a wide range of results. There is not a consistent main root cause of failures among all systems, nor a consistent MTBF or mean time to repair. However, by looking at the overall view including all systems there are several observations to be made.

On average, for the biggest Petascale systems, there are failures of any type once every 7–10 h, while the systems suffer system-wide outages (SWO), in general once every week. The MTBF has continued to decrease from one failure every couple of months in very small systems (LANL systems in the Table) to a failure every several hours for the Blue Waters system. Moreover, the time to restart the machine and the applications after a system-wide outage is taking longer times for larger machines. The frequency of failures and the system complexity is making the task of failure detection and prediction much harder.

Table 2.1 Types of failures

Hardware failures	Software failures
Failures that affect group of nodes	
Switch; Power supply	Scheduler; FS; Cluster Management Software
Individual node failure	
Processor; Mother Board; Disk	OS; Client Daemon

Table 2.2 Different failure characterization studies and their results

System	MTBF	Root cause analysis	Citation
-A cluster of 12 SGI Origin 2000 (1500 CPUs) -A PC cluster (1000 CPUs) -A cluster of 162 Itanium dual CPUs	MTTI of 1 day, less than 1 h and about 6 h respectively	Software was at the origin of most outages (59–84 %)	Lu et al. [51]
Blue Gene/L during 6 months	More than 10 h	–	Leangsuksun et al. [45]
- Blue Gene/L (131k CPUs) - Red Storm (11k CPUs) - Thunderbird (9k CPUs) - Spirit (1k CPUs) - Liberty (512 CPUs)	–	Software caused 64 % of failures, while hardware only 18 %	Oliner et al. [57]
22 different systems at LANL, mostly large clusters of SMP and NUMA nodes, over a period of 10 years	–	Hardware is the main cause of failures with percentages ranging from 30 % to more than 60 %. Software is the second largest contributor, with percentages from 5 to 24 %	Schroeder et al. [68]
Blue Gene/P	The job level MTBF is about three times larger than that system level MTBF	–	Zheng et al. [87]
Blue Gene/L, Blue Gene/P, SciNet, Google	–	A large fraction of DRAM errors can be attributed to hard errors	Hwang et al. [38]
Beowulf-style PC clusters: Platium, Titan	6 h	Software represents the cause of most outages with 84 % for Platium and 60 % for Titan	Lu et al. [50]
Blue Waters	6–8 h	Software is the cause of almost all storage failures, more than 50 % of the node failures and more than 50 % of system wide outages	Di Martino [53]

2.4.1 Location Propagation

In general on all systems, over 15 % of failures affect more than one node (without considering system wide outages). For example failures in the voltage converter module, or problems with the cabinet controller on the Blue Waters system, affect a whole blade consisting of four nodes.

Large-scale systems contain a large number of nodes that are organized in a hierarchy. For example for the BlueGene systems, nodes are gathered into mid-planes and multiple mid-planes form a rack. The propagation path for different error types follows closely the way components are connected in the system. For example, if a fan breaks, all nodes sharing the same rack will be affected. In general, sequences of non faulty events, like warnings, following a failure do not propagate on different locations and if they propagate they appear on a small number of nodes: only around 22 % for Mercury and 25 % for Blue Gene/L show some kind of propagation. This phenomenon is consistent with all the systems presented in Table 2.2.

A very small number of failures appear on locations that do not follow the topology of the machine. An example of such a failure can be seen on the Mercury machine, when it experiences NFS (Network File System) problems. The event "`rpc: bad tcp reclen d+ (nonterminal)`" indicates network file system unavailability to any requests for a machine. In applications using the network file system this could cause file operations to fail and the application to quit. Also all nodes from which the application tries to access the network file system will be affected by this problem. This failure usually occurs nearly simultaneously on a large number of nodes depending on where the application was running.

In general some failure types are more likely to create cascade failures than others. This is the case, for example, of network and filesystem failures. In general, errors in memory or processor caches do not show the same behavior.

2.4.2 Failure Statistics

We divided all failures in 5 main categories that can be encountered in all systems: hardware, software, network, facility and unknown. Table 2.3 presents the percentage of failures representing each considered type for several HPC systems.

Table 2.3 Percentage of different failure types

Category	Blue Waters (%)	Blue Gene/P (%)	LANL systems (%)
Hardware	43.12	52.38	61.58
Software	26.67	30.66	23.02
Network	11.84	14.28	1.8
Facility/Environment	3.34	2.66	1.55
Unknown	2.98	–	11.38
Heartbeat	12.02	–	–

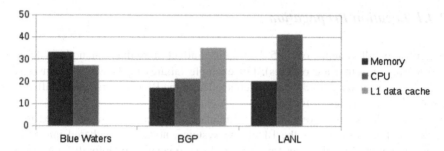

Fig. 2.3 Percentage of main hardware failures

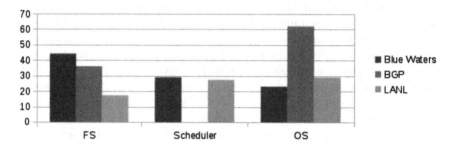

Fig. 2.4 Percentage of main software failures

Hardware represents the majority of failures for all systems, with the lowest percentage of 43.12 % for the Blue Waters system and 61.58 % for the LANL systems. As shown in Fig. 2.3, the majority of hardware failures were memory and processor errors. Moreover, failures with hardware root causes were limited to a single node in 96 % of the cases, or a single blade consisting of 4 service nodes in 99.3 % of the cases.

Software errors represent over 30 % of total failures for the Blue Waters system, while for the LANL system they represent only 23 %. In general, as the system increases in size and complexity, the number of software failures has continued to increase while the hardware failures represent a smaller percentage.

Figure 2.4 presents the main causes of software failures. The main ones are filesystem problems (Lustre for the Blue Waters system, GPFS for BGL and several for LANL: Cluster File System, Parallel File System, NFS, Scratch FS and Vizscratch FS), failures of the job scheduler and operating system problems. On the Blue Waters system, 12 % of total software failures caused system-wide outages (SWO) and represent over 75 % of all causes that triggered SWO. Moreover, software failures, when not causing an SWO, propagate to more than one node in 15 % of the cases.

Environmental failures include power-outages, failures related to temperature, cooling hardware problems and others. Table 2.4 presents the main failure types for each main category for each system.

Table 2.4 Main specific failure types

Blue Waters	Blue Gene/P	LANL systems
Hardware		
RAM 33.12%	L1 data cache parity error 35.27%	CPU 41.35%
CPU 27.04%	CPU 21.81%	DIMM 20.08%
	Memory 16.72%	
Software		
FS (Lustre) 27.2%	OS 62.11%	Other software 21.89%
Scheduler 18.9%	FS 36.02%	OS 20.99%
		DST 21.02%
		FS 12.33%

2.4.3 Additional Information

Some of the studies give additional information to what is presented in the Table 2.2. Schroeder et al. [68] demonstrate that the number of failures per socket in different systems is rather stable from 1996 to 2004. They also find that average failure rates differ wildly across systems, ranging from 20–1000 failures per year.

In a paper from 2011, Zheng et al. [87] analyze Blue Gene/P at Argonne National Lab and by using both RAS and job logs, they filter out failures that do not affect any jobs. By characterizing only the failures that lead to application crashes, they make a couple of interesting observations which might influence the fault-tolerant protocol used by different applications. Another observation their study reveals is that the probability of job interruption is high if there exist historical records of application-related interruptions. Moreover, most errors due to bugs in the application tend to be reported in the first hour. Therefore it is not recommended to introduce checkpointing early in the execution period if the job has historical records of application-related interruptions. Another interesting find is that the MTBF after filtering all failures that do not lead to application crashes is about three times larger than that without applying the filtering.

Tsai et al. [77] present a study that uses data collected from a population of over 50,000 customer deployed disk drives to examine the relationship between disk soft errors and failures, in particular failures manifested as hard errors. They observe that soft errors alone cannot be used as a reliable predictor of hard errors. However, in the cases where soft errors do accurately predict hard errors, sufficient warning time exists for preventive actions. Disk failures will be inspected in more detail in the next section.

In [38], Hwang et al. analyze data on DRAM errors collected on a diverse range of production systems in total covering nearly 300 terabyte-years of main memory. The authors provide a detailed analytical study of DRAM error characteristics, including both hard and soft errors.

Table 2.5 presents high-level statistics on the frequency of correctable and uncorrectable errors per year of operation, broken down by the type of hardware platform.

Table 2.5 Memory errors per year

Platform	Technology	Time (days)	Scrubbers	Corrected errors	Uncorrected errors
Google—platform 1	DDR1	2.5 years	no	45.80%	0.17%
Google—platform 2	DDR1	2.5 years	yes	22.30%	2.15%
Google—platform 3	DDR2	2.5 years	yes	19.60%	2.65%
Google—platform 4	Fully-Buffered DIMM	2.5 years	no	N/A	0.27%
Blue Gene/L	N/A	214	no	5.32%	N/A
Blue Gene/P	N/A	583	no	3.55%	1.34%
SciNet	N/A	211	no	2.51%	N/A

For the Google platforms the last two columns represent the percentage of machines affected by at least one error, while for the HPC system the percentage refers to nodes. As we see, memory errors are not rare events. For example, about a third of all machines at Google experience at least one uncorrectable memory error per year and the average number of correctable errors per year is over 22,000. These numbers vary across platforms, with some system experiencing nearly 50% of correctable errors, while in others it represents only 15–30%. For the platforms with a low percentage of machines affected by correctable errors, the average number of correctable errors per machine per year is the same or even higher than for the other platforms.

In general, we have seen that in all studies that analyze memory failures, the number of errors is highly variable depending on the machine under study. Some systems develop a very large number of correctable errors compared to others. The overall image of memory errors shows that for all platforms, 20% of the nodes with errors make up more than 90% of all observed errors for the corresponding system. This can be explained by the observation that memory errors are highly correlated.

While correctable errors, depending on their number, typically do not have an immediate impact on applications, uncorrectable errors usually result in a crash. Uncorrectable errors are less common than correctable errors (as seen in Table 2.5); however, they do happen at a significant rate. Memory failures will be inspected in more detail in the next section.

Recently, systems have started to have heterogeneous nodes, which may have different failure rates. The study from 2013 [75] has shown that even homogeneous nodes present different rates both in the failure rate as well as the reliability.

2.4.3.1 Time to Repair

There are several studies that analyze the mean time to repair [45, 53, 68, 69]. In general, the studies conclude that hardware type has a major effect on repair times. While systems of the same hardware type exhibit similar mean and median time to repair, repair times vary significantly across systems of different type. All studies make an analysis of the relative impact that different types of failures have on the

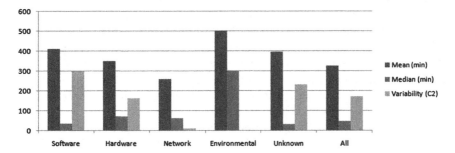

Fig. 2.5 Time to repair for different failure types

total number of node repair hours (hours required to repair failures due to the same root cause, multiplied by the number of nodes involved in the failure). Figure 2.5 shows the median and mean of time to repair as a function of the root cause, and as an aggregate across all failure records. The figure summarizes the information given by all studies presented in Table 2.2.

Both the median and the mean time to repair vary significantly depending on the failure type that is being analyzed. The mean time to repair ranges from less than 3 h for failures caused by human error, to nearly 10 h for failures due to environmental problems, while the repair time across all failures (independent of root cause) has an average close to 6 h. The most frequent type of failures affecting the systems are hardware and software.

After merging the results from all the studies, we observed that failures with software root causes are responsible for the largest percentage of the total node repair hours. This is surprising considering that hardware failures represent the majority of failures for all analyzed systems. Together, hardware and software node repair hours represent over 90 % of the total downtime. For larger system this percentage is even higher. For example, for the Blue Waters system, over 98 % of the total node repair hours are represented by software and hardware problems. The rest of 2 % represents network, facility, environmental, and unknown failures.

In general, hardware problems are closely monitored and are well managed by the vendor management system. In addition, hardware is easier to diagnose than software. There is also a difference in the distribution of the number of nodes involved in failures with different types of root causes. Failures with hardware root causes propagate outside the boundary of the smallest unit of node aggregation in a very small percentage of cases. Conversely, software failures propagate in a much larger percentage of the cases (for the Blue Waters system this number is 20 times more often than the hardware propagation percentage).

One reason for the high variability in repair times of software and hardware failures might also be the diverse set of problems that can cause these failures. For example, the root cause information for hardware failures spans on a very large list of different categories, compared to only two (power outage and A/C failure) for environmental problems on the LANL smaller systems. Breaking down the hardware problems in specific categories, the average time to repair for each still presents a high variability. Moreover, even the same hardware problem can have different times to repair depending on when it occurred. For example, the variability for repair times of CPU, memory, and node interconnect problems, as expressed with the squared coefficient of variation, are 36, 87, and 154, respectively on the LANL system. This indicates that there are other factors as well contributing to the high variability of hardware failures. Software failures have a similar behavior.

Another important observation from these studies is that the time to repair for all types of failures is extremely variable, except for environmental problems. These failures are better understood and the solution policies are already in place and it usually requires changing a component in the system, which explains the long average times to repair such problems.

2.4.4 Silent Errors

All the characteristics and statistical properties extracted so far are covering only a subset of all actual faults, namely those that can be detected by software and hardware monitoring systems. These are the faults that have a fail-stop behavior or degrade the performance of the systems and/or applications. Silent data corruptions (SDC) are undetected faults that are usually materialized as bit flips in storage (both volatile memory and nonvolatile disk) or even within processing cores.

In general, a single bit flip in memory can be detected and even corrected if the system is using an Error Correcting Code (ECC). Double bit flips, however, even though they are detected, often force an instant reboot of the node since ECC cannot correct them. For smaller systems, the frequency of double bit flips was considered to be seemingly low. Current studies [24] on the Oak Ridge National Laboratory's Cray XT5 system have shown that the density of DIMMs causes uncorrected but detected errors to occur on a daily basis (at a rate of one per day for 75,000+ DIMMs).

Single bit flips in the processor cores mostly remain undetected since few processor structures are protected. For this reason, the sensitivity of register files and ALUs for SDC is significantly higher. The Blue Gene/L architecture uses an unprotected L1 cache. Due to the high number of silent corruptions, the next generation system, the Blue Gene/P implemented ECC in L1. However, since hardware redundancy still remains extremely costly, not all storage systems afford to implement ECC.

It is believed that in today's systems, the frequency of bit flips is no longer dominated by single-events caused by radiation from space but it is increasingly attributed to fabrication miniaturization and aging of silicon given the increasing likelihood of repeated failures in DRAM after a first failure has been observed [38].

Bit-flips in stale data or in the instruction caches do not impact the system and application's execution. Only those in active data or the system's code can have extremely big effects and potentially render computational results invalid without ever being detected. This creates a severe problem for today's science that relies increasingly on large-scale simulations.

There are several solutions for overcoming the effects of silent corruption on systems and applications. Redundant computing is the method of choice when the frequency of silent errors is not extremely high since it can detect silent data corruption that impacts the results. Detection requires dual redundancy, while correction requires at least triple redundancy. Such high levels of redundancy are costly. However, with the current systems and depending on the application, it might be preferable to flawed scientific results. For the Exascale era redundancy at the process level, as currently defined, would not be feasible. Thus, the state of research for HPC requires urgent investigation to level the path for the Exascale computing.

The research direction in this field spans two separate directions: (i) more efficient application level redundancy and (ii) application level detection.

The study in [38], analyzes the potential for redundancy to detect and correct soft errors in MPI message-passing applications. For this purpose, the authors investigate the challenges inherent to detecting soft errors within MPI applications by providing transparent MPI redundancy. Their theoretical model assumes that corruption in application data manifests itself by producing differing MPI messages between replicas. With this model, the authors study the best suited protocols for detecting and correcting corrupted MPI messages.

In scientific applications that involve dense matrices, checksum encodings have yielded "Algorithm-Based Fault Tolerance" (ABFT) in the event of data corruption from either hard or transient (soft) errors in the hardware. The second research direction in dealing with SDC for Exascale systems deals with optimizing or finding new ways of detecting the corruption at the application level. In [70], the authors developed a new sparse checksum encoded algorithm that can be applied to all the key operations in the Preconditioned Conjugate Gradient method (PCG), including sparse matrix-vector multiplication, vector operations and the application of a preconditioner through sparse triangular solution. Their method detects a single error in the matrix and vector elements and in the metadata representing the sparse matrix row or column indices.

In [17] the authors convert the detection problem to a one-step look-ahead prediction issue. By modeling the values taken by different variables used by an algorithm based on their history, one can predict the future state of each of them. A running HPC application often iteratively operates on a set of data, whose values thus change over time. As shown in Fig. 2.6, at each iterative time-step, the detector dynamically predicts the possible range for the next-step data value. The detector will consider a

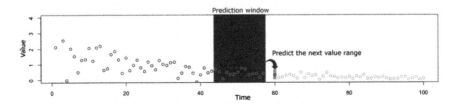

Fig. 2.6 Value prediction model for SDC detection

value as an outlier if it falls outside this range. The study explores the most effective prediction methods for different HPC applications (the Auto Regressive model, the Auto Regressive Moving Average Model, Linear Curve Fitting, and Quadratic Curve Fitting). Experiments show that this method can obtain an F-measure around 80 % for silent bit-flip errors.

2.5 Modeling

Failures and downtime intervals have a severe impact on the performance of applications in large scale HPC environments. Research efforts have been deployed to understand the failure behavior on such computing systems. Failure modeling is an important research direction used in guiding reliability-aware resource allocation and optimizing fault-tolerant protocols in order to minimize the performance loss due to failures. The failure's distribution is also used by fault-tolerant protocols, like checkpointing, to decide an advantageous trade-off between frequently creating checkpoints, which takes resources away from completing execution of the application but reduces the amount of lost calculation, and infrequent checkpoints, which diverts less resources but incurs greater losses when a fault occurs. Prediction can also be built when knowing the failure distribution and later used for marking suspicious components and monitoring them more frequently.

In general, studies that analyze failures on different HPC systems, like the ones presented in the previous section, are also extracting the failure distribution. Most research characterizes an empirical distribution by using three import metrics: the mean, the median, and the squared coefficient of variation (C^2). The squared coefficient of variation is a measure of variability and is defined as the squared standard deviation divided by the squared mean which has the advantage of allowing comparison of variability across distributions with different means.

Another used method is the empirical cumulative distribution function (CDF) that studies how well the data is fit by several probability distributions commonly used in reliability theory, like the exponential, the Weibull, the gamma, and the log-normal distributions. The method uses the maximum likelihood estimation to parameterize the distributions and evaluate the goodness of fit either by visual inspection and the negative log-likelihood test. The primary problem with using goodness-of-fit measures is that usually they do not take into account the number of free parameters in a model; with enough free parameters, any model can precisely match any dataset. For example, a phase-type distribution with a high number of phases would likely give a better fit than any of the above standard distributions, which are limited to one or two parameters. The standard distribution is preferred whenever the quality of fit allows it.

There are several goodness-of-fit tests that are currently used from Anderson-Darling and Kolmogorov–Smirnov to the chi-square test. The chi-square goodness-of-fit test can be applied to discrete distributions such as the binomial and the Poisson, while the Kolmogorov–Smirnov and Anderson–Darling tests are restricted to continuous distributions.

2.5.1 Randomness Testing

Before investigating the distribution fitting for failures, several tests of randomness need to be run in order to identify whether the generated failures have a truly random behavior. A random data series exhibits trends of periodicity, autocorrelation, or nonstationarity. Fitting probability distribution to nonrandom data is not statistically relevant since data with such properties do not respect the basic assumption of all standard statistical tests. For this purpose, when fitting distribution to a dataset, randomness needs to be tested first.

The methodology used in the literature focuses on two classes of randomness tests: parametric and nonparametric tests. In the first case the algorithm has information about the distribution of the data and only parameters need to be found. The second class refers to tests where the distribution of the observed data is unknown. In this category, there are tests of randomness based on runs or trends, such as the Mann-Kendall test, the Bartels' rank test, or the Wald-Wolfowitz test, as well as tests based on entropy estimators. A commonly used example of run-based randomness test is the Wald-Wolfowitz test in which each interval of time is compared with the mean in order to determine the mutually independent property of the intervals. Another frequently used test is the autocorrelation that is used to discover repeated patterns that differ only by a lag in time.

The autocorrelation function describes the correlation between values in the data at different times. Plotting the autocorrelation values makes it easy to visualize the lags that offer correlation. Examples of the auto-correlation function, for a periodic and random data set, can be seen in Fig. 2.7. Random data sets have only one peak for lag 0, which means that the signal has a high similarity only with itself. Periodic data sets have multiple peaks, visible in Fig. 2.7b. Thresholds can be chosen to decide when a data set is periodic either by setting it manually or in automatic using different heuristics [29].

Since no method is perfect, in general it is recommended to run multiple tests on the same data set and compare the results. The runs test and the up/down test return one value called a probability value and noted p-value. This value is used to either reject the null hypothesis about the randomness of the data, if the p-value is smaller than or equal to the significant threshold, or to confirm that the data is truly random otherwise. Depending on how many samples are available in the data set, the statistical threshold might have different values, between 1 and 15 %. For large data sets, a significance level is chosen before data collection and is usually set by all research to 5 %. Other significance levels, for example 1 %, may be used, depending on the field of study. For the autocorrelation function, in general a confidence interval of 95 % is considered. Thus, if the value of the autocorrelation test is out of this confidence interval, the sample contains a high correlation of order 1, which implies the nonrandomness of the data.

In general, most HPC systems pass the randomness test. By analyzing studies that looked at different HPC systems from several national laboratories, around 85 % of systems presented a random time between failures. Interestingly, after filtering out

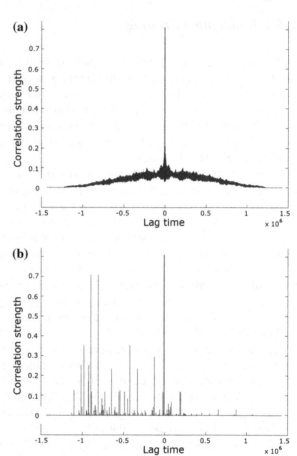

Fig. 2.7 Auto-correlation plots for different signals. **a** Random signals. **b** Periodic signals

the failures that can be predicted with a failure prediction method, the remaining failures pass the randomness test for 75 % of the systems that had nonrandom time intervals initially. This means that, now, fault-tolerant protocols can be optimized for these systems as well, since failures can be fitted to a distribution.

2.5.2 Fitting Distributions

The principle behind fitting a data set with a distribution function is to find the type of distribution (for example, exponential, normal, log-normal, gamma) and the value of its parameters (mean, variance, etc.) that give the highest probability of producing the observed data. The objective of a fitting distribution algorithm in the fault tolerance field is to find the mathematical model that best describes the inter-arrival time between failures. As mentioned before, only traces with a truly random behavior can be used to find a good probability distribution that is a good match to the empirical distribution.

2.5.2.1 Fitting Methodology

There are several available methods that can be used to fit the empirical data to probability distribution functions. In general, the methodology used has three distinct steps: (1) select a set of candidate distributions either by domain knowledge about the given data set or by including as many distribution functions as possible; (2) estimate the parameter values for each empirical distribution; and (3) choose the best fit with the most likely similarity either by manual inspection or by using automatic thresholds.

Many distributions could be used as input candidates in the first step of the fitting methodology, but in general, in the HPC community there are several commonly used distribution functions to model failures [9, 39] namely, the Poisson, exponential, Weibull, log-normal, normal, and gamma distributions.

The second step of the methodology deals with computing the best parameter values for each candidate distribution. Specifically, the maximum likelihood estimates (MLE) method is being used [46] to chose the values that are the most likely to fit the empirical data. Several older studies also use the moment matching method. However, since this methods has been shown to be sensitive to outliers [23], recent work has mainly used the MLE method. This method aims to maximize the logarithm of the likelihood function that corresponds to the closest distance between the empirical distribution and the samples resulting from the distribution function with certain parameters. The negative log likelihood value produced by the MLE is being used to rank different distributions. This method will give a list of ordered distributions, however, without giving an indication of how good the distributions actually fit the empirical data.

In order to check if a distribution is actually a good model for the given data, we must check also the goodness-of-fit between the data sample and the synthetic sample generated by the distribution. In most cases, the number of inspected distributions is relatively low so a visual inspection of the fitting is desirable. Figure 2.8 shows

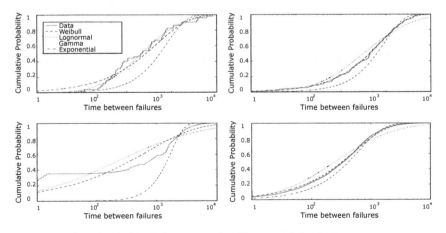

Fig. 2.8 The CDF, fitted with a Poisson, normal and log-normal distribution

four cumulative distribution functions of four separate data sets that represent the measured number of failures per compute node in different year intervals, with four different distributions fitted to it: the Weibull, log-normal, gamma, and exponential distributions. It is visible that, in general, the distribution between failures is well modeled by a Weibull or gamma distribution, for some year intervals better than for others. Both distributions create an equally good visual fit and the same negative log-likelihood. In general, the simpler exponential distribution is a poor fit. This can also be seen by looking at its C^2, which is equal to 1 for the exponential distribution, for example for the first two figures. This value is significantly lower than the data's C^2 of 1.9. Using the C^2 goodness-of-fit method, choosing the distribution can be made in an automatic way.

There are many goodness-of-fit tests in the literature, but most of them are not used in practice in the HPC community. The Kolmogorov–Smirnov test is a nonparametric test and one of the most popular methods to date. Through this test, samples are being compared with a reference probability distribution. The Kolmogorov–Smirnov test quantifies a distance between the empirical distribution function of the sample and the cumulative distribution function of the reference distribution. What makes this test attractive is that it also rejects the true randomness hypothesis.

Another popular method for visualizing the fitness of a distribution is the standard probability-probability plot (P-P plot) and quantile-quantile plot (Q-Q plots). In a Q-Q Plot, the cumulative distribution function associated with the empirical measure of the sample and the CDF from the theoretical distribution are plotted against one another. If the extracted distribution would fit exactly the given data observations, the resulting Q-Q plot would be very nearly a line intersecting the origin and having slope of 1. Figure 2.9 presents examples of Q-Q plots for failure data sets from two wide-area distributed computing environments after fitting them on a Weibull distribution.

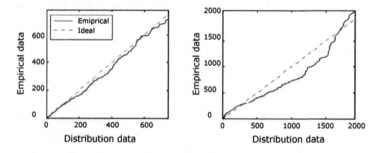

Fig. 2.9 The Q-Q plot for two data sets and Weibull distribution samples

2.5.2.2 Failure Distribution for HPC System

There are several studies that analyze different HPC systems. Some even investigate the relationship between the distribution of failures without considering prediction and the probability distribution of the false negative alerts when prediction is performed.

Table 2.6 presents the results obtained by different studies when fitting the failure distributions for several HPC systems.

There was an assumption in the past that the failure rate at all nodes follows a Poisson process with the same mean. In this case the distribution of failures across nodes would be expected to match a Poisson distribution. The large majority of studies for current HPC systems have found that the Poisson distribution is a poor fit, the Weibull and log-normal distributions being a much better fit, visually as well as measured by the negative log-likelihood.

Overall, we observed that the failure rate varies widely across systems, ranging from as low as 17 failures per year to about 1200 failures per year depending on the size and the architecture used by each system. In fact, variability in the failure rate is high even among systems of the same hardware type. Moreover, the same system might experience different distributions depending on when in the system's lifecycle the failures are inspected. For example, some systems present a complete different distribution that best fits their results when analyzing the first on second half of the system life.

For failure inter-arrival distributions, it is useful to know how the time since the last failure influences the expected time until the next failure. This notion is captured by the distribution's hazard rate function. An increasing hazard rate function predicts that if the time since a failure is long then the time lag until the next failure will be short. And a decreasing hazard rate function predicts the reverse, i.e., not seeing a failure for a long time decreases the chance of seeing one in the near future. In general, the analyzed systems are well-fit by a Weibull distribution, in most cases with a shape parameter of less than 1, indicating that the hazard rate function is decreasing.

Table 2.6 Failure distribution for several systems

System	Failure distribution	Citation
20 systems at LANL	Weibull distribution with decreasing hazard rate	Schroeder et al. [68]
O2K	Weibull distributions	Lu et al. [50]
Titan	Exponential distribution	Lu et al. [50]
Platium	Exponential distribution	Lu et al. [50]
Blue Gene/L	Weibull distribution	Taerat et al. [74]
Blue Gene/P	Weibull distribution	Harper et al. [34]

Another important overall observation after inspecting the studies, is that some nodes experience a significantly higher number of failures than other nodes in the same cluster. In some cases, it was observed that nodes that make up only about 5 % of the entire cluster account for over 20 % of all the failures. One explanation is that these nodes run different workloads. For example, nodes that are used for visualization, as well as computation, thus resulting in a more varied and interactive workload compared to the other nodes, experience a higher failure rate. Similar observations are made for other systems as well. For example, it was observed that front-end nodes, which run a more interactive and varied workload have a different, higher failure rate than all other nodes. Another interesting observation is that, while the whole system best fits the Weibull and gamma distributions, individual node failures are best fit by the log-normal distribution, followed by the Weibull and the gamma distribution.

As the failure distribution varies depending on the node's workload, as well as for other reasons, it is important to characterize how this phenomenon influences the applications running on a system. The study from [75] uses the failure trace obtained from prominent HPC platforms to study and compare different distributions, Exponential, Weibull, and Log-Normal for fitting the failures that affect applications running on k nodes. Their results indicate that Weibull distribution results in the better reliability model in most of the cases for the given data.

Table 2.7 shows the distribution that fits the time to repair for several systems. The distributions, as well as their parameters are very different depending on the system. This is to be expected since the process is dependent on the failover mechanisms offered by the vendor, the policy of the center and the policies used by system administrators.

Not all failures propagate to the application level and crash the job. Indeed, as seen in the previous section, a large percentage of failures either do not affect the application at all or degrade its performance without causing it to crash. All failure distribution functions presented thus far deal with modeling all system level failures without making a distinction between the fail-stop ones. Moreover, job-related redundancy is not negligible. Some studies have shown that over half of the resubmitted jobs were allocated to the same failed nodes by the scheduler. When analyzing application crashes it is important to separate the redundant failures, either due to the same faulty node or because users keep submitting the same buggy code, thereby leading to the same type of application errors at different locations.

Table 2.7 Failure distribution for several systems

System	Time to repair distribution	Citation
20 systems at LANL	Log-Normal distribution	Schroeder et al. [68]
O2K	Inverse normal distribution	Lu et al. [50]
Titan	Gamma distribution	Lu et al. [50]
Platium	Truncated Weibull distribution	Lu et al. [50]

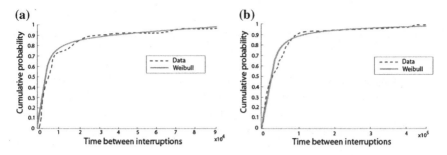

Fig. 2.10 Experimental CDF for inter-arrival times of interruptions. **a** Due to system failures and **b** due to application errors

In general, application interruption distribution can be fitted by a Weibull distribution with a shape parameter of less than 1, indicating that the hazard rate function is decreasing as in the case of system failures. In general, over 65 % of job interruptions are caused by system failures, and the rest are caused by application errors. Figure 2.10 presents the distribution fitting results of interruption inter-arrival times for the Blue Waters system at the National Center for Supercomputing Applications for a period of 4 months. Weibull distribution still gives a best fit for both interrupts caused by system failures and interrupts caused by application errors, having a shape parameter of less than 1. Another observation is that the hazard rate of interruptions caused by system failures is less than the one for interruptions caused by application errors. The main reason is that application errors generally need more time for fixing, whereas some system failures can be easily solved by rebooting.

2.5.3 Including Prediction

The distribution of failures is in general known for most of the current HPC system. However, when prediction is used, and preventive actions are taken for the failures that are known in advance, the distribution of the predicted failures, as well as that of the false negatives is, in general, unknown. In other fields [52], the analysis of false negatives has given new insights into combining a predictor with fault tolerance actions, but since prediction in the HPC field is still an area of research that is rarely used in practice, such studies are rare.

In [9], the authors present such a study for several systems that were deployed at the Los Alamos National Laboratory in the past several years. Firstly, they check the randomness hypothesis of failure intervals and identify whether it is possible to fit the entire failure set to any classic probability distribution functions. After which, by studying the randomness of the false negatives alerts, they investigate the impact of the prediction process on the randomness of the data and on the final distribution fit of the predicted failures.

Interestingly, the authors report several systems with nonrandom behavior that become random after prediction is applied and only false negatives are analyzed. In this case, after prediction, the unpredicted failures can be fitted by one of the classical distribution functions. In general, the best fit distributions for each system is different, with most systems having a Weibull, other log-normal and rarely some systems are fitted by exponential distributions, all with different parameters. While no pattern can be extracted, the results are still important since after prediction fault-tolerant protocols can be optimized for the rate of false negative alerts.

Table 2.8 shows the best fitted distribution for all failures and false negatives for several systems. The figure investigates the relationship between the initial failure distribution and the false negative distribution function. In general, it was found that the best fitted distribution for the false negative alerts is the same as for the initial failure data set, but having different parameters. Interestingly, this means that a failure predictor does not change the initial distribution and affects only the scale parameters of the initial distribution.

Moreover, when the best fitted distribution is exponential, the ratio between the parameter μ_u for the initial distribution and the parameter μ_y for the false negative distribution is given by $\mu_y/\mu_u \approx 1 - r$, where r represents the recall value for the given predictor. Similarly for the Weibull distribution, the systems that fit this distribution have approximately the same shape parameter for both distributions, and the scale parameters follow the same pattern as before: $a_u/a_y \approx 1 - r$. Basically, the failure prediction mechanism acts as a scaling filter affecting only the time scale. Therefore the distribution of the false negative alerts can be estimated from the initial failure distribution, by using the recall value to scale the parameters.

The analysis of false negatives and the impact of its distribution on fault-tolerant protocols is a current area of research that has been applied on a relatively small number of systems. Moreover, there is no study on current Petascale systems that have shown more complexity and different behaviors than previous generation machine. Thus, the results of false negative distribution analysis still need to be validated on current HPC system in order for a general conclusion to be drawn.

Table 2.8 Distribution fitting for all failures and for false negatives

System	All failures		False negatives		Param ratio
	Dist. fit	Parameters	Dist. fit	Parameters	
Blue Gene/L	Exp.	$\mu_x = 62431.3$	Exp.	$\mu_y = 113289$	0.55
LANL system 3	Exp.	$\mu_x = 215705$	Exp.	$\mu_y = 393538$	0.54
LANL system 4	Exp.	$\mu_x = 204544$	Exp.	$\mu_y = 371218$	0.54
LANL system 5	Exp.	$\mu_x = 197671$	Exp.	$\mu_y = 382671$	0.51
LANL system 6	Exp.	$\mu_x = 1007800$	Exp.	$\mu_y = 1912690$	0.54
LANL system 23	Weib.	$a_x = 509380$ $b_x = 0.84$	Weib.	$a_y = 895274$ $b_y = 0.85$	0.56

2.5.4 Per Component Failure Distribution

Large-scale applications using large numbers of processors and memory in parallel are relatively more sensitive to individual component failures. It is important to understand the failure distribution of each component and how this might affect the application's tolerance to failures. Application-centric models provide more accurate reliability estimates compared to general models. This can, for example, improve the efficiency of fault-tolerant algorithms by tuning the application checkpointing strategies to the tailored model.

The components (such as memory, CPU, disk, and the network) of current HPC systems have different failure dynamics in terms of time and space. Some components fail randomly and frequently while others fail in a correlated manner. Moreover, modern computing systems use multiple heterogeneous types of processors, networks or storage systems. This makes the failure dynamics even more diverse.

As mentioned previously, the failure data of the entire system is best fitted by the log-normal and Weibull distributions with p-values of around 0.4–0.5 respectively, by using the standard method of maximum likelihood estimation of the Kolmogorov–Smirnov test to evaluate these distributions.

In [36], the authors analyze the distribution of the most frequent failures on a particular HPC system (Mercury at the National Center for Supercomputing Applications). They found that the best fit for each failure type is still a Weibull or log-normal distribution, but depending on the analyzed type the parameters that describe the distribution might vary considerably.

For certain storage failures the distribution that best fits the failure rate is the log-normal and Weibull with average p-values around 0.52–0.61. The Network File System failures have irregular sharp spike shapes and are in general difficult to model. However, the data still fits reasonably to an exponential, log-normal or Weibull distribution (p-values 0.33, 0.22 and 0.12 respectively). Network availability failures are best fitted by a log-normal distribution with p-values around 0.38. Memory and processor cache errors are the most similar and are both fitted by log-normal distributions (p-values 0.82 and 0.68), Weibull distributions (p-values 0.68 and 0.58) and exponential distributions (p-values around 0.55).

The parameters for these distributions are shown in Table 2.9. In addition to having heterogeneous scale and shape parameters, the hazard rates are different as well depending on the failure type. As mentioned previously, a decreasing hazard rate means that when a component has been without failure for longer, the probability of the component failing in the future becomes lower. A shape parameter with value less than 1 indicates a decreasing hazard rate.

Storage and network errors show a clear decreasing hazard rate since the shape parameter is below one. For memory and processor cache errors the shape parameter has values above or slightly below 1, which means that the hazard rate is relatively constant. Overall the hazard rate is decreasing for the dominant failures in the system, namely storage, memory, and filesystem. This indicates that the presence of a failure in the system, in most cases, is followed by a period of increased failure rate.

Table 2.9 Failure distribution parameters for NCSA's Mercury system

Inter-event time in days

Failure type	Distribution fit	Time interval 1	Time interval 2	Time interval 3
All failures	Weibull	$\lambda = 0.26$	$\lambda = 0.16$	$\lambda = 0.17$
		$k = 0.66$	$k = 0.61$	$k = 0.58$
Storage errors (F1)	Weibull	$\lambda = 3.16$	$\lambda = 7.57$	$\lambda = 10.68$
		$k = 0.84$	$k = 0.55$	$k = 0.65$
NFS errors (F2)	Weibull	$\lambda = 1.83$	$\lambda = 13.08$	$\lambda = 8.07$
		$k = 0.53$	$k = 0.92$	$k = 1.41$
Network unavailability (F3)	Log normal	$\mu = -1.71$	$\mu = -2.76$	$\mu = -2.62$
		$\sigma = 2.03$	$\sigma = 2.24$	$\sigma = 2.12$
Memory errors (F4)	Weibull	$\lambda = 7.18$	$\lambda = 5.5$	$\lambda = 2.27$
		$k = 0.84$	$k = 0.85$	$k = 0.7$
Processor cache errors (F5)	Weibull	$\lambda = 2.52$	$\lambda = 4.78$	$\lambda = 4.54$
		$k = 1.32$	$k = 1.45$	$k = 1.09$

The differences between the characteristics of different failure types are also visible when looking at the effect of node failure distributions on job failure probability. The weighted sum of the probabilities that the failure affects one or more of the nodes the job is running on can be used in order to compute the node failure distribution. We are assuming that all nodes are equally likely to experience a given failure.

Figure 2.11 shows this function for different error types. The results are for the Mercury system, but are general and they describe the behavior of other HPC systems. When looking at all failures at once, the distribution that fits the data is Weibull. For

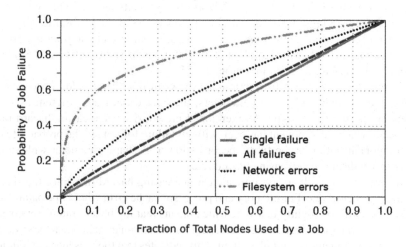

Fig. 2.11 Effect of node failure distribution on job failure probability. More details in [36]

network failures the exponential distribution offers the best fit, while for filesystem failures, the log-normal fits better. The black solid line represents the case when failures affect a single node. In this case the job failure probability rises linearly because the probability of a single node failure affecting a job rises linearly with the number of nodes used by the job. When looking at different failure types that are likely to affect a large numbers of nodes, the probability of having a job failure is high even when using a few nodes. When looking at all failures at once, the job failure probability is similar to that of a single node, but slightly higher due to the chance of a failure affecting multiple nodes (depending on the system this number can be as low as 10%).

These differences in characteristics show that, when analyzing a system, it is important to also look at individual failure types and extract their distribution and probabilities.

Recently, studies [6, 38, 77] have focused on certain components and analyzed them separately and extensively. This is the case with DRAMs on today's large-scale systems, since main memory is currently one of the leading hardware causes for application crashes. Designing, evaluating, and modeling systems that are resilient against memory errors requires a good understanding of the underlying characteristics of errors in DRAM in the field. Studies in this field provide a detailed analytical study of DRAM error characteristics, including both hard and soft errors.

In general, failure rates are observed to increase with age, even when the early stage of the disk's lifecycle is included. The analysis of the failure distribution shows moderate degree of spatial correlation and a high temporal correlation between successive failures. Depending on the study, between 40 and 80% of errors arrive within one minute of the previous error. This is visible when inspecting the arrival-rate distributions that present very long tails in all the studies. This observed locality implies that the errors are detected close in time, even though they may have developed long before they were detected.

The data collected by these studies cover over 1 million disks that were analyzed and their behavior investigated. Patterns were extracted by using latent sector errors, for both *nearline* storage and enterprise class disks. The correlation between latent sector errors (hard errors) and recovered errors (soft errors) have also been analyzed.

In general, about one-third of all hard errors occur within a very short time, less than one second, after an initial soft error. However, a large portion of hard errors has a time delay between the soft and hard error of over one hour, sometimes reaching even several days. In the study that gives the most pessimistic results, only about one third of the hard errors have a lead time of one hour or greater, while in the most optimistic results this number is closer to 80%. These results are very promising considering that such a lead time could allow preventive actions to finish before the hard errors occur.

2.6 Prediction

Over the years, approaches on prediction have been developed in a variety of fields, from astrology, meteorology to stock analysis and politics to computer science and engineering. High reliability and availability are important requirements for many systems, such as switches and track circuits for railroad networks [18], liquid storage tanks during earthquakes [58] or routers and batteries for self adapting sensor networks [1]. For this purpose, several methods have been used depending on the environmental variables gathered by each system. For example, multivariate statistical models have been developed to improve the ability to predict the occurrence of broken rails; Data mining classification models are used to create predictive models on a combination of hourly temperature readings with fire reports in order to build the context and model environmental variability for sensor networks.

Fatigue prediction has also been an area of increasing research in several fields. The term "Fatigue" refers to a failure of a component as a result of cyclic stress and it occurs following the same patters no matter the system analyzed. There are three phases that characterize fatigue failures: initiation, propagation, and catastrophic overload failure. The duration of each of these three phases depends on many factors including fundamental raw material characteristics, magnitude, and orientation of applied stresses or processing history. In the past, predicting fatigue life has been one of the most important problems in design engineering for reliability and quality. Holmgren et al. present an overview in their 1996 paper [37] of different methods for fatigue life prediction. Even though their study focuses on bogie beams, the presented methods are general and can give a good background for understanding the evolution of aging predictors in computer systems.

The fatigue life prediction methods use, in general, three primary steps. First, a theoretical or constitutive equation is defined, which forms the basis for modeling. Depending on the type of system that needs to be modeled, appropriate assumptions need to be made in constructing the constitutive equation. Second, the constitutive equation is translated into a model. The model considers the predicted stress–strain values for the system under study and returns stress values for the simulated conditions. Third, the model is tested and validated by measurement data.

In general, environmental metrics are used as the input data that creates the model and triggers predictions. Out of all environmental metrics, the most used is the load history, since it is uniaxial and proportional. Fatigue can then be evaluated with the S–N curve, also known as the Wöhler curve, which represents a graph of the stress amplitude against the logarithmic scale of cycles to failure. In many applications we deal with multi-axiality and non-proportional loading. In this instance, the S–N curve is insufficient for fatigue prediction. Weibull distribution and modified Goodman diagram modeling are two other frequently used methods. Moreover, critical plane models examine stress state in different orientations in space and can therefore incorporate some effects of multi-axiality and non-proportionality. Because they can accurately predict the fatigue failure phenomenon for many structural applications, they have gained a wide acceptance among the engineering community.

In computer science, prediction methods are used in various areas. For example, branch prediction in microprocessors tries to prefetch instructions that are most likely to be executed; similarly, memory or cache prediction tries to forecast what data might be required next. In the fault-tolerance community, the focus is on predicting computer system failures, a topic that has attracted interest for more than 30 years. However, what is understood by the term "failure prediction" varies among research communities and has also changed over the decades. In reliability theory, the goal of reliability prediction is to assess the future reliability of a system from its design or specification. For clouds [31, 80], run-time information can be used to identify the current execution state, and to check whether the design-time model will satisfy a set of wanted/unwanted properties in the future.

As computer systems are growing more and more complex, they are also changing dynamically due to the mobility of devices, updates and upgrades, changing execution environments, online repairs, the addition and removal of system components and the systems/networks complexity itself. Classical prediction methods do not work online and are therefore not capable to reflect the dynamics of run-time systems and failure processes. Such methods are typically useful in design for long term or average behavior predictions and comparative analysis. New methods have been developed specifically for short-term predictions on rapidly changing computing systems.

These new methods are almost entirely based on data mining, by using either classification, for predicting the outcome from a set of finite possible values; regression, for predicting a numerical abnormal value; clustering, for summarizing data and identifying groups of similar data points; association analysis, for finding relationships between attributes; or deviation analysis, for finding exceptions in major trends or structures.

Over the years, approaches on failure prediction in computer science have been developed in relation to reliability theory and preventive maintenance [30, 56, 68], by using the lifetime distribution or the component aging rate. Models evolved by trying to incorporate several factors into the distribution, for example the manufacturing process [79] or code complexity [22]. As the methods from other fields, these solutions are tailored to long-term predictions and, in general, do not work appropriately for online failure prediction. We will look at these methods next and in the following sub-section we will analyze methods for short term prediction.

2.6.1 Long-Term Prediction

In general long-term predictions use deterministic models for approximating the aging indicators. In addition some studies use an automated procedure for statistical testing of their correctness in order to find the optimal rejuvenation schedule under utility functions [2]. Aging in this context of long-term prediction refers, in general to software aging. The software aging phenomenon is defined by an increase in the failure rate or in the performance degradation of a system, which can be induced due to unreleased resources, the accumulation of errors in the system state, filesystem degradation or to the consumption of resources such as physical memory.

While a priori the most straightforward solutions to fix the software bugs that cause software aging, in practice this can be rarely applied due to many reasons from application complexity to budget constraints.

Software rejuvenation is the most used solution for the aging phenomenon, by restarting the software continuously after certain time frames or at a specific performance degradation level. This approach has been investigated in context of application replication in order to avoid service outages [2].

In general, in order to apply software rejuvenation techniques effectively, the aging process needs to be modeled. These models allow to estimate the current or future progress of the performance degradation or failure state. Moreover, they can facilitate the schedule of optimal rejuvenation times or the management of system administrator tasks, such as alerting operators of anticipated crashes. These approaches are known as adaptive or proactive software rejuvenation.

In [14], the authors present a comprehensive analysis of Software Aging and Rejuvenation literature, by reviewing almost 500 papers that were published in the fields of software engineering and software dependability. The aim of the paper is to provide an overall picture of the state of the art in this field. For this purpose, the paper surveys relevant studies that have been used to forecast the software aging phenomenon and to apply software rejuvenation. It also presents a study of the kind of systems and aging symptoms that have been studied, and the techniques that have been proposed to rejuvenate complex software systems.

In general, the work addressing software rejuvenation, though rich in research studies, often lacks experimentation on real systems. Most of the studies are validated by numerical examples and by simulations instead of real scenarios. In order for these studies to be useful for practical scenarios it is not enough to study these techniques in simulated environments since the models presented in the survey make assumptions about the system being modeled, which can be validated only by comparing the actual behavior of the system with the prediction of models, and because the deployment of software rejuvenation on real systems can reveal practical issues that would be neglected otherwise.

A lot of research has been focused on software rejuvenation techniques as well, however most studies do not specify the particular rejuvenation technique analyzed. This phenomenon denotes that the focus of the community is more on finding the theoretical optimal scheduling rather than on the design of the actual rejuvenation action. Rejuvenation techniques are useful to reduce the performance decrease by keeping the cost of software rejuvenation low. Most studies analyze approaches that are independent from the application and that involve a restart of the corresponding software. At the same time, other approaches are application-depended by using specific features of the system to improve the availability and efficiency of the software rejuvenation technique. The best results in the literature are given by these application-specific methods, especially in the context of embedded systems and distributed systems. Another interesting result of current research is given by rejuvenation techniques that are selective and can restart parts of the system instead of the whole software.

Another long-term prediction method category is using failure models in order to predict the reliability of a component in the system. Most studies use the Weibull and other failure distributions discussed in the previous section in order to define the state of components. This prediction can be combined with short-term predictions in order to increase their coverage [9].

2.6.2 Short-Term Prediction

If the system knows about a critical situation in advance, it can try to trigger preventive measures in order to mitigate the effect of a failure, or it can prepare repair mechanisms for the upcoming failure in order to reduce time-to-repair. For this purpose, short-term predictions need to be accurate and leave enough time for the preventive measure to be taken.

Recent methods for short-term failure prediction are typically based on run-time monitoring as they take into account a current state of the system. The taxonomy we will be using is presented in Fig. 2.12.

There are two levels of online failure prediction in the literature: component level and system level failure prediction. The first level assumes methods that observe components (hard drive, mother board, DRAM, etc.) with their specific parameters and domain knowledge and define different approaches that give best prediction results for each [38]. One example of this type of approach is to compare the execution

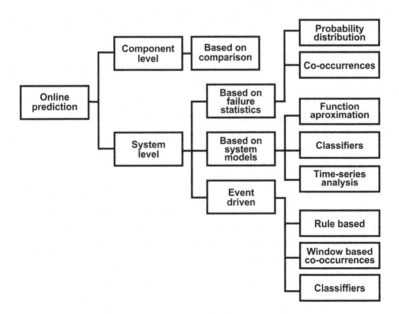

Fig. 2.12 Online failure prediction taxonomy

of good components with failed ones. A couple of studies from different fields that fit in this category are [8, 60]. For the HPC community, one example is [88] in which matrices are used to record system performance metrics at every interval. The algorithm afterwards detects outliers by identifying the nodes that are far away from the majority. One example is [44], where the authors implement their own data collection module that gathers relevant data across the system and assembling them into a uniform format. In the second step they apply two feature extraction techniques: PCA and ICA to generate matrices with lower dimensionality and in the last step the nodes that are far away from the majority are determined and considered potential anomalies. Their data mining algorithm is specifically designed for HPC systems and the results are different than previous studies.

The second level is represented by system level failure prediction, in which monitoring daemons observe different system parameters (system log, scheduler logs, performance metrics, etc.) and investigate the existence of correlations between different events. In the last couple of years, a significant number of papers have been proposed that focus on providing predictions by analyzing different HPC systems. However, most predictors are able to use the information extracted in the training phase for only short prediction span after which a new training phase is required. For example, [89] is using almost 3 months of training for predicting only half a month of execution. When dealing with real long time execution of a HPC system, the results of this type of prediction are unknown and can become unusable for real large-scale applications.

System level failure prediction has several categories:

Prediction Based on Failure Statistic The basic idea of failure prediction based on failure statistics is to draw conclusions about upcoming failures from the aggregated occurrences of previous failures. This may include the time of occurrence as well as the types of failures that have occurred. The two sub-categories includes: Probability Distribution Estimation and Co-Occurrence. Prediction methods belonging to the first category try to estimate the probability distribution of the time to the next failure from the previous occurrence of failures. The second type of failure predictors use the fact that system failures can occur close together either in time or in space (e.g., at proximate nodes in a cluster environment). This can be exploited to make an inference about failures that might come up in the near future.

Prediction Based on System Models The motivation for analyzing periodically measured system variables such as the amount of free memory or CPU usage in order to identify an imminent failure is the fact that some types of errors affect the system even before they are detected. The key notion of failure prediction based on monitoring data is that some errors can be grasped by their side-effects on the system such as exceptional memory usage, CPU load, disk I/O, or unusual function calls in the system. These side-effects are called symptoms. Symptom-based online failure prediction methods frequently address non-fail-stop failures, which are usually more difficult to grasp. The following subcategories are included in this category: function approximation that refers to mimicking a target value, which is supposed to be the output of an unknown function of measured system variables as input data (this includes stochastic models, regression and machine learning); classifiers where failure prediction is achieved by classifying whether the current situation is failure-prone or not (this includes for example Bayesian networks); and time-series analysis where sequences of monitored system variables are treated as time series and time-series analysis is used in order to predict outlier moments in the series.

Event-Driven Failure Prediction Failure prediction approaches that use error reports as input data have to deal with event-driven input data. This is one of the major differences to system model failure prediction that uses symptom monitoring-based approaches, which in most cases operate on periodic system observations. Furthermore, symptoms are in most cases represented by metrics that are real-valued while error events are mostly discrete, categorical data such as event IDs, component IDs, log messages, etc.

2.6.2.1 Prediction Based on Failure Statistics

The predictors in this category assume the correlation between failures either in time or in space. In order to estimate the probability distribution of the time to the next failure, nonparametric methods as well as Bayesian predictors have been applied. Figure 2.14 presents how failures occur in a cluster, where the horizontal axes represents time and $M_1, M_2, \ldots, M_J$ denote the J nodes in use. ALL_{CL} refers to the set of failure event times over the entire cluster. ALL refers to the set of unique failure event times over the entire cluster. Several papers [36, 38, 84] study the failure inter-arrival time distribution as well as the failure co-occurrence properties for different systems. Their results are later used by predictors in this category so it is important to understand their results. The next section gives a brief overview of their most important findings that can later be used by failure predictors.

Cumulative Distribution of Failure Inter-Arrivals

Figure 2.13 presents a cumulative distributions (CDF) graph of the time between failures for an entire cluster over different epochs of time. This study by Heien et al. [36] analyzes a large scale system from the National Center for Supercomputer application, but their results are general and have been seen on other systems as well. In this figure, on the horizontal axes F_1, F_2 represent different failure types, while on the vertical axes each failure is analyzed for different time intervals. The solid lines indicate inter-event times for the cluster as a whole.

These time intervals between consecutive events correspond to the time line labeled ALL_{CL} in our previous figure (Fig. 2.14). It is visible that some failures are correlated and so occur nearly simultaneously on multiple machines. Current studies have shown that network or filesystem failures are a good example of such a spatial co-occurrence. In the logs generated by the cluster, a correlated failure appears as sequence of time-clustered events of the same failure type across machines. For this purpose, different studies assume different time thresholds for determining when

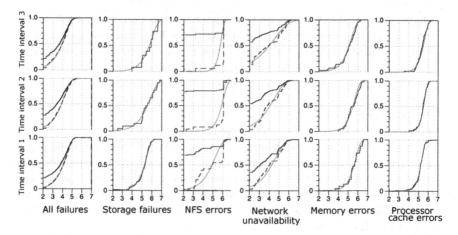

Fig. 2.13 Failure inter-event cumulative distributions for different epochs (*Solid line* cluster as a whole, *dotted line* discounts simultaneous failures, *red line* best fit)

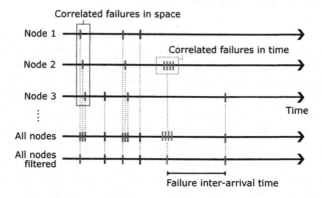

Fig. 2.14 Example of failure co-occurrence

failures are correlated. Depending on the system administrators expertise, one needs to assume that a correlated failure occurs when the time separating two consecutive failure events is less than a given threshold. Usual values for this threshold can range between tens of seconds, 1 min and several minutes.

After merging all simultaneous failures as a single failure, the resulting cumulative distribution is shown as the dotted line in Fig. 2.13. These time intervals between failures correspond to the time line labeled ALL in our previous figure (Fig. 2.14). The red line indicates the line of best fit. The best fit is found with any of the methods presented in Sect. 2.5.

Failure Correlation. All studies have found that some of the failures are correlated across different machines while other do not present this behavior. The failures that have a strong space correlation are shown in Fig. 2.13 as a large cumulative distribution at the start of the plot. Depending on the study, between 10 and 40 % of all the failures on different machines occur within 30 s of each other. One example of this phenomenon is when multiple nodes are being restarted simultaneously after a shared power failure.

Besides space correlation between failures, it is important to understand and investigate whether different failure types are correlated, such as memory and processor failures being correlated due to common causes like overheating or motherboard failure. To determine whether different failures are time-correlated between machines, current research is dividing the life interval of a machine into time intervals of different lengths and noting the existence of failures in each period. Thus, the value of each period is 1 if a specified failure type occurred and 0 otherwise. In general, each study chooses the time interval length of a couple of hours since most failure events are separate by several hours. The cross-correlation is then calculated between all combinations of failure types, with a cross-correlation of 1 indicating exact correlation (at some time delta) between two failure patterns. When looking at the entire lifetime of a cluster and on broad failure categories, studies have shown that the average cross-correlations between different failures over all time intervals ranged from 0.04 to 0.11, none of which indicate strong positive or negative correlation. However, when analyzing shorter periods of time and/or specific categories, there is a visible correlation between different failure types. For this purpose, we present the results of a study that investigates the daily failure probability of certain general types of failures following another failure.

The authors looked at the probability that a node will fail within 24 h following a failure of a particular type. At the same time they are also looking at the percentage of cases when a node failure of any type follows a particular type of failure within an hour window. The percentage of cases when a failure of a particular type follows any failure within a one hour time window is also investigated. The results are presented in Fig. 2.15.

Figure 2.15a shows what types of failures are good precursors for other failures and Fig. 2.15b shows the types of failures that have precursors. In general, many failures seem to follow environmental and network failures. Also, by looking at Fig. 2.15c, we observe that these failures in general affect a large number of nodes which suggest they propagate not only in time but also space. All the results seem to indicate a strong correlation in space and time for failures affecting a cluster. These studies have encouraged the development of predictors based on failure statistics and co-occurrences.

Statistical Prediction

In order to estimate the probability distribution of the time to the next failure, non-parametric methods as well as Bayesian predictors have been applied. In [20], the authors investigate reliability prediction by analyzing a decade of field data made available by Los Alamos National Lab. They focus on investigating the impact of factors such as the power quality, temperature, fan and chiller reliability, system usage and utilization, and external factors, such as cosmic radiation, on system reliability. They observed that some types of failures increase the likelihood of follow-up failures more than others and that this information can be used for creating effective failure prediction models based on root cause distribution.

(a)

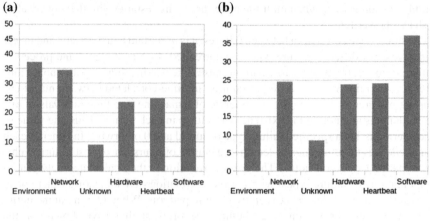

(b)

(c)

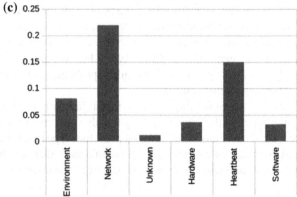

Fig. 2.15 Correlations between failures. **a** Probability of having a failure of any type after a failure of type X. **b** Probability of failure of type X following another failure of any type. **c** The probability that any node-failure follows a failure of type X

Bayesian failure prediction has the goal of estimating the probability distribution of the next time of failure by benefiting from the knowledge obtained from previous failure occurrences in a Bayesian framework [15, 35]. In [35], the authors use a mixture model of naive Bayes clusters trained by using expectation-maximization algorithm in order to predict disk failures.

Another paper [59] uses Bayesian statistics to develop an anomaly detection/ prediction system that employed naive Bayesian networks to perform intrusion detection on traffic bursts. Their model has the capability to potentially detect distributed attacks in which each individual attack session is not suspicious enough to generate an alert.

Due to sharing of resources, system failures can occur close together either in time or in space (at a closely coupled set of components or computers). As mentioned in the previous subsection, it has been observed several times that failures occur in clusters in a temporal as well as in a spatial sense. Liang et al. [48] choose such an approach to predict failures of IBM's Blue Gene/L from event logs containing reliability, availability and serviceability data. The key to their approach is data pre-processing employing first a categorization and then temporal and spatial compression: temporal compression combines all events at a single location occurring with inter-event times lower than some threshold, and spatial compression combines all messages that refer to the same location within some time window. Prediction methods are rather straightforward: using data from temporal compression, if a failure of type application I/O or network appears, it is very likely that a next failure will follow shortly. If spatial compression suggests that some components have reported more events than others, it is very likely that additional failures will occur at that location. Fu and Xu [25] further elaborate on temporal and spatial compression and introduce a measure of temporal and spatial correlation of failure events in distributed systems. A different approach is given in [49, 82], where the authors investigate parameter co-occurrences between different application log messages for extracting dependencies among system components. The authors mine dependencies from the tuple-form representations of the log messages looking for patterns that could indicate a failure in the system that prevented tasks from completing.

Statistical methods do not have extremely good results when looking at the entire set of failures affected by a system, however, they have proven a great insight for some particular failures. The study that uses these types of prediction and that has the best results shows 50 % precision and 48 % recall on a 350 node-based cluster [82]. Location prediction is one of the limitations of these methods, so it is no surprise that the same methods applied on larger and more complex systems give lower results. However, these types of method can be used to predict a state of instability for a node, without being able to give an exact time when the failure will occur. When using large time windows for when the node might fail, the methods give better results.

Another example of using the statistical failure predictor is by combining it with a rule-based method. In [64], the authors use a meta-learning predictor to choose between a rule-based method and a statistical method depending on which one gives better predictions for a corresponding state of the system and their results show that they can obtain better predictors, showing a 90 % prediction and 70 % recall on small clusters.

2.6.2.2 Prediction Based on System Models

One frequently used method is represented by regression techniques where parameters of a function are adapted such that the curve best fits the measurement data, for example by minimizing the mean square error. The simplest form of regression one can use in order to develop a predictor is curve fitting of a linear function. For this purpose, system administrators analyze either performance metrics or the count of particular events in order to predict future failures. The usual performance metrics that represent the input for regression models are temperature and CPU usage. A steady increase in temperature until exceeding a given threshold for a given cabinet could indicate future problems in several components on the corresponding cabinet. At the same time, an increase in the count or frequency of correctable memory errors usually can indicate a future occurrence of an uncorrected memory error.

In [2], the authors apply deterministic function approximation techniques such as splines to characterize the functional relationships between the target function and input data. Deterministic modeling offers a simple and concise description of system behavior with few parameters. They consider the problem of automated modeling in server-type applications whose performance degrades depending on the "work" done since last rejuvenation. Their input data is for example the number of served requests by a system. In this case a failure can be seen as a type of performance degradation caused mostly by resource depletion. This is common for filesystems for example where failures on the metadata or object target server propagate at the application level as performance diminishes.

Pattern recognition techniques operate on sequences of error events trying to identify patterns that indicate a failure-prone system state. The most used method for pattern recognition is by far the Markov chain model. The approach is based on the assumption that failure-prone system behavior can be identified by characteristic patterns of errors. The most used technique by current system administrators as well as research in this field is to use a hidden Markov model.

An HMM is mathematically equal to a stochastic finite automaton defined by 5 tuples $(Q, \sum, \Delta, \pi, O)$, where $Q = \{q_1, q_2, \ldots, q_N\}$ is a finite set of states , $\sum$ is an alphabet of output symbols, $\Delta = \{a_{ij} \text{ with } 1 \leq i, j \leq N, \sum_{j=1}^{N} a_{ij} = 1\}$ is a state transition probability distribution and $\pi = \{\pi_i \text{ with } 1 \leq i \leq N, \sum_{j=1}^{N} \pi_i = 1\}$ is an initial state distribution and O is the set $\{e_j(x) \text{ with } 1 \leq j \leq N, \sum_{j=1}^{N} e_j(x) = 1\}$ of output symbol probabilities.

HMMs are called hidden because only the outputs can be observed from outside and the actual state q_i is hidden from the observer. The objective of this type of predictors, as before, is to assess the risk of failure for some time in the future. Similar to the previous methods, here failures are predicted by analysis of error events that have occurred in the system by using the property of systems that the frequency of error occurrence increases before a failure occurs. Given a sequence of observations (events generated by the system in the past) a Hidden Markov Model is successfully developed from a probabilistic finite state automata. The overall probability of the given sequence can be afterwards found by sequence likelihood.

Building a failure predictor from a sequence of error events takes two steps: (i) firstly the number of states in the automata needs to be fixed and (ii) secondly the probability that leads to failure needs to be computed. All methods used in the literature construct the best automaton governing the given data. There are a number of variations on HMM problems depending on how many states a system administrator would take. The simplest model has one state, the most complex model has a state for each and every symbol of the data but certainly neither extreme is justified. Figure 2.16 presents two models, one with two states and one with five that both fit the input data AAACACBBBCCBBAACAAACB. The number of states for

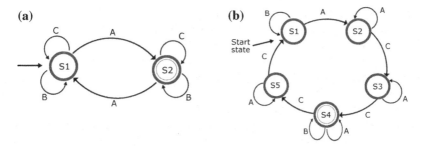

Fig. 2.16 State models. **a** Two states automaton. **b** Five states automaton

the Markov model is either chosen empirically based on the previous knowledge of the failure and its behavior or by a hierarchical algorithm that fits the data for increasing number of states and the best one is chosen. In general, current work is using between 5 and 15 states depending on the analyzed failure and system.

In [66] the authors propose to use hidden semi-Markov models (HSMM) in order to add one additional level of flexibility to the theoretical method. Two HSMMs are trained from previously recorded log data: one for failure and one for non-failure sequences. Online failure prediction is then accomplished by computing the likelihood of the observed error sequence for both models and by applying Bayes decision theory to classify the sequence (and hence the current system status) as failure-prone or not.

The second step implemented by [33] uses two semi Markov models that quantify the reliability of a node in the overall system. In the process the method identifies nodes that tend to be the source of a large number of failures and predicts the reliability of these nodes. The first discrete-time semi-Markov model is built for each system where state transitions are driven by functions derived from the distributions fitted to the result of the neural-gas filtering analysis. The second semi-Markov process computes transaction probabilities and event arrival rates directly from event observations.

Other methods include covariance models with an adjustable timescale to quantify the temporal correlation and a stochastic model to describe the spatial correlation. In [25], the authors build a neural network to approximate the number of failures in a given time interval. The set of input variables consists of a temporal and spatial failure correlation factor together with variables, such as CPU utilization or the number of packets transmitted by a computing node.

Support vector machines (SVM) are another popular modern pattern recognition and regression algorithm that is used by current research for offering model-based failure prediction. The principle of the SVM classifier is to project the data into a higher dimensional space where the classes are separated by a linear hyperplane which is defined by a small set of support vectors. There are open source or free tools already developed that implement a wide range of SVM algorithm, like the MySVM package developed by Ruping [63]. Other researchers develop their own, depending on the needs of their analysis. Murray et al. [55] have used the MySVM package on their data, gathered with the Self-Monitoring and Reporting Technology (SMART) system in order to predict failures of hard disk drives.

Depending on the analyzed data set, predictors based on system models can be accurately used in diverse systems. For disk drives, the predictor detects around 50 % of failures with only a 0.6 % false alarms. If even lower false alarm rates are needed, studies have shown that changing the combination of attributes can offer a prediction for 25.0 % of failures with no measured false alarms. For disk drives it is important to have a very low false alarm rate since it reduces the number of returned good drives, thus lowering costs to manufacturers of implementing improved SMART algorithms.

2.6.2.3 Event-Driven Prediction

Failure prediction methods in this category analyze the events generated by the system and derive a set of rules/patterns/correlations between different events. In general, the rules express temporal ordering of events in the form "if errors A and B occur within x seconds, then error C occurs within y seconds with probability P." Several parameters such as the maximum length of the data window, types of error messages, and ordering requirements have to be prespecified.

The predictors follow the same work flow (Fig. 2.17): (1) the input data is pre-processed so that it fits the standards required by the analyzing modules; (2) data mining algorithms are applied on the preprocessed data in order to extract patterns and rules between events; and (3) the system is monitored and predictions are triggered based on the extracted rules. In the preprocessing phase, failures are divided into different classes after which rules are being extracted for each class. In general, the groups used are either general groups of failures, like network, hardware or software,

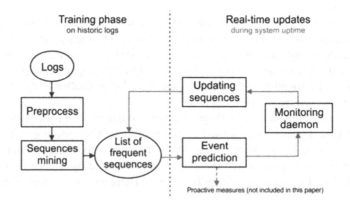

Fig. 2.17 Prediction methodology

or more specific, like memory ECC errors, cpu cache error and so on. Most of the research in this field uses a predefined number of classes, that are usually defined by the system administrator. For analyzing general types of events a predefined number of classes is enough. However, when moving the analysis at a more specific layer, a more flexible method of defining events is necessary. Since systems can change during the course of their lifetime, novel events may appear, thus the number of classes may need to change in time. In [33] the authors propose a clustering method that groups events automatically. The method proposed in [27] is another way of automatically extracting all events generated in the past by a system and keeping them accurate by monitoring everything that is generated by the system. Table 2.10 presents examples of event types found on different systems with their corresponding regular expressions.

Extensive research has been focusing on using system logs, scheduling logs, performance metrics or usage logs in order to extract a correlation between events generated by a system. There are numerous methods, starting with simple brute force extraction of rules between nonfatal events and failures [67] and going to more sophisticated techniques. Event logs are a rich source of information for analyzing the cause of failures in cluster systems. However, the size of these files has continued to increase with the ever growing size of supercomputers, making the task of analyzing log files a hard and error prone process when handled manually.

Table 2.11 presents the number of events generated by systems throughout time, starting with LANL systems from 1996 to the current Blue Water system from 2014. The number of events generated by the Blue Waters system is two orders of magnitude larger than previous generation systems and it keeps increasing as new components

Table 2.10 Examples of event types

System	Template	Event type
BGL	Failed to configure resource mgmt subsystem err = d+	Processor cache error
Blue Waters	* panic - * syncing: *	LBUG
Blue Waters	Lustre: * @ @ @ Request sent has failed due to network error: n+	MDT failure
BGQ	Component state change: component * is in the * state *	Info notification
BGQ	ECC-correctable single symbol error: DDR Controller d+, failing SDRAM address *, BPC pin *, n+	DDR single symbol error

Table 2.11 Log file statistics

System	Events/Day	Total event types
Blue Waters	11.5 GB (90 mil events)	10,495
Blue Gene/P	8.12 MB (120,000 events)	252
Blue Gene/L	5.76 MB (25,000 events)	186
Mercury	152.4 MB (1.5 mil events)	563
LANL systems	433,490 in 5 years	53

are added in the system. Filtering both in time and space is frequently necessary for several data mining algorithms in order to reduce the input data set and to cluster all failures that belong to the same problem.

Filtering Failures. Failure events from the same location often occur in bursts or clusters of notifications. Some clusters are homogeneous, with their failures having the same type, while others are heterogeneous and their failures usually report different attributes of the same event. For example, a memory failure cluster is heterogeneous, in the sense that every consecutive entry reports the same memory error referring to the same unique system state. Filtering failures in time requires to coalesce a cluster into a single failure record. Identifying such bursts from the log, requires sorting/grouping all the failures according to the associated general/specific category and location. Studies that focus on the application level might also consider the job ID. Failures that occur within the same subsystem and that are reported by the same location (and/or the same job), are filtered into a single entry if the gaps between them are less than a specified threshold. Figure 2.18 presents the usual number of remaining failure records after using this filtering technique with different threshold values. In general, the threshold is chosen by the system administrator and is a fix value between 5 and 10 min depending on the study.

Bursts of messages can occur on multiple locations as well, especially since HPC systems host parallel jobs. For example, all the tasks from a job will experience the same I/O failure if they access the same directory. At the same time, a network or a file system failure is very likely to be detected by multiple locations. As a result, many studies consider it essential to filter across locations. Spatial filtering removes failures that occur closer in time than a given threshold for the same event type and/or from the same job, but from different locations. Similarly, Fig. 2.19 presents the number of remaining failure records after spatial filtering with different threshold values. The values presented in the figure are for the Blue Gene/L system but the filtering percentage is similar for other systems as well.

In general, choosing the appropriate value for spatial filtering is easier than the temporal filtering, seen in Figs. 2.18 and 2.19 by the fact that the resulting failure count is not very sensitive to the threshold value for spatial filtering. Most studies chose the same threshold as for the temporal filtering, between 5 and 15 min.

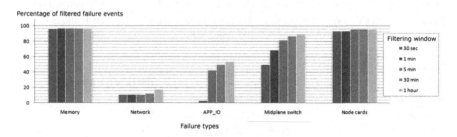

Fig. 2.18 Percentage of failures filtered with temporal filtering for different thresholds

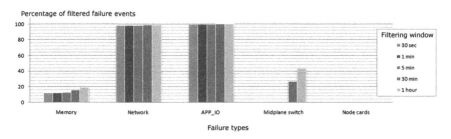

Percentage of filtered failure events

Fig. 2.19 Percentage of failures filtered with spatial filtering for different thresholds

Extracting Events' Behavior. Large scale systems experience a large variety of events during their lifetime and they output notifications for each of them. Once an error is triggered for one component, either software or hardware, there is not a consistent way of registering how the system will behave. For example, in case a node experience a network failure and is incapable of generating log messages, the failure is announced in the log files by a lack of generated messages. Conversely, some component failures may cause logging a large numbers of notifications. For example, memory failures can result in a single faulty component generating hundreds or thousands of messages in less than a day.

At the same time, some errors are notified by a single message. For example on NCSA's Mercury system, NFS related errors that indicate unavailability of the network file system for a machine, need a single instance of the generated message to notify a potentially fatal failure to an application using this resource. However, this is not always the case. Memory errors, for example, are often correctable by the ECC capabilities, so only when the system generates a large numbers of these errors in a short time span, it is likely to have a permanent failure of a component.

Each failure type behaves differently and affects the systems differently. An alternative to the methods that simply apply the same data mining algorithm on the pre-process data is to model the normal behavior of the system for each event that might be generated. By characterizing the way a failure affects these models the input data can be transformed into a unified format that can be used as an input in the data mining algorithm. Such an example is [29], where the authors extract all the event types and then plot the number of occurrences per time step for each event type into separate signals. Each event type has occurrences at different times in a system lifespan. By choosing a sampling rate and mapping the number of messages generated by the system in each sampling slot and for each event type, time series of number of occurrences for each event type can be extracted. The obtained time series are regarded as signals and can be analyzed with signal processing modules. The sampling rate is chosen differently depending on the characteristics of each signal.

Extracting all the signals for different systems has shown that there are three types of events: periodic, silent, and noisy. An example of each of the three types can be seen in Fig. 2.20. Usually, periodic signals are generated by daemons or by events that deal

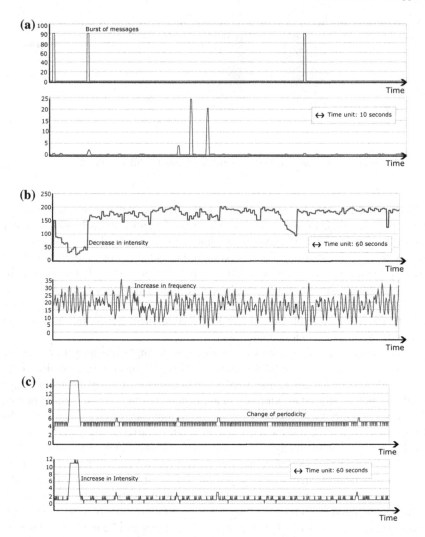

Fig. 2.20 Different signals generated by HPC systems. **a** Silent signals. **b** Noisy signals. **c** Periodic signals

with monitoring information. Examples of these signals are presented in Fig. 2.20c. We call the second type silent signals because most of the signal is a flat line around the zero value, and only from time to time there is a burst of messages. This type is presented in Fig. 2.20a and is usually characteristic for error messages, for example in case of PBS (Portable Batch System) errors. Noisy signals are chatty signals that send notifications very often. Two examples of such events are presented in Fig. 2.20b. These types of signals are usually warning messages that are generated both in case of normal behavior and failures, usually preceding error messages or when a problem

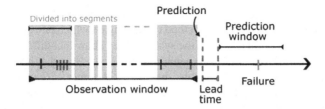

Fig. 2.21 Rule extraction methodology

is corrected. We observed that even some failure events can experience this behavior, for example in the case of memory errors that could be corrected by ECC. Anomaly detection can be applied for each signal and a unified model can be created for each.

Prediction

Event-driven predictors are in general characterized by two main methods: (1) period-based approaches and (2) rule-based approaches with a few variations. Figure 2.21 shows the methodology for the first type of method. In general, there is an observation window, a lead time and a prediction window. The observation window is usually composed of a set of consecutive time intervals $I = I_1, I_2, \ldots, I_n$, either of the same size (like in [64]) or dynamically adjusted (like in [26]). The observation window is used to collect evidence that determines whether a failure will occur within the prediction window. *A priori* based data mining algorithms are used in order to find correlated events that occur frequently together in the same time interval.

The second method is called rule-based prediction and it uses an observation window in order to extract rules between different failures and between failures and events. This is done either by brute force (for smaller systems) by investigating all to all correlations or with more advanced data mining solutions, like the Grite algorithm [40].

The period-based method has been applied in [85]. The authors are using a three-phase failure predictor for the Blue Gene/L systems: event preprocessing where the raw RAS log is cleaned and categorized; the base prediction phase where different base learning methods are applied on the preprocessed log to identify fault patterns and correlations. Similarly, [64] uses a period-based predictor close to [85] in that it uses a fixed time window for creating the time intervals. The method consists of two steps: (1) a preprocessing step that converts sys-logs into a data set that is appropriate for running classification techniques by extracting a set of features. These features can accurately capture the characteristics of failures. (2) In the next step the method applies different classifiers besides the rule-based method in order to compare the results (a rule-based classifier and a Bayesian network that combines both methods). A different approach on the classical period-based prediction is presented in [26] where the time intervals are not fixed, but they are rather defined by the data. For each failure type, the time intervals are defined by the time between two consecutive failures. The same algorithm is then applied to find frequently occurring events for each failure.

In [43], the authors are using a meta-learning predictor to choose between a rule-based method and a statistical method depending on which one gives better predictions for a corresponding state of the system. They show that different prediction methods capture different failures so combining predictors together is beneficial and has the potential to increase the results drastically. Another successful rule-based approach is presented in [28] where the authors combine signal analysis with data mining in order to extract the rules. A modified version of the gradual item set mining algorithm is used for extracting patterns of the form "the more/less $X_1, \ldots,$ the more/less X_n". In our case the algorithm gives us rules of the form "X_1 has anomalies, $\ldots, X_n$ has anomalies" where X are events in the system. This method has the advantage of extracting multiple event correlations instead of only pairs. The results indicate that this type of prediction has good results on its own and gives enough lead time for different preventive actions.

Table 2.12 presents the prediction results for the most successful studies in literature up to date. While some of the results seem extremely good, at a closer look it is clear that some are obtained either by using long training phases for only a couple of days of prediction, or not considering the lead time between when the prediction is done and when the failure occurs. Moreover, most of the presented methods do not provide any location information. This makes it impossible for proactive methods to know which application processes should be migrated. At the same time, predictions with location information will enable checkpointing data only on those failure-prone components, thereby avoiding system-wide checkpointing which is significantly time consuming. When filtering out the predictor's that have a small lead time or do not offer location prediction, the results are slightly lower, the best study offering around 50 % recall and 80 % precision.

For the second problem, Yu et al. [85] offers a study of the influence that the observation window has on the prediction's results. They look at both period-based and event-based approaches. The accuracy achieved by the period-based approach is growing with the increase of the observation window, while for the event-driven approach it is the opposite. The period-based approach achieves its best performance, in general, when the observation window is bigger than a couple of days, while the

Table 2.12 Prediction results for different state of the art-related work

System	Method	Precision	Recall	Lead time (s)	Citation
BGL	Rule-based method	0.7	0.3–0.4	5 min	[48]
BGL	Statistical method	0.5	0.48	–	[64]
BGL	Multiple methods	0.9	0.7	–	[64]
BGP	Rule-based method	0.4/0.4/0.35	0.8/0.7/0.6	0/300/600	[89]
BGP	Rule-based method	0.5	0.5	30 min	[85]
LANL systems	Signal analysis	0.9	0.5	10 s	[29]
BGP	Signal analysis with rule-based method	0.7	0.6	10 s	[28]

event-driven one reaches its peak when the observational window is as low as a couple of hours. Looking at the entire results, the event-driven approach outperforms the period-based approach significantly.

In general, thresholds, like the observation window, greatly influence the results. Figure 2.22 presents the prediction results for several methods when varying one of the parameters. Event-driven approaches are in general sensitive to the events that occur shortly before a failure. That is the reason why, in general, event-driven predictors will achieve their best performance with small observation windows. On the contrary, the period-based approaches takes more benefits from past statistical information which makes them achieve their best results with larger observation windows. There are also thresholds, like the correlation threshold in the first graph from Fig. 2.22, that offer trade-offs between precision and recall. Depending on how the predictions are used, a larger coverage or a higher accuracy might be desirable.

A failure in the system most of the time does not seem to impact a massive numbers of jobs. At a closer analysis, we observed that only around 44 % of the failures lead to at least one application crash, out of which the most surprising were filesystem failures.

Filesystem failures are one of the main reasons for a low recall for current large-scale system. For example, on the Blue Waters system, the Lustre Metadata failures have very few precursors since most of them occur at the same time with the actual failure. Metadata servers for the Lustre filesystem store namespace metadata, such as filenames, directories, access permissions, and file layout. When applications detect an MDT (Meta Data Target) failure, they connect to the the backup MDT and continue their execution. Just in less than 17 % of the cases, applications having trouble connecting to the back-up MDT fail. During an OST (Object Storage Target) failure, when applications attempt to do I/O to a failed Lustre target, these are blocked waiting for OST recovery. An application does not detect anything unusual, except that the I/O may take longer to complete. Rarely, when an OST is marked as inactive, the file operations that involve the failed OST will return an IO error and the application might be terminated.

In general, prediction from the application's point of view is more complex and differs in results compared to the one for system failures. Moreover, when analyzing the prediction results from the application's perspective, the online methodology is

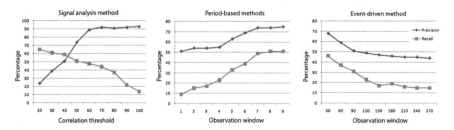

Fig. 2.22 Evaluation of different approaches using different thresholds

highly sensitive to the lead time offered by each prediction. The lead time represents the time interval between when the prediction is triggered and when the failure actually occurs. Location prediction gets a slightly new meaning as well when application crashes need to be predicted instead of system failures. If a given method predicts a failure correctly in time, but the failure occurs on a different node, all methods will give a false negative and a false positive in the final results. However, if an application was running on multiple nodes, one of which corresponds to the predicted node, and the application takes global preventive actions, the mis-predicted failure could be masked. Depending on the fault avoidance strategy, a predictor that only looks at applications as a whole and not as a set of running nodes could increase the recall significantly. By taking the lead time and the new definition of location prediction into consideration one can recompute the results for different methods. We observed that the prediction results have, in general slightly better values than when applying the same method for system failures. For the method presented in [28] and when analyzing the Blue Waters system, the application crash prediction has a higher recall value with 5 % for the same precision.

2.6.3 Checkpointing Challenges

We consider that the prediction performance presented in the previous section has the potential to be used in the future in order to reduce the effects of failures on application. For this purpose, failure prediction is useful only when coupled with a proactive failure management that tries to apply countermeasures. The decision to actually trigger a countermeasure may follow a complex process involving (i) cost of the actions, (ii) the confidence in the prediction and (iii) the effectiveness and complexity of the actions. These promising advances in failure prediction precision and recall open the possibility to reduce drastically the rework time by actually checkpointing right before the failure; a technique know as proactive checkpointing.

However, proactive checkpointing alone, cannot systematically avoid re-executing the application from scratch if failures are not perfectly predicted. Since executions on large scale HPC systems are very expensive (in time and energy), taking the risk of long (potentially near to full) re-executions is unacceptable. Therefore, failure prediction and proactive checkpointing should be combined with periodic checkpointing. Nevertheless, little is known about the benefits of failure prediction and proactive checkpointing when combined with periodic checkpointing.

Most predictions offer small lead time windows for proactive actions to be taken. To be consistent with short term prediction, fault tolerance strategies require significantly improvement. One promising direction is multi-level checkpointing. There are currently two environments providing multi-level Checkpoint/Restart: SCR (Scalable Checkpoint/Restart) [54] and FTI (Fault Tolerance Interface) [7]. Recent results show that a process context of 1GB can be saved in 2–3 s in local SSD (i.e. 2 SSD mounted in RAID0). Such checkpoint speed is orders of magnitude faster than checkpointing on remote file system which requires tens of minutes in current petascale systems and may require many hours in projected exascale systems. An experiment with FTI on a large scale execution (1/2 million GPU cores) of an earthquake simulation on a hybrid system composed of CPU and GPUs demonstrates very low overhead on the execution time (i.e., less than 10 %) when using such checkpoint strategy compared to no fault tolerance. Other research results demonstrate that checkpointing on remote node memory is even faster than on local HDD or SSD [86]. These results demonstrate that proactive checkpoints can be taken even with a few seconds before the predicted failure happens. However, proactive checkpointing introduces a whole new dimension with several challenges:

- To decrease the checkpoint size and maximize efficiency many applications rely on user-guided checkpointing, in which users specify points in the code where to checkpoint, so that the amount of data that need to be saved is minimal. However, upon a failure prediction, the checkpoint is triggered by the prediction runtime and the application may be in the middle of a complex kernel execution that requires a high memory footprint. Thus, new ways of combining user-guided checkpointing with proactive checkpointing need to be found.
- Furthermore, it is important to remember that the application needs to restart after the failure and still produce correct results. This is the classic checkpointing coordination problem that may imply the use of a fault-tolerant protocol. In application level checkpointing, the coordination is implicit, while in system level checkpointing capturing the state of the execution is explicit and relies on a fault-tolerant protocols. If the approach relies on coordinated checkpointing or on hierarchical fault-tolerant protocols [32], the coordination (global or partial) needs to be fast enough to store the state of the application before the failure occurs.

Any proactive checkpointing implementation that does not provide high performance solutions for these two problems will not be able to work efficiently in combination with a prediction strategy. There are several theoretical studies that propose to combine classic periodic checkpointing with proactive fault tolerance actions in order to study the theoretical benefit of such approaches in case such an implementation will work on future systems. One such example is presented in Aupy et al. [3], where the authors propose a fault-tolerant strategy that uses the prediction alerts to compute an optimal checkpointing interval. In their follow-up work [4], the authors assume that the fault-prediction systems that do not provide exact prediction dates, but instead time intervals during which faults are predicted to strike, with different probabilities at each moment of time. Li et al. [47] consider a different model of prediction mechanism that provides a probability of failure when the application ask for a prediction. Moreover, they consider a specific application model where proactive checkpoints or migration can be performed at a predefined location during the execution. Cappello et al. [12] proposed two proactive fault tolerance strategies, both relying on a perfect prediction mechanism. The perfect prediction mechanism is supposed to have a 100 % recall, 100 % precision and enough lead time to perform either checkpointing or migration. Even though the scenario is not realistic since there is no prediction method that can offer these results, it can show the trade-off of combining prediction either with checkpointing or with migration.

In [10], the authors combine an existing multi-level checkpointing strategy with a short-term event driven prediction. The results show that the benefit of such an implementation can give a decrease in the waste of the checkpointing strategy of around 10 % for the Blue Waters system and over 20 % for Blue Gene/L. Considering the extra 3–4 % overhead induced by this hybrid approach, the overall benefit becomes over 15 % for smaller system, and around 7 % for current Petascale computing.

Long-term predictions can be combined with checkpointing in order to reduce the I/O overhead and compute resource wastage induced by current checkpointing strategies. In [76], the authors propose a couple of methods that place checkpoints by taking advantage of the temporal locality in failures, instead of naively taking periodic checkpoints.

Acknowledgments Ana Gainaru's work is supported by the Blue Waters sustained-Petascale computing project, funded by the National Science Foundation (award number OCI 07-25070) and the state of Illinois. This chapter is build on material from publications co-authored with numerous colleagues. The authors would like to thank Leonardo Bautista-Gomez, Mohamed Slim Bouguerra, Jeremy Enos, Joshi Fullop, Eric Heien, Derrick Kondo, and William Kramer.

References

1. Anaya IDP, Simko V, Bourcier J, Plouzeau N, Jézéquel J-M (2014) A prediction-driven adaptation approach for self-adaptive sensor networks. In: Proceedings of the 9th international symposium on software engineering for adaptive and self-managing systems, SEAMS 2014. ACM, New York, pp 145–154
2. Andrzejak A, Silva L (2007) Deterministic models of software aging and optimal rejuvenation schedules. In: 10th IFIP/IEEE international symposium on integrated network management, IM'07, pp 159–168
3. Aupy G, Robert Y, Vivien F, Zaidouni D (2012) Impact of fault prediction on checkpointing strategies. Rapport de recherche RR-8023, INRIA
4. Aupy G, Robert Y, Vivien F, Zaidouni D (2013) Checkpointing strategies with prediction windows. In: 2013 IEEE 19th Pacific Rim international symposium on dependable computing (PRDC), pp 1–10
5. Avizienis A, Laprie J-C, Randell B, Landwehr C (2004) Basic concepts and taxonomy of dependable and secure computing. IEEE J Dependable Secur Comput 1:11–33
6. Bairavasundaram LN, Goodson GR, Pasupathy S, Schindler J (2007) An analysis of latent sector errors in disk drives. In: Proceedings of the 2007 ACM SIGMETRICS international conference on measurement and modeling of computer systems, SIGMETRICS'07. ACM, New York, pp 289–300
7. Bautista-Gomez L, Tsuboi S, Komatitsch D, Cappello F, Maruyama N, Matsuoka S (2011) FTI: high performance fault tolerance interface for hybrid systems. In: Proceedings of 2011 international conference for high performance computing, networking, storage and analysis, pp 1–32
8. Bolander N, Qiu H, Eklund N, Hindle E, Rosenfeld T (2009) Physics-based remaining useful life predictions for aircraft engine bearing prognosis. In: Conference of the prognostics and health management society
9. Bouguerra MS, Gainaru A, Cappello F (2013) Failure prediction: what to do with unpredicted failures? In: 28th IEEE international parallel and distributed processing symposium

10. Bouguerra MS, Gainaru A, Cappello F, Gomez LB, Maruyama N, Matsuoka S (2013) Improving the computing efficiency of HPC systems using a combination of proactive and preventive checkpointing. In: Proceedings of IEEE IPDPS 2013. IEEE Press

11. Cappello F, Geist A, Gropp B, Kale L, Kramer W, Snir M (2009) Toward exascale resilience. Int J High Perform Comput Appl 23:374–388

12. Cappello F, Casanova H, Robert Y (2010) Checkpointing versus migration for post-petascale supercomputers. In: 2010 39th international conference on parallel processing (ICPP), pp 168–177

13. Chen MY, Accardi A, Kıcıman E, Lloyd J, Patterson D, Fox A, Brewer E (2004) Path-based failure and evolution management. In: Proceedings of the international symposium on networked system design and implementation, NSDI'04, pp 309–322

14. Cotroneo D, Natella R, Pietrantuono R, Russo S (2014) A survey of software aging and rejuvenation studies. J Emerg Technol Comput Syst 10(1):8:1–8:34

15. Csenki A (1990) Bayes predictive analysis of a fundamental software reliability model. IEEE Trans Reliab 39:177–183

16. DeBardeleben N, Daly J, Scott S, Harrod W (2009) High-end computing resilience: analysis of issues facing the HEC community and path forward for research and development. National HPC workshop on resilience

17. Di S, Berrocal E, Bautista-Gomez L, Heisey K, Gupta R, Cappello F (2014) Toward effective detection of silent data corruptions for HPC applications. In: Proceedings of the 28th ACM international conference on supercomputing, SC'14

18. Dick T, Barkan C, Chapman E, Stehly M (2000) Predicting the occurrence of broken rails: a quantitative approach. In: Proceedings of the American railway engineering and maintenance of way association annual conference

19. Dongarra J, Beckman P, Moore T, Aerts P, Aloisio G, Andre J-C, Barkai D, Berthou J-Y, Boku T, Braunschweig B, Cappello F, Chapman B, Chi X (2011) The international exascale software project roadmap. Int J High Perform Comput Appl 25(1):3–60

20. El-Sayed N, Schroeder B (2013) Reading between the lines of failure logs: understanding how HPC systems fail. In: 2013 43rd annual IEEE/IFIP international conference on dependable systems and networks (DSN), pp 1–12

21. Elnozahy E, Bianchini R, El-Ghazawi T, Fox A, Godfrey F, Hoisie A, McKinley K, Melhem R, Plank J, Ranganathan P et al (2008) System resilience at extreme scale. Technical report for the defence advanced research project agency

22. Farr W (1996) Software reliability modeling survey. Handbook of software reliability engineering. McGraw-Hill, New York, pp 71–117

23. Feitelson DG (2002) Workload modeling for performance evaluation. Performance evaluation of complex systems: techniques and tools. Springer, Berlin, pp 114–141

24. Fiala D, Mueller F, Engelmann C, Riesen R, Ferreira K, Brightwell R (2012) Detection and correction of silent data corruption for large-scale high-performance computing. In: Proceedings of the international conference on high performance computing, networking, storage and analysis, SC'12. IEEE Computer Society Press, Los Alamitos, pp 78:1–78:12

25. Fu S, Xu C (2007) Quantifying temporal and spatial fault event correlation for proactive failure management. In: IEEE proceedings of symposium on reliable and distributed systems

26. Gainaru A, Cappello F, Fullop J, Trausan-Matu S, Kramer W (2011) Adaptive event prediction strategy with dynamic time window for large-scale HPC systems. In: Managing large-scale systems via the analysis of system logs and the application of machine learning techniques, SLAML'11. ACM, New York, pp 4:1–4:8

27. Gainaru A, Cappello F, Trausan-Matu S, Kramer W (2011) Event log mining tool for large scale HPC systems. In: Proceedings of the 17th international conference on parallel processing—volume part I, Euro-Par'11. Springer, Berlin, pp 52–64

28. Gainaru A, Cappello F, Snir M, Kramer W (2012) Fault prediction under the microscope: a closer look into HPC systems. In: Proceedings of 2012 international conference for high performance computing, networking, storage and analysis. IEEE Press

29. Gainaru A, Cappello F, Kramer W (2012) Taming of the shrew: modeling the normal and faulty behavior of large-scale HPC systems. In: Proceedings of IEEE IPDPS 2012. IEEE Press
30. Gertsbakh I (2000) Reliability theory: with applications to preventive maintenance. Springer, Berlin
31. Guan Q, Zhang Z, Fu S (2011) Ensemble of Bayesian predictors for autonomic failure management in cloud computing. In: 20th international conference on computer communications and networks, pp 1–6
32. Guermouche A, Ropars T, Snir M, Cappello F (2012) HydEE: failure containment without event logging for large scale send-deterministic MPI applications. In: 2012 IEEE 26th international parallel and distributed processing symposium (IPDPS), pp 1216–1227
33. Hacker T, Romero F (2009) An analysis of clustered failures on supercomputing systems. J Parallel Distrib Comput 69:652–665
34. Hacker TJ, Romero F, Carothers CD (2009) An analysis of clustered failures on large supercomputing systems. J Parallel Distrib Comput 69:652–665
35. Hamerly G, Elkan C (2001) Bayesian approaches to failure prediction for disk drives. In: Proceedings of the eighteenth international conference on machine learning, pp 202–209
36. Heien E, Kondo D, Gainaru A, LaPine D, Kramer B, Cappello F (2011) Modeling and tolerating heterogeneous failures in large parallel systems. In: Proceedings of 2011 international conference for high performance computing, networking, storage and analysis. ACM, p 45
37. Holmgren M (1996) Comparison between different methods for fatigue life prediction of bogie beams. Rakenteiden Mekaniikka, vol 29
38. Hwang A, Stefanovici I, Schroeder B (2012) Cosmic rays don't strike twice: understanding the nature of DRAM errors and the implications for system design. SIGARCH Comput Archit News 40(1):111–122
39. Javadi B, Kondo D, Vincent J-M, Anderson D (2011) Discovering statistical models of availability in large distributed systems: an empirical study of SETI@home. IEEE Trans Parallel Distrib Syst 22(11):1896–1903
40. Jorio D, Laurent A, Teisseire M (2009) Mining frequent gradual itemsets from large databases. In: International conference on intelligent data analysis
41. Kharbas K, Kim D, Hoefler T, Mueller F (2012) Assessing HPC failure detectors for MPI jobs. In: Proceedings of the 2012 20th euromicro international conference on parallel, distributed and network-based processing, pp 81–88
42. Kiciman E, Fox A (2005) Detecting application-level failures in component-based internet services. IEEE Trans Neural Netw 16(5):1027–1041
43. Lan Z, Gu J, Zheng Z, Thakur R, Coghlan S (2010) Dynamic meta-learning for failure prediction in large-scale systems: a case study. J Parallel Distrib Comput 6:630–643
44. Lan Z, Zheng Z, Li Y (2010) Toward automated anomaly identification in large-scale systems. IEEE Trans Parallel Distrib Syst 21:147–187
45. Leangsuksun C, Ostrouchov G, Scott SL (2008) Using log information to perform statistical analysis on failures encountered by large-scale HPC deployment. In: Proceedings of the 2008 high availability and performance computing workshop
46. Lehmann EL, Casella G (1998) Theory of point estimation, vol 31. Springer, New York
47. Li Y, Lan Z (2006) Exploit failure prediction for adaptive fault-tolerance in cluster computing. In: Sixth IEEE international symposium on cluster computing and the grid, CCGRID 06, vol 1
48. Liang Y (2006) Blue Gene/L failure analysis and prediction models. In: Proceedings of the international conference on dependable systems and networks, pp 425–434
49. Lou J (2010) Mining dependency in distributed systems through unstructured logs analysis. ACM Spec Interes Group Oper Syst (SIGOPS) 44
50. Lu C-D (2013) Failure data analysis of HPC systems. Technical report CoRR abs/1302.4779
51. Lu C-D, Reed DA (2005) Scalable diskless checkpointing for large parallel systems. Technical report, Ph.D. dissertation, University of Illinois at Urbana-Champain
52. Mane SV (2008) False negative estimation: theory, techniques and applications. ProQuest, Ann Arbor

53. Martino CD, Baccanico F, Fullop J, Kramer W, Kalbarczyk Z, Iyer RK (2014) Lessons learned from the analysis of system failures at petascale: the case of Blue Waters. In: IEEE/IFIP international conference on dependable systems and networks (DSN 2014)
54. Moody A, Bronevetsky G, Mohror K, de Supinski BR (2010) Design, modeling, and evaluation of a scalable multi-level checkpointing system. In: Proceedings of the 2010 ACM/IEEE international conference for high performance computing, networking, storage and analysis, pp 1–11
55. Murray J, Hughes G, Kreutz-Delgado K (2003) Hard drive failure prediction using non-parametric statistical methods. In: Proceedings of ICANN/ICONIP
56. Nassar FA, Andrews DM (1985) A methodology for analysis of failure prediction data. In: IEEE real-time systems symposium, pp 160–166
57. Oliner A, Stearley J (2007) What supercomputers say: a study of five system logs. In: IEEE international conference on dependable systems and networks
58. Panigrahi PK, Dwivedi M, Khandelwal V, Sen M (2003) Prediction of turbulence statistics behind a square cylinder using neural networks and fuzzy logic. J Fluids Eng 125:385–387
59. Papadogiannakis A, Polychronakis M, Markatos EP (2010) Improving the accuracy of network intrusion detection systems under load using selective packet discarding. In: Proceedings of the third European workshop on system security, EUROSEC'10. ACM, New York, pp 15–21
60. Patra A, Bidhar S, Kumar U (2010) Failure prediction of rail considering rolling contact fatigue. Int J Reliab Qual Saf Eng 17(3):167–177
61. Rani S, Leangsuksun C, Tikotekar A, Rampure V, Scott S (2006) Toward efficient failure detection and recovery in HPC. In: Proceedings of high availability and performance workshop
62. Ricoux P (2013) European exascale software initiative EESI2—towards exascale roadmap implementation. In: 2nd IS-ENES workshop on high-performance computing for climate models
63. Ruping S (2000) MySVM manual. Technical report, University of Dortmund, CS Department, AI Unit
64. Sahoo RK, Oliner AJ, Rish I, Gupta M, Moreira JE, Ma S, Vilalta R, Sivasubramaniam A (2003) Critical event prediction for proactive management in large-scale computer clusters. In: Proceedings of the ninth ACM SIGKDD international conference on knowledge discovery and data mining, KDD'03. ACM, New York, pp 426–435
65. Salfner F (2006) Modeling event-driven time series with generalized hidden semi-Markov models. Technical report 208, Department of Computer Science, Humboldt University
66. Salfner F, Malek M (2007) Using hidden semi-Markov models for effective online failure prediction. In: Symposium on reliable distributed systems, pp 161–174
67. Salfner F, Lenk M, Malek M (2010) A survey of online failure prediction methods. Comput Surv 42:1–42
68. Schroeder B, Gibson G (2010) A large-scale study of failures in high-performance computing systems. IEEE Trans Dependable Secur Comput 7(4):337–350
69. Schroeder B, Gibson GA (2007) Understanding failures in petascale computers. J Phys: Conf Ser 78:012022
70. Shantharam M, Srinivasmurthy S, Raghavan P (2012) Fault tolerant preconditioned conjugate gradient for sparse linear system solution. In: Proceedings of the 26th ACM international conference on supercomputing, ICS'12. ACM, New York, pp 69–78
71. Snir M, Wisniewski RW, Abraham JA, Adve SV, Bagchi S, Balaji P, Belak J, Bose P, Cappello F, Carlson B, Chien AA et al (2013) Addressing failures in exascale computing. Argonne report ANL/MCS-TM-332
72. Stearley J (2005) Defining and measuring supercomputer reliability, availability and service-ability (RAS). In: Proceedings of the Linux cluster institute conference
73. Stearley J, Oliner AJ (2008) Bad words: finding faults in spirit's syslogs. In: The eighth IEEE international symposium on cluster computing and the grid, pp 765–770
74. Taerat N, Naksinehaboon N, Chandler C, Elliott J, Leangsuksun C, Ostrouchov G, Scott S, Engelmann C (2009) Blue Gene/L log analysis and time to interrupt estimation. In: International conference on availability, reliability and security, ARES'09, pp 173–180

75. Thanakornworakij T, Nassar R, Leangsuksun CB, Paun M (2013) Reliability model of a system of k nodes with simultaneous failures for high-performance computing applications. Int J High Perform Comput Appl 27(4):474–482
76. Tiwari D, Gupta S, Vazhkudai S (2014) Lazy checkpointing: exploiting temporal locality in failures to mitigate checkpointing overheads on extreme-scale systems. In: 2014 44th annual IEEE/IFIP international conference on dependable systems and networks (DSN), pp 25–36
77. Tsai T, Theera-Ampornpunt N, Bagchi S (2012) A study of soft error consequences in hard disk drives In: IEEE international conference on dependable systems and networks, pp 1–8
78. US Department of Energy (2012) Fault Management Workshop. http://shadow.dyndns.info/publications/geist12department.pdf. Accessed July 2013
79. Vilalta R, Apte C, Hellerstein J, Ma S, Weiss S (2002) Predictive algorithms in the management of computer systems. IBM Syst J 41:461–474
80. Wang C, Talwar V, Schwan K, Ranganathan P (2010) Online detection of utility cloud anomalies using metric distributions. NOMS. IEEE, pp 96–103
81. Workshop, I-A (2012) HPC resilience at extreme scale. http://institute.lanl.gov/resilience/docs/Inter-AgencyResilienceReport.pdf. Accessed July 2013
82. Xu W, Huang L, Fox A, Patterson D, Jordan M (2009) Online system problem detection by mining patterns of console logs. In: Proceedings of the 2009 ninth IEEE international conference on data mining, ICDM'09. IEEE Computer Society, Washington, pp 588–597
83. Yamanishi K (2005) Dynamic syslog mining for network failure monitoring. In: Proceedings of the 11th ACM SIGKDD, international conference on knowledge discovery and data mining. ACM Press, pp 499–508
84. Yigitbasi N, Gallet M, Kondo D, Iosup A, Epema D (2010) Analysis and modeling of time-correlated failures in large-scale distributed systems. In: 2010 11th IEEE/ACM international conference on grid computing (GRID), pp 65–72
85. Yu L, Zheng Z, Lan Z, Coghlan S (2011) Practical online failure prediction for Blue Gene/P: period-based versus event-driven. In: IEEE conference on dependable systems and networks workshops, pp 259–264
86. Zheng G, Shi L, Kale L (2004) FTC-Charm++: an in-memory checkpoint-based fault tolerant runtime for Charm++ and MPI. In: 2004 IEEE international conference on cluster computing, pp 93–103
87. Zheng Z, Yu L (2011) Co-analysis of RAS log and job log on Blue Gene/p. In: Proceedings of the 2011 IEEE international parallel and distributed processing symposium, pp 840–851
88. Zheng Z, Li Y, Lan Z (2007) Anomaly localization in large-scale clusters. In: IEEE international conference on cluster computing, pp 322–330
89. Zheng Z, Lan Z, Gupta R, Coghlan S, Beckman P (2010) A practical failure prediction with location and lead time for Blue Gene/P. In: IEEE conference on dependable systems and networks workshops, pp 15–22

Chapter 3
Fault-Tolerant MPI

Aurélien Bouteiller

Abstract As supercomputers are entering an era of massive parallelism where the frequency of faults is increasing, the MPI standard remains distressingly vague on the consequence of failures on MPI communications. In this chapter, we present the spectrum of techniques that can be applied to enable MPI application recovery, ranging from fully automatic to completely user driven. First, we present the effective deployment of most advanced checkpoint/restart techniques within the MPI implementation, so that failed processors are automatically restarted in a consistent state with surviving processes, at a performance cost. Then, we investigate how MPI can support application-driven recovery techniques, and introduce a set of extensions to MPI that allow restoring communication capabilities, while maintaining the extreme level of performance to which MPI users have become accustomed.

3.1 Introduction

High Performance Computing, as observed by the Top 500 ranking,[1] has exhibited a constant progression of the computing power by a factor of two every 18 months for the last 15 years and the pace of progress has been only slightly disturbed by the financial turmoil in 2008. Following the long-term trend, the Exaflops milestone should be reached as soon as 2022. The International Exascale Software Project (IESP) [27] proposes an outline of the characteristics of an Exascale machine, based on the foreseeable limits of the hardware and maintenance costs. A machine in this performance range is expected to be built from gigahertz processing cores, with thousands of cores per computing node (up to 10^{12} flops per node), thus requiring millions of computing nodes to reach the Exascale. Software will face the challenges of complex hierarchies and unprecedented levels of parallelism.

---

[1] http://www.top500.org/.

A. Bouteiller (✉)
Department of EECS, University of Tennessee,
1122 Volunteer Boulevard, Knoxville, TN 37996, USA
e-mail: bouteill@icl.utk.edu

© Springer International Publishing Switzerland 2015
T. Herault and Y. Robert (eds.), *Fault-Tolerance Techniques
for High-Performance Computing*, Computer Communications and Networks,
DOI 10.1007/978-3-319-20943-2_3

145

One of the major concerns is reliability. If we consider that failures of computing nodes are independent, the reliability probability of the whole system (i.e., the probability that all components will be up and running during the next time unit) is the product of the reliability probability of each of the components (see Chap. 1, and especially Sect. 1.3.2.1 for the discussion on the MTBF of large machines). A conservative assumption of a ten-year mean time to failure translates into a probability of 0.99998 that a node will still be running in the next hour. If the system consists of a million of nodes, the probability that at least one unit will be subject to a failure during the next hour jumps to $1 - 0.99998^{10^6} > 0.99998$. This probability being disruptively close to 1, one can conclude that many computing nodes will inevitably fail during the execution of an Exascale application.

Fault-tolerant algorithms have a long history of research. Only recently, since the practical issue has been raising, High Performance Computing (HPC) software has been adapted to deal with failures. As most HPC applications are using the Message Passing Interface (MPI) [87] to manage data transfers, introducing failure recovery features that can support MPI applications is paramount to maintain productivity on future systems. One of the most popular fault-tolerance technique deployed today, coordinating checkpointing, builds a consistent recovery set [54, 73]. As today's HPC users are facing occasional failures, they can still dismiss as an inconvenience the slow, inefficient recovery procedure, involving restarting all the computing nodes even when only one has failed. Considering future systems will endure higher fault frequency, recovery time could become another gap between the peak performance of the architecture and the effective performance users can actually harvest from the system.

In this chapter, we present a range of additions and fault-tolerance constructs that can be added to an MPI library in order to support more advanced recovery strategies. Advanced fault-tolerance techniques have the potential to prevent full-scale application restart and therefore lower the cost incurred for each failure, but they demand from MPI the capability to detect failures and resume communications afterward. In Sect. 3.2, we first consider the case of automatic fault tolerance, that is a fault tolerant recovery procedure that does not require actions from the application, so that legacy codes are rendered fault tolerant without modifications. This goal can be achieved with the integration of an uncoordinated checkpointing mechanism into the MPI library, that ensures that the application can be restored to an equivalent state as before a failure struck. Then, we turn our attention to interfaces that permit deploying arbitrary fault tolerance techniques, in applications and/or middleware that are enhanced to integrate their own recovery mechanism. We will discuss what type of user fault tolerance is possible in the current iteration of the MPI standard in Sect. 3.7, and then, in Sect. 3.8, present additional APIs that permit restoring a consistent communication environment, under the control of an application recovery mechanism, which can then benefit from a full post-failure communication capability to perform the necessary application and dataset recovery operations.

3.2 Automatic Uncoordinated Fault Tolerance in MPI

Automatic fault-tolerant algorithms, which can be provided either by the operating system or the middleware, remove some of the complexity in the development of applications by masking failures and the ensuing recovery process. The most common approaches to automatic fault tolerance are replication, which consumes a high number of computing resources, and rollback recovery. Rollback recovery stores system-level checkpoints of the processes, enabling rollback to a saved state when failures happen. Consistent sets of checkpoints must be computed, using either coordinated checkpointing or some variant of uncoordinated checkpointing with message logging (for brevity, in this chapter, we use indifferently message logging or uncoordinated checkpointing). Coordinated checkpointing minimizes the overhead of failure-free operations, at the expense of a costly recovery procedure involving the rollback of all processes. Conversely, message logging requires every communication to be tracked to ensure consistency, but its uncoordinated recovery procedure proves more efficiency in failure prone environments. See Sect. 1.4.2 for discussions on replication, and Sects. 1.2.2 and 1.3.2 for coordinated checkpointing. Uncoordinated checkpointing was briefly presented in Sect. 1.2.3. In this chapter, we will detail its implementation in an actual MPI library.

Because message logging does not rely on a full restart, it is able to recover faster from failures. From previous results [54], it is expected that a typical application makespan will be better than coordinated checkpoint when the MTBF is less than 9 h while coordinated checkpoint will not be able to progress anymore for a MTBF less than 3 h. Still, message logging suffers from a high overhead on communication performance. Moreover, the better the latency and bandwidth offered by newer high performance networks, the higher the relative overhead. Those drawbacks need to be addressed to provide a resilient and fast fault tolerant MPI library to the HPC community. In this section, we will present refinements of the classical message logging theoretical concepts as well as practical effective implementation issues designed to mitigate the high cost on communication of message logging, and thereby improve its practical effectiveness in HPC production systems.

In the first part of this section, we first describe a refinement of the classical model of message logging, closer to the reality of high-performance network interface cards, where message receptions are decomposed in multiple dependent events (Sect. 3.3). We better categorize message events allowing (1) the suppression of intermediate message copies on high performance networks and (2) the identification of deterministic and nondeterministic events, thus reducing the overall number of messages requiring latency disturbing management. We demonstrate how this refinement can be used to reduce the fault-free overhead of message logging protocols by implementing it in Open OMPI [35]. Its performance is compared with the previous reference implementation of message logging MPICH-V2. Results outline a several orders of magnitude improvement of the fault-free performance of pessimistic message logging and a drastic reduction in the overall number of logged events. Then, we discuss the impact of the event logging protocol employed, and we observe by comparing experimentally the optimistic and pessimistic event logging strategies, which are situated at both extremes in terms of event logging synchrony (Sect. 3.4). With the observation that message payload copy is a major contributor to the overall cost of message logging, we then focus our attention on reducing this overhead. First, we present varied strategies to increase the throughput of message exchanges over the network by overlapping, as much as possible, the copy of the message payload and its sending over the network (Sect. 3.5). Then, we investigate how a combination between uncoordinated and coordinated checkpointing has the potential to drastically reduce both the volume and the overhead on message bandwidth of payload logging (Sect. 3.6).

3.2.1 Rollback Recovery Execution Model

Message logging is defined in the more general model of message passing distributed systems. Communications between processes are considered explicit: processes explicitly request sending and receiving messages; and a message is considered as delivered only when the receive operation associated with the data movement is complete. Additionally, from the perspective of the application each communication channel is FIFO, but there is no particular order on messages traveling along different channels. The execution model is pseudo-synchronous; there is no global shared clock among processes but there is some (potentially unknown) maximum propagation delay of messages in the network. A formal interpretation is to say the system is asynchronous and there is an *eventually reliable* failure detector. Failures can affect both the processes and the network. Usually, network failures are managed by some CRC mechanism and message reemission provided by the hardware or low-level software stack and do not need to be considered in the model. Therefore, the considered failure model is definitive crash failures, where a failed process completely stops sending any subsequent message.

In the following section, we define our execution model. We consider a distributed execution, with explicit message passing. Any process may be subject to permanent (fail-stop) failures. After a failure, a process will be replaced and rejoin the distributed execution by loading a checkpoint image saved by the failed processes prior to the failure.

3.2.1.1 Events and States

Each computational or communication step of a process is an event. An execution is an alternate sequence of events and process states, with the effect of an event on the preceding state leading the process to the new state. As the system is basically asynchronous, there is no direct time relationship between events occurring on different processes. However, Lamport defines a causal partial ordering between events with the *happened-before* relationship [53]. It is noted $e \prec f$ when event f is causally influenced by e.

Events can be classified into two categories. An event is *deterministic* if, in a given state, no other event can apply. On the contrary, if in a given state multiple events can apply and lead to different outcome states, these events are considered nondeterministic. The arrival of a network packet is a notorious example of a nondeterministic event: the ordering of packet arrival depends on network jitter between independent channels, resulting in an uncertain matching between packets and posted receptions (see [11] for a classification of MPI reception events).

3.2.1.2 Recovery Line

Rollback recovery addresses mostly fail-stop errors: a failure is the loss of the complete state and actions of a process. A checkpoint is a copy of a past state of a particular process stored on some persistent memory (remote node, disk, ...), and used to restore the process in case of failure. The recovery line is the configuration of the entire application after some processes have been reloaded from checkpoints. If the checkpoints can happen at arbitrary dates, some messages can cross the recovery line [14]. Consider the example execution of Fig. 3.1. When process P_1 fails, it rolls back to checkpoint C_1^1. If no other process rolls back, messages m_3, m_4, m_5 are crossing the recovery line. A recovery set is the union of the saved states (checkpoint, messages, events) and a recovery line.

3.2.1.3 In-transit Messages

Messages m_3 and m_4 are crossing the recovery line from the past, they are called *in-transit* messages. The *in-transit* messages are necessary for the progression of the recovered processes, but are not available anymore, as the corresponding send operation is in the past of the recovery line. For a recovery line to form a complete recovery set, every *in-transit* message must be added to the recovery line.

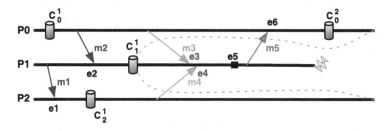

Fig. 3.1 Recovery line based on rollback recovery of a failed process

3.2.1.4 Orphan Messages

Message m_5 is crossing the recovery line from the future to the past; such messages are referred to as *orphan* messages. By following the happened-before relationship, the current state of P_0 depends on the reception of m_5; by transitivity, it also depends on events e_3, e_4, e_5 that occurred on P_1 since C_1^1. Since the channels are asynchronous, the reception of m_3 and m_4, from different senders, can occur in any order during re-execution, leading to a recovered state of P_1 that can diverge from the initial execution. As a result, the current state of P_0 depends on a state that P_1 might never reach after recovery. Checkpoints leading to such inconsistent states are useless and must be discarded; in the worst case, a domino effect can force all checkpoints to be discarded.

3.2.2 Building a Consistent Recovery Set

Two different strategies can be used to create consistent recovery sets. The first one is to create checkpoints at a moment in the history of the application where no *orphan* messages exist, usually through coordination of checkpoints. The second approach avoids coordination, but instead saves all *in-transit* messages to keep them available without sender rollback, and keep track of nondeterministic events, so that *orphan* messages can be regenerated identically. We focus our work on this second approach, deemed more scalable.

3.2.2.1 Coordinated Checkpoint

Checkpoint coordination aims at eliminating *in-transit* and *orphan* messages from the recovery set. Several algorithms have been proposed to coordinate checkpoints, the most usual being the Chandy–Lamport algorithm [20] and the blocking coordinated checkpointing [15, 64, 73], which silences the network. In these algorithms, waves of tokens are exchanged to form a recovery line that eliminates *orphan* messages and detects *in-transit* messages. Coordinated algorithms have the advantage of having almost no overhead outside of checkpointing periods, but require that every process, even if unaffected by failures, rolls back to its last checkpoint, as this is the only recovery line that is guaranteed to be consistent.

3.2.2.2 Message Logging

Message Logging is a family of algorithms that attempt to provide a consistent recovery set from checkpoints taken at independent dates. As the recovery line is arbitrary, every message is potentially *in-transit* or *orphan*. Event Logging is the mechanism used to correct the inconsistencies induced by *orphan* messages, and nondeterministic events, while Payload Copy is the mechanism used to keep the history of *in-transit* messages. While introducing some overhead on every exchanged message, this scheme can sustain a much more adverse failure pattern, which translates to better efficiency on systems where failures are frequent [54].

Event Logging: In event logging, processes are considered *Piecewise deterministic*: only sparse nondeterministic events occur, separating large parts of deterministic computation. Event logging suppresses future nondeterministic events by adding the outcome of nondeterministic events to the recovery set, so that, during recovery, it can be forced to a deterministic outcome (identical to the initial execution). In message logging, the network, more precisely the order of reception, is considered the unique source of nondeterminism. The relative ordering of messages from different senders (e_3, e_4 in Fig. 3.1), is the only information necessary to be logged. For a recovery set to be consistent: then, no unlogged nondeterministic event can precede an *orphan* message.

Payload Copy: When a process is recovering, it needs to replay any reception that happened between the last checkpoint and the failure. Consequently, it requires the payload of *in-transit* messages (m_3, m_4 in Fig. 3.1). Several approaches have been investigated for payload copy, the most efficient one being the sender-based copy [66]. During normal operation, every outgoing message is saved in the sender's volatile memory. The surviving processes can serve past messages to recovering processes on demand, without rolling back. Unlike events, sender-based data do not require stable or synchronous storage (although this data is also part of the checkpoint). Should a process holding useful sender-based data crash, the recovery procedure of this process replays every outgoing send and thus reproduces the missing messages.

3.2.3 Short Survey of Related Works

Though fault tolerance can be fully managed by the application [21, 71], the software engineering cost prevents a large number of applications from benefiting of the entire capacity of modern clusters. FT-MPI [31, 32] aims at helping an application to express its failure recovery policy by taking care of rebuilding internal MPI data structures (communicators, rank, etc.) and triggering user provided callbacks to restore a coherent application state when failures occur. Though this approach is very efficient to minimize the cost of failure recovery techniques, it still adds a significant level of complexity to the design and implementation of parallel applications.

The next step toward easing application development is automatic fault-tolerant MPI libraries, where failures are completely hidden from the application, thus avoiding any modification of the user's code. Consistent recovery can be achieved automatically by building a coordinated checkpoint set where no-orphan message exists (with the Chandy and Lamport algorithm [20, 54, 83], or blocking the application until channels are empty [49, 72, 73]). Communication Induced Checkpoint (CIC) [52] is another approach that aims at constructing a consistent recovery set, but without coordination. The CIC algorithm maintains the dependency graph of events and checkpoints to compute *Z-paths* as the execution progresses. Forced checkpoints are taken whenever a Z-path would become a consistency breaking *Z-cycle*. This approach has several drawbacks: it adds piggyback to messages, and is notably not scalable because the number of forced checkpoints grows uncontrollably [3]. In all coordinated checkpoint techniques, the only consistent recovery set is when every process, including non-failed ones, restart from a checkpoint.

Another approach, that allows for faster recoveries according to [54], is to use message logging. Manetho [29], Egida [67] and MPICH-V [9] feature the main flavors of message logging (optimistic, pessimistic, and causal). Optimistic message logging protocols, such as [23, 80, 81, 85], delay the storing of determinants to the stable storage and keeps them in the process memory. As a consequence, they are more subject to creating orphan processes and to piggyback more determinants with messages. Active optimistic message logging protocol [69] copes with this drawback by aggressively saving determinants to the stable storage as soon as possible. Because they are based on the classic message logging model, when implemented in MPI, all these protocols face difficulties to distinguish between deterministic and nondeterministic events and introduce extra memory copies leading to a performance penalty on high-throughput networks.

3.3 Message Logging and Zero-Copy MPI Communication

Though the classical model has been used successfully in many implementations of message logging in the past, it is unable to capture the full complexity of MPI communications, resulting in nonoptimal performance. This was left unaddressed as long as the performance gap between network and memory bandwidth was hiding the ensuing overhead. But as the performance of network interface cards progressed it became clear that extra memory copies on the critical path of messages were the source of significant performance penalties.

The discrepancy between the model and the reality of MPI communication basically roots in the existence of non-blocking communications. Those are intended to maximize opportunities for overlapping communication and computation, by having the application post in advance its intention to communicate, compute while the communication actually takes place, and wait for completion of the communications later. One of the most important optimizations for a high-throughput communication library is *zero-copy*: the ability to send and receive directly into the application's user space buffer without intermediary memory copies. Because the classical message logging model assumes that the message reception is a single atomic event, it cannot catch the complexity of zero-copy MPI communications involving distinct matching and completion events, as is the case with non-blocking communication.

3.3.1 Understanding Non-blocking MPI Communication

To better understand the difficult match between advanced MPI communication operations and the classical message logging model, one has to first understand the steps of a MPI communication when it is managed internally by the MPI implementation. To illustrate, Fig. 3.2 shows the basic steps of two concurrent non-blocking zero-copy communications between three MPI processes.

3.3.1.1 Request Post

At the receiver side, the application posts the intent for a message to be received, specifying the message source, tag, and reception buffer. The postoperation creates a *request*. As an example, the post-event $Post_{r1}^{any}$ creates the request $r1$ to receive a message from any sender.

3.3.1.2 Fragments

Every message is divided into a number of network fragments when it is transferred over the network, the number depending on its length. Though MPI enforces a FIFO semantic for messages from a particular sender, at the lowest network level, there is no particular order between fragments. Consequently, when receiving two different messages m_1 and m_2, the first fragment of m_1 coming first does not imply that the last fragment of m_1 arrives before the last fragment of m_2. Therefore, unlike in the classic model, with MPI communications the reception order of a message cannot be fully described by a single event denoting *message reception*, but rather depends on the relative ordering of the multiple fragments composing the messages. Although

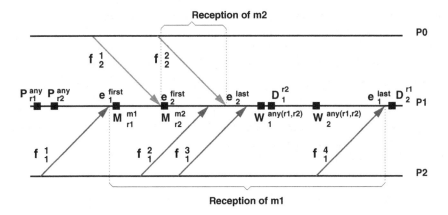

Fig. 3.2 Consequence of message fragmentation on event ordering of non-blocking MPI receives

there is a very large number of such network nondeterministic events, only the order of events denoting the first and last fragments of a message are actually meaningful to the application, as described in the next paragraphs.

3.3.1.3 Matching

In order to correctly dispatch incoming fragments to the reception buffers, the MPI implementation needs to match a reception request with the incoming fragments. When the first fragment of a message is delivered by the network, requests are considered in order by the matching logic; the first request (according to the post-ordering) with a matching source and tag is associated with the incoming message fragments. All upcoming fragments of this message are then delivered directly into the request's reception buffer. If no request matches, the message is *unexpected*; it is copied into an internal buffer until it matches an upcoming posted request.

In the example of Fig. 3.2, $M_{r_1}^{m_1}$ is the matching determinant between the request created by the any source non-blocking receive $P_{r_1}^{any}$ and the first fragment reception event e_1^{first}. Though the relative order of the fragments from the network is always nondeterministic, the FIFO by channel MPI semantic allows for most of the matching determinants to be deterministic. The only nondeterministic ones are promiscuous receptions, i.e., when a request can match a message coming from any source. Those promiscuous matching determinants are the only events that need to be logged in order to replay a correct matching during recovery.

3.3.1.4 Completion of Requests

When using non-blocking communications, several requests can concurrently progress while the application is computing. When the next computation needs to access the buffer involved in an ongoing communication operation, the application has to wait for the completion of the corresponding requests. All the functions allowing the application to check the status of a request (like MPI_Wait) are a *completion test*. The most commonly employed completion test operations are always deterministic, namely the MPI_Recv, MPI_Send, MPI_Wait, and MPI_Waitall functions. However, for MPI_Waitany, as an example, the outcome of the MPI call depends on the ordering between the arrival date of the last fragments of messages associated to the input parameter requests. MPI_Waitsome, MPI_Test, MPI_Testany, MPI_Testsome, and MPI_Iprobe add to the previous source of nondeterminism a dependency between the arrival date of the last fragments and the date of the completion test.

3.3.1.5 Consequences on Message Logging

By not discriminating between matching and completion, the legacy model of atomic *message reception event*, prevents messages from being directly delivered into the application buffer. The only software layer where the MPI matching can be delayed is the very low level interface with the network. Implementing message logging at this level has two severe limitations. First, the message logging mechanism cannot easily take advantage of the optimized network drivers and second, at this level it is impossible to make a distinction between deterministic and nondeterministic delivery determinants. As an example, in MPICH, only the lowest blocking point-to-point transport layer called the *device* matches the classical model, explaining why previous state-of-the-art message logging implementations, such as MPICH-V, replace the low level device with the MPICH- V fault-tolerant one (see Fig. 3.3a). This device has adequate properties regarding the hypothesis of message logging: (1) messages are delivered in one single atomic step to the application (though message interleave is allowed inside the MPICH- V device), (2) intermediate copies are made for every message to fulfill this atomic delivery requirement, so that the matching is delayed to the delivery time, (3) as the message logging mechanism replaces the regular low level device, it cannot easily benefit from zero-copy and OS bypass features of modern network cards, and (4) because it is not possible to distinguish the deterministic events at this software level, every message generates an event (which is obviously useless for deterministic receptions). We will show in the performance analysis section how these strong model requirements lead to dramatical performance overhead in an MPI implementation when considering high-performance interconnects.

3.3.2 A Split Model for Matching and Delivery Events

By relaxing the strong model described previously, it is possible to interpose the event logging mechanism inside the MPI library. Then it is only necessary to log the communication events at the library level and the expensive events generated by the lower network layer can be completely ignored. This requires consideration of the particularity of the internal MPI library events, but allows to use the optimized network layers provided by the implementation. The remainder of this section describes this improved model.

3.3.2.1 Network Events

From the lower layer come the packet-related events: let m denote a message transferred in $length(m)$ network packets. We note that r_m^i equals the ith packet of message m, where $1 \leq i \leq length(m)$. Because the network is considered reliable and FIFO, we have $\forall 1 \leq i \leq length(m) - 1, r_m^i \prec r_m^{i+1}$. We denote $tag(m)$ the tag of message m and $src(m)$ its emitter. Packets are received atomically from the network layer.

3.3.2.2 Application Events

From the upper layer comes the application-related events. We note that $Post(tag, source)$ is a reception post, $Probe(tag, source)$ is the event of checking the presence of a message, and $Wait(n, \{R\})$ is the event of waiting n completions of the request identifier set $\{R\}$. Because the application is considered piecewise deterministic, we can assign a totally ordered sequence of identifiers to upper layer events. Let r_0 be a request identifier obtained by the $Post_0(tag_0, source_0)$ event. Since posting is the only way to obtain a request identifier, if $r_0 \in \{R\}$, $Post_0(tag_0, source_0) \prec Wait_0(n, \{R\})$. There is at most one event $Post$ per message and at least one $Wait$ event per message. If $r_{m_0}^1 \prec Probe_0(tag_0, source_0) \prec Post_0(tag_0, source_0)$, then $Probe_0(tag_0, source_0)$ must return true. Otherwise, it must return false. The main difference between $Probe$ and $Post$ is that in case $r_{m_0}^1$ precedes one of these events, $Probe_0(tag_0, source_0)$ will not discard $r_{m_0}^1$, while $Post_0(tag_0, source_0)$ will always do so.

3.3.2.3 Library Events

The library events are the result of the combination of a network-layer event and an application-layer event. There are two categories of library events: (1) Matching (denoted by M) and (2) Completing (denoted by C). Matching binds a network communication with a message reception request; completing checks the internal state of the communication library to determine the state of a message (completed or not).

(1) To build a Matching event from a reception event and a Post-event, we define a reception-matching pair of events: r_m^1 and $Post_0(tag_0, source_0)$ match for reception if and only if $(source_0 = src(m) \vee source_0 = ANY) \wedge (tag_0 = tag(m) \vee tag_0 = ANY)$. The Matching event built from the reception-matching events is causally dependent from the two elements of the matching pair: $Post_0(tag_0, source_0) \prec M_0$ and $r_m^1 \prec M_0$. The reception-matching pair is determinist if and only if $source_0 \neq ANY$. Additionally, based on the same rules, we can build a Matching from a Probe event and a reception event. In this case, the result of the Matching M_0 is successful if and only if $r_m^1 \prec Probe_0(tag_0, source_0)$. Otherwise, the Matching event takes a special value (undefined source). Because the order between r_m^1 and $Probe_0(tag_0, source_0)$ is nondeterministic, all probe-matching pair events are nondeterministic.

(2) Similarly, to build a Completing event from a reception event and a Wait event, we define a completion-matching pair of events: $r_m^{length(m)}$ and $Wait(n, \{R\})$ match for completion if and only if there is a matching event M_0 built from r_m^1 containing the request identifier r_0 and $r_0 \in \{R\}$. The Completing event built from the completion-matching events is causally dependent on the two elements of the matching pair: $Wait(n, \{R\}) \prec C_0$ and $r_m^{length(m)} \prec C_0$. All the r_m^i events are nondeterministic per definition. Thus, every $Wait(n, \{R\})$ event is nondeterministic, because the result of these events depends upon the internal state of the library,

which depends upon the $r_m^{length(m)}$ events. However, according to the matching and completion rules, if $r_m^{length(m)}$ and $Wait(n, \{R\})$ is a completion-matching pair, the Completing event built is deterministic if and only if $n_0 = |R_0|$ (case of Wait, WaitAll, Recv).

Although this refinement generates supplementary events, most of them are deterministic and do not need to be logged. Only nondeterministic events (nondeterministic Matching due to ANY sources; nondeterministic Matching due to probe-matching events; nondeterministic completion due to WaitSome, WaitAny, TestAll, Test, TestAny and TestSome) are logged and force a synchronization with the event logger.

3.3.3 A Generic Framework for Message Logging in Open MPI

The Open MPI architecture is a typical example of the modern generation MPI implementation. Figure 3.3b summarizes the Open MPI software stack dedicated to MPI communication. Regular components are summarized with plain lines, while the additional fault-tolerant components are dashed. At the lowest level, the BTL exposes a set of communication primitives appropriate for both send/receive and RDMA interfaces. A BTL is MPI semantics agnostic; it simply moves a sequence of bytes (potentially noncontiguous) across the underlying transport. Multiple BTLs might be in use at the same time to strip data across multiple networks. The PML implements all logic for point-to-point MPI semantics including standard, buffered, ready, and synchronous communication modes. MPI message transfers are scheduled by the PML based on a specific policy according to short and long protocol, as well as using control messages (ACK/NACK/MATCH). Additionally, the PML is in charge of providing the MPI matching logic as well as reordering the out-of-order fragments. All remaining MPI functions, including some collective communications, are built on top of the PML interface. While in the current implementation of the fault-tolerant components only point-to-point based collectives are supported, other forms

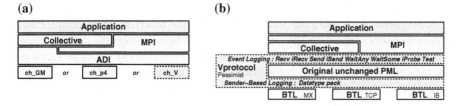

Fig. 3.3 Comparison between the MPICH and the Open MPI architecture and the interposition level of fault tolerance (fault-tolerant components are dashed). **a** MPICH-1.2/MPICH-V. **b** Open MPI/Open MPI-V

of collective communication implementations (such as hardware based collectives) can also be supported as they are deterministic.

In order to integrate the fault-tolerance capabilities in Open MPI, we added one new class of components, the VPROTOCOL (dashed in the Fig. 3.3b). A VPROTOCOL component is a symbiotic shell component enveloping the default PML. Each is an implementation of a particular fault-tolerant algorithm; its goal is not to manage actual communications but to extend the PML with message logging features. As all of the Open MPI components, the VPROTOCOL module is loaded at runtime at the user's request. When it is loaded, it replaces some of the interface functions of the PML with its own. Once it has logged or modified the communication requests according to the needs of the fault-tolerant algorithm, it calls the real PML to perform the actual communications. This modular design has several advantages compared to the MPICH-V architecture: (1) it does not modify any core Open MPI component, regular PML message scheduling and device optimized BTL can be used, (2) expressing a particular fault-tolerant protocol is easy, it is only focused on reacting to some events, not handling communications and (3) the best-suited fault-tolerant component can be selected at runtime.

3.3.4 Pessimistic Message Logging Implementation

The VPROTOCOL pessimist is the first implementation based on our refined model. It provides four main functionalities: sender-based message logging, remote event storage, event logging for any source receptions, and event logging for nondeterministic deliveries. Each process has a local Lamport clock, used to mark events; during Send, iSend, Recv, iRecv, and Start, every request receives this clock stamp as a unique identifier.

3.3.4.1 Sender-Based Logging

The improvements we propose to the original model still rely on a sender-based message payload logging mechanism. We integrated the sender-based logging to the data type engine of Open MPI. The data type engine is in charge of packing (possibly noncontiguous) data into a flat format suitable for the receiver's architecture. Each time a fragment of the message is packed, we copy the resulting data in an *mmaped* memory segment. Because the sender-based copy progresses at the same speed as the network, it benefits from cache reuse and releases the send buffer at the same date. Data is then asynchronously written from memory to disk in background to decrease the memory footprint.

3.3.4.2 Event Logger Commits

Nondeterministic events are sent to event loggers processes (EL). An EL is a special process added to the application outside the `MPI_COMM_WORLD`; several might be used simultaneously to improve scalability. Events are transmitted using nonblocking MPI communications over an intercommunicator between the application process and the event logger. Though asynchronous, there is a transactional acknowledgment protocol to ensure that every event is safely logged before any MPI send can progress.

3.3.4.3 Any Source Receptions

Any source logging is managed in the iRecv, Recv, and Start functions. Each time an any source receive is posted, the completion function of the corresponding request is modified. When the request is completed, the completion callback logs the event containing the request identifier and the matched source. During recovery, the first step is to retrieve the events related to the MPI process from the event logger. Then every promiscuous source is replaced by the well specified source of the event corresponding to the request identifier. Because the MPI matching is FIFO per channel, enforcing the source is enough to replay the original matching order.

3.3.4.4 Nondeterministic Deliveries

In MPI, several completion operations have the potential to generate a nondeterministic outcome (iProbe, WaitSome, WaitAny, Test, TestAll, TestSome, and TestAny functions). Every nondeterministic completion test is assigned a unique clock according to the local Lamport clock. When the operation returns, a delivery event containing the list of all requests completed by the operation is created for this clock. During replay, when the completion operation's clock is equal to the clock of the first event, the corresponding requests are completed by waiting specifically for each of them.

Should the outcome of the completion test be that no request completed, to avoid the creation of a large number of events for consecutive unsuccessful completion test, we use lazy logging; only one event is created for all the consecutive operations. If a completion test succeeds, any pending lazy event is discarded. During recovery, any completion test whose clock is lower than the first event in the log has to return that no request completed.

3.3.5 Performance of MPI Message Logging

The full experimental conditions are described in [11]. Because the proposed approach does not change the recovery strategy used in previous works, we only focus on failure-free performance.

3.3.5.1 Benefits from Event Distinction

One of the main differences of the refined model is the split of message receptions into two distinct events. In the worst case, this might lead to logging twice as many events compared to the model used in other message logging implementations. However, the closer match between the new model and MPI internals allows for detecting (and discarding) deterministic events. Table 3.1 characterizes the amount of nondeterministic (actually logged in Open MPI-V) events compared to the overall number of exchanged messages. Though we investigated all the NPB kernels (BT, MG, SP, LU, CG, FT) to cover the widest spectrum of application patterns, we detected nondeterministic events in LU and MG only. In all other benchmarks, Open MPI-V does not log any event, thanks to the detection of deterministic messages. On both MG and LU, the only nondeterministic events are any source messages; there are no nondeterministic deliveries or probes. In MG, two-thirds of the messages are deterministic, while in LU less than 1 % are using the any source flag, outlining how the better fitting model drastically decreases the overall number of logged events in the most usual application patterns. As a comparison, MPICH-V2 logs at least one event for each message (and two for rendezvous messages). According to our experiments, the same results hold for class A, C, and D of the NAS. The ratio of logged events does not correlate with the number of computing processes in LU and decreases when more processes are used in MG, meaning that the fault-tolerant version of the application is at least as scalable as the original one. Separate research efforts have confirmed that most events are deterministic in a wide range of applications [44].

Avoiding logging of some events is expected to lower the latency cost of a pessimistic protocol. Figure 3.4 presents the overhead on Myrinet round trip time of enabling the pessimistic fault-tolerant algorithm. We normalize Open MPI-V pes-

Table 3.1 Percentage of nondeterministic events to total number of exchanged messages on the NAS Parallel Benchmarks (Class B)

	BT	SP	LU					
#processors	All		4	32	64	256	512	1024
%nondeterministic	0	0	1.13	0.66	0.80	0.80	0.75	0.57
	FT	CG	MG					
#processors	All		4	32	64	256	512	1024
%nondeterministic	0	0	40.33	29.35	27.10	22.23	20.67	19.99

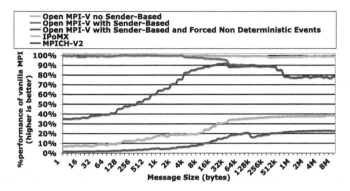

Fig. 3.4 Myrinet 2000 ping-pong performance of pessimistic message logging in percentage of non-fault-tolerant MPI library

simist (labeled Open MPI-V with sender-based in the figure) according to a similar non-fault-tolerant version of Open MPI, while we normalize the reference message logging implementation MPICH-V2 according to a similar version of MPICH-MX; in other words, 100 % is the performance of the respective non-fault-tolerant MPI library. We deem this as reasonable as (1) the bare performance of Open MPI and MPICH-MX is close enough that using a different normalization base introduces no significant bias on the comparison between fault-tolerant protocols, and (2) this ratio reflects the exact cost of fault tolerance compared to a similar non-fault-tolerant MPI implementation, which is exactly what needs to be outlined. IPoMX performance is also provided as a reference to support our discussion on the overhead breakdown of MPICH-V2.

In this ping-pong test, all Recv operations are well-specified sources and there is no WaitAny. As a consequence, Open MPI-V pessimist does not create any event during the benchmark, and reaches exactly the same latency as Open MPI (3.79 µs). To measure the specific cost of handling nondeterministic events in Open MPI-V pessimist, we modified the NetPIPE [82] benchmark code; every Recv has been replaced by the sequence of an any source iRecv and a WaitAny. This altered code generates two nondeterministic events for each message. The impact on Open MPI-V pessimist latency is a nearly three time increase in latency. The two events are merged into a single message to the event logger; the next send is delayed until the acknowledgment comes back. This corresponds to the expected cost on latency of pessimistic message logging. Still, the better detection of nondeterministic events removes the message logging cost for most common types of messages.

Because MPICH-V does not discard deterministic events from logging, there is a specific overhead for every message (the 40 µs latency increases to reach 183 µs), even on the original deterministic benchmark. This specific overhead comes on top of the cost of supplementary memory copies.

3.3.5.2 Benefits from Zero-Copy Receptions

Figure 3.4 shows the overhead of MPICH-V. With the pessimistic protocol enabled, MPICH-V reaches only 22 % of the MPICH-MX bandwidth. This bandwidth reduction is caused by the number of memory copies in the critical path of messages. Because the message logging model used in MPICH-V assumes delivery is atomic, it cannot accommodate the MPI matching and buffering logic; therefore, it does not fit the intermediate layer of MPICH (similar to the PML layer of Open MPI). As a consequence, the event logging mechanism of MPICH-V replaces the low level *ch_mx* with a TCP/IP based device. The cost of memory copies introduced by this requirement is estimated by considering the performance of the NetPipe TCP benchmark on the IP emulation layer of MX: IPoMX. The cost of using TCP, with its internal copies and flow control protocol, is as high as 60 % of the bandwidth and increases the latency from 3.16 to 44.2 µs. In addition, the MPICH- V device itself needs to make an intermediate copy on the receiver to delay matching until the message is ready to be delivered. This is accountable for the 20 % remaining overhead on bandwidth and increases the latency to 96.1 µs, even without enabling event logging.

On the contrary, in Open MPI-V the model fits tightly with the behavior of MPI communications. The only memory copy comes from the sender-based message payload logging; there are no other memory copies. As a consequence, Open MPI-V is able to reach a typical bandwidth as high as 1570 Mbit/s (compared to 1870 Mbit/s for base Open MPI and 1825 Mbit/s for MPICH-MX). The difference between Open MPI-V with or without sender-based logging highlights that, although the cost is not completely eliminated, the cache reuse effect from progressing with the packing convertor is beneficial. While the sender-based copy fits in cache, the performance overhead of the extra copy is reduced to 11 % and jumps to 28 % only for messages larger than 512 kB.

3.3.5.3 Sender-Based Impact

While the overall number of memory copies has been greatly reduced, the sender-based message payload copy is mandatory and can't be avoided. Figure 3.5 explains the source of this overhead by comparing the performance of Open MPI and Open MPI-V pessimist on different networks. As the sender-based copy is not on the critical path of messages, there is no increase in latency, regardless of the network type. On Ethernet, bandwidth is unchanged as well, because the time to send the message on the wire is much larger than the time to perform the memory copy, thus a perfect overlap.

Counterintuitively, Open MPI bandwidth for the non-fault-tolerant version is better on Myrinet 10G than on shared memory: the shared memory device uses a copy-in copy-out mechanism between processes, producing one extra memory access for each message (i.e., physically reducing the available bandwidth by two). Adding a third memory copy for handling sender-based logging to the two involved in regular

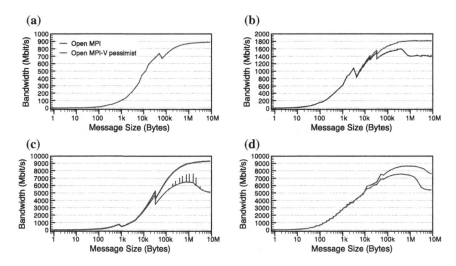

Fig. 3.5 Ping-pong performance comparison between Open MPI and Open MPI-V pessimist on various networks. **a** TCP gigabit ethernet. **b** MX Myrinet 2G. **c** MX Myrinet 10G. **d** Shared memory

shared-memory transfer has up to 30 % impact on bandwidth for large messages, even when this copy is asynchronous. This is the expected result considering that the performance bottleneck for shared-memory network is the pressure on memory bus bandwidth.

As the sender-based message logging speed depends on memory bandwidth, the faster the network, the higher the relative copy time becomes. Myrinet 2G already exhibits imperfect overlap between memory copies and network transmission, though when the message transfer fits in cache, the overhead is reduced by the memory reuse pattern of the sender-based mechanism. With the faster Myrinet 10G, the performance gap widens to 4.2 Gbit/s (44 % overhead). As the pressure on the memory subsystem is lower when using Myrinet 10G network than when using shared memory, one could expect sender-based copy to be less expensive in this context. However the comparison between Open MPI-V on Myrinet 10G and shared memory shows a similar maximum performance on both media, suggesting that some memory bandwidth is still available for improvements from better software engineering. Similarly, the presence of performance spikes for message sizes between 512 kB and 2 MB indicates that the cache reuse strategy does not fit well with the DMA mechanism used by this NIC.

3.3.5.4 Application Performance and Scalability

Figure 3.6a presents the performance overhead of various numerical kernels on a Myrinet 10G network with 64 nodes. Interestingly, the two benchmarks exhibiting

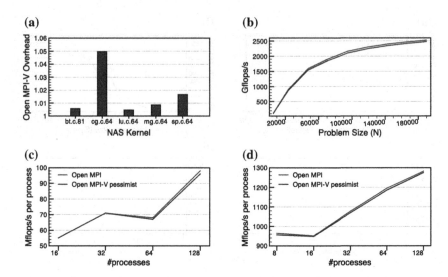

Fig. 3.6 Application behavior comparison between Open MPI and Open MPI-V pessimist on Myrinet 10G. **a** NAS normalized performance (Open MPI = 1). **b** Weak scalability of HPL (90 procs, 360 cores). **c** Scalability of CG Class D. **d** Scalability of LU Class D

nondeterministic events suffer from a mere 1 % overhead compared to a non-fault-tolerant run. The more synchronous CG shows the highest performance degradation, a moderate 5 % increase in execution time. Because there are no nondeterministic events in CG, overhead is solely due to sender-based payload logging.

Figure 3.6b compares the performance of a fault-tolerant run of HPL with regular Open MPI on 90 quad-core processors connected through Myrinet 10G, one thread per core. While the performance overhead is limited, it is independent of the problem size. Similarly, for CG and LU (Fig. 3.6c, d), the scalability when the number of processes increase follows exactly the same trend for Open MPI and Open MPI-V. For up to 128 nodes, the scalability of the proposed message logging approach is excellent, regardless of the use of nondeterministic events by the application.

3.3.6 Concluding Remarks

The model of message logging could be refined to match the reality of high-performance network interface cards, where message receptions are decomposed in multiple interdependent events. From a finer decomposition of events impacting the life cycle of a MPI message, the need for intermediate message copies impacting bandwidth on high-performance networks is lifted; deterministic and nondeterministic events are clearly discriminated, allowing a reduction of the overall number of messages requiring latency disturbing management. One question that naturally arises is the effect of this new development on the historical knowledge and understanding about the importance of the event logging technique, and it's synchronicity.

3.4 Comparing Event Logging Strategies

Pessimistic message logging, as we discussed so far, is the most synchronous event logging technique. It ensures the always no-orphan condition: all the previous nondeterministic events of a process must be logged before a process is allowed to impact the rest of the system. Therefore, any process has to ensure that every event is safely logged before any MPI send can proceed. Since no-orphan process can be created, only the failed processes have to restart after a failure. In order to improve latency, the no-orphan condition can be relaxed. Over the years, different versions of message logging have been proposed to address the issue of high latency associated with synchronous logging of events to a stable storage [2, 9, 30]. Causal message logging [8, 67] piggybacks unlogged events on outgoing messages, so that any process depends only on events either logged or known locally. Optimistic message logging [69] pushes one step further; nondeterministic events are buffered in the sender process memory and logged asynchronously, while message sending is never delayed. The consequence is that a message sent by a process may depend on an unlogged event and may become orphaned. Thus a recovery protocol is needed to detect orphan messages and to recover the application in a consistent global state after a failure. To be able to detect orphan messages, dependencies between nondeterministic events need to be tracked during the entire execution; dependency information needs to be piggybacked on application messages.

As we described earlier, the model of message logging was recently refined to match the reality of high-performance network interface cards, where message receptions are decomposed in multiple interdependent events. In this section, we present two implementations of message logging, pessimistic and optimistic message logging, respectively, being the most and the least synchronous possible paradigms, based on that same generic failure recovery framework, implementing the finer message logging model in a production MPI implementation [11]. Then we present a comprehensive experimental comparison of those two approaches using microbenchmarks and exploring their behavior on a range of scientific kernels. Results demonstrate how improvements targeted at adapting message logging to high-performance networks have dramatically altered the knowledge acquired in previous work about the impact of synchronicity on event logging performance [30].

3.4.1 Active Optimistic Message Logging

In order to compare the impact of the synchronicity of event logging, we need to consider two best performing strategies in their class. Although there are little possible variations in the pessimistic event logging protocol, optimistic logging can take more varied forms. Optimistic message logging traditionally suffers from two main drawbacks. First, it is less efficient than pessimistic message logging on recovery because orphan processes may be created. In the event of a failure, a recovery protocol must be executed to detect orphan processes and these orphan processes must rollback, in addition to the failed processes. Second, to track dependencies between processes during failure-free execution, dependency information must be piggybacked on application messages, adding overhead on communications [76].

A new optimistic message logging solution, called *active optimistic message logging* [69] (O2P), has been recently proposed to mitigate some inefficiencies in existing optimistic message logging protocols. In the standard model of optimistic message logging, determinants are buffered is the process memory and logged asynchronously. O2P is an active optimistic message logging protocol, i.e., it logs nondeterministic determinants on stable storage as soon as possible to reduce the probability that a message depends on an unlogged determinant when it is sent. Thus it reduces the risk for a failure to produce orphan messages. In addition, it has been proved that to be able to detect orphan messages, only dependencies to unlogged nondeterministic determinants have to be tracked [23]. Since active optimistic message logging maximizes the probability that previous nondeterministic determinants are logged when a message is sent, it also reduces the amount of data that needs to be piggybacked on application messages.

3.4.1.1 Dependency Tracking

To track dependencies between application processes, O2P uses a dependency vector. A dependency vector is an n-entry vector, n being the number of processes in the application. Entry j of process p_i dependency vector is the last unlogged nondeterministic event of process p_j that the current state of p_i depends on. If entry j is empty, it means that p_i doesn't depend on any unlogged nondeterministic event from the process p_j. When a process sends a message, it piggybacks its dependency vector on the message. The process receiving that message updates its own dependency vector with the piggybacked vector.

When a nondeterministic event occurs at process p_i, it sends the event to the EL and saves it in entry i of its dependency vector. This entry is emptied by the process when it receives the acknowledgment from the EL. To limit the piggybacked data size, dependency vectors are implemented as described in [79]. Only nonempty entries that have changed since the last message sent to the same process are piggybacked on the message.

3.4.1.2 Event Logger

In order to make a process aware of the events saved by other processes, the EL maintains a n-entry vector that we call the stable vector. Entry k of the stable vector is the last event of process p_k received by the EL. The stable vector is included in the acknowledgements sent by the EL. When a process delivers an acknowledgment from the EL, it updates its dependency vector according to the stable vector received. This mechanism contributes to reduce the size of the piggybacked data.

3.4.1.3 Piggyback Mechanisms

Piggyback mechanisms have a significant impact on O2P failure-free performance. Due to active optimistic message logging, most of the time, there is no data to piggyback on application messages. That is why we have implemented a solution that optimizes this case. Piggybacked data are sent in a separate message. An additional flag is included in the application message header to make the destination process aware of the presence of piggybacked data. Thus an additional message is sent only if there is data to piggyback.

3.4.2 Optimistic Versus Pessimistic: Experimental Evaluation

In this section, we compare the performance obtained by the optimistic and pessimistic protocols taking into account the impact of the new message logging model. The VPROTOCOL framework enables the implementation of message logging protocols in the Open MPI library. It is based on the refined model presented in the previous section. The two protocols compared in this paper are implemented in this same framework, allowing for a fair and equitable comparison. The full description of the experimental conditions can be found in [10].

3.4.2.1 Ping-Pong Performance

For this set of experiments, NetPIPE is deployed on two Dell PowerEdge 1950 servers while a third one hosts the Event Logger. The results of Fig. 3.7a show a regular Gigabit Ethernet ping-pong for the two protocols. With the default options, there is no nondeterministic event in this benchmark. Therefore, thanks to the optimizations introduced by the refined model, there is no event to log and the latency overhead is unnoticeable. As a consequence, both protocols exhibit very similar behavior.

Impact of nondeterministic events To investigate the impact of event logging, we force any source receptions in the NetPIPE benchmark. According to the pattern of communication in this benchmark, a nondeterministic event is created by each recep-

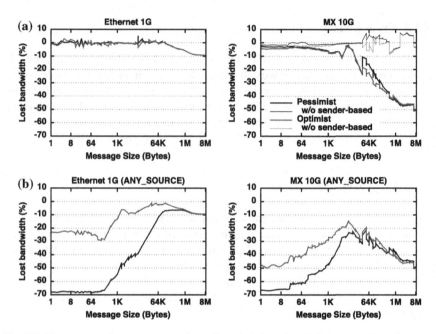

Fig. 3.7 Ping-pong performance comparison of pessimistic and optimistic protocols. **a** Common case without nondeterministic events. **b** With forced nondeterministic events

tion and is immediately followed by a send, forcing the pessimistic protocol to log an event synchronously before allowing the send to proceed. Figure 3.7b illustrates that the consequence is a threefold increase in latency. The overhead induced by the optimistic protocol is much smaller: while the event is still sent to the Event Logger immediately, the next send does not need to be delayed until the reception of the acknowledgment. The impact of piggybacked data management is very small: as the application has only two processes, the maximum number of events to piggyback is at most one.

High-performance networks Focusing on the Myrinet 10G network results from Fig. 3.7a, the very low latency of both protocols illustrates that without nondeterministic events, the cost of event logging is well contained on high-performance networks. As seen in Fig. 3.7b, the relative overhead of managing nondeterministic events is not modified; the pessimistic protocol still endures a threefold increase in latency while the optimistic one sees a milder degradation. However, the performance penalty associated with sender-based payload logging, a shared characteristic of all message logging protocols, sees its share increase as the network becomes faster. The Myrinet network is fast enough that even being asynchronous, the extra memory copy generated by the sender-based payload logging drains more memory bandwidth than available. Experiments where the sender-based mechanism is disabled, depicted in Fig. 3.7a, further support that explanation, with no bandwidth degradation compared to non-fault tolerant MPI.

3.4.2.2 Application Benchmarks

Figure 3.8 presents the performance of all the NAS kernels for 64 processes on the Myrinet network. Every kernel is evaluated with or without the sender-based mechanism being active. While it is a required component for a successful recovery, deactivating the sender-based overhead reveals the performance differences imputable to the event logging protocols. Among the NAS kernels, only two generates nondeterministic events: MG and LU. As expected, the performance of event logging exhibits almost no differences between the protocols on the benchmarks where there is no nondeterministic events. Even on those with nondeterministic events, the performance varies only by less than 2 %, which is close to the error margin of measurements. On this fast network, the sender-based overhead clearly dominates the performance and flattens any performance difference coming from the synchronicity of the event logging.

3.4.3 Concluding Remarks

Pessimistic and optimistic message logging are, respectively, the most and the least synchronous message logging solutions. Optimistic message logging exchanges the ability to delay logging of determinants with the need to rollback some non-failed processes during recovery. As outlined by NetPIPE, this allows optimistic logging for reaching a twofold better latency in that case. However, in many application kernels, the performance degradation due to synchronous message logging is very limited. When the application actually uses nondeterministic communication patterns, a five to six percent difference can be measured between the two protocols. From a broader

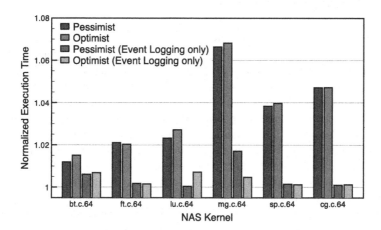

Fig. 3.8 Normalized performance of the NAS kernels on the Myrinet 10G network (Open MPI = 1)

perspective, from the combined effects of the event logging model improvements, and the increase in the network interface performance, sender-based payload copy has become the dominant overhead of message logging.

3.5 Optimizing Sender-Based Message Logging

The previous sections demonstrated that, with a new careful modeling, the cost of event logging and has tremendously decreased [10], leading the once negligible overhead incurred by in-transit messages to now dominate. The technique considered as the most efficient today to replay in-transit messages is sender-based message logging: the sender keeps a copy of every outgoing message. Although sender-based logging requires only a local copy, done in memory, and could theoretically be overlapped by actual communication over the network, it has appeared experimentally to remain a significant overhead. The bandwidth overhead of the sender-based copy is now standing alone in the path of ubiquitous automatic and efficient fault-tolerant software. In this article, we consider and compare multiple approaches to reduce or overlap this cost to a nonmeasurable overhead in the Open MPI implementation of message logging: Open MPI-V [11].

Many works have recently considered the more general issue of copying memory regions in multicore systems using specific hardware [41, 90], or how the memory management can play a significant role in the communication performance [40, 84]. However, the interactions between simultaneously transferring the data to the Network Interface Card and obtaining an additional copy in the application space has not been addressed.

3.5.1 Strategies for Sender-Based Copies

In the message logging fault-tolerant protocol, two mechanisms are used: event logging and sender-based message logging. The event logging mechanism defuse the threat on recovery consistency posed by orphan messages, those who carry a dependency between the nondeterministic future of the recovering processes and the past of the survivors. The outcome of every nondeterministic event is stored on a stable remote server; upon recovery, this list is used to force the replay to stay in a globally consistent state. In this section, we focus our efforts on improving the second mechanism, message payload copy, thus we do not further discuss the event logging method. The necessity of the sender-based message logging comes from in-transit messages, i.e., messages sent in the past of the survivors but not yet received by the recovering processes. Because only the failed processes are restarted, messages sent in the past from the survivors can not be regenerated. The sender-based message logging approach keeps a memory copy of every outgoing message on the sender, so that any in-transit message is either regenerated (because the sender also failed and therefore is replaying the execution as well), or is readily available.

There are mostly two parameters governing the payload logging: (1) the backend storage system, and (2) the copy strategy from the user memory to the backend storage system. We have designed three backend storages: (a) a file that is mapped in memory, (b) heap memory as backend, allocated using memory mapping of private anonymous memory, and (c) a dummy backend storage, that does not implement message logging, but provides us a mean to measure the overhead due only to the copy itself. We have also designed three copy methods: (a) a pack method, that copy the message in one go into the backend space, (b) a convertor method, that chops the copy of the message according to the Open MPI pipeline, and (c) a thread method, that creates an independent thread responsible of doing the copies. In the following, we describe with more details these strategies.

3.5.2 Backend Storages

Memory Mapped file: It should be noted that there is no necessity for the log to be persistent: if a process crashes, it will restart in its own history, and recreate the messages that have been logged after the last checkpoint (still, messages preceding the last checkpoint must be saved with the checkpoint image, because they are part of the state of the process). However, a file backend is natural, because the volume of message to be logged can be significant, and this should not reduce the amount of memory available for the application. Mapping the backend file into memory is the most convenient way of accessing it.

We designed this backend file as a growing storage space, on which we open a moving window using the mmap system call. When the window is too small to accept a new message (we use windows of 256 MB, unless some message exceeds the size of the window), we wait that all messages are logged (depending on the copying method, described later), make the file grow if necessary, and move the window entirely to a free area of the file.

Heap Memory: If the amount of memory available on the machine is large enough to accept at the same time the application and the copy of the messages payload (up to garbage collection time), then the payload logging can be kept in memory. This second method uses anonymous private memory allocated with the mmap system call to create such a backend for our message logging system.

Dummy Storage: In order to measure independently the overhead introduced by the copy method itself, we also designed a Dummy Storage that does not really implement message logging: after a message is logged, the pointer to store the message payload is moved back to the beginning of the same memory area, reallocated if the size of the message is larger than the largest message seen until the call. When messages are sent often, the pages related to this area will most likely be present in the TLB, and for very short messages, it is even possible that the area itself remains in the CPU cache between two emissions. Though this storage cannot be considered as a backend storage for message logging, it helps us evaluate the overheads of the copy methods themselves, without considering other parameters like TLB misses and pages fault.

3.5.3 Copy Methods

Pack.: The Pack method consists in copying the payload of the message using the memcpy libc call, from the user space to the backend storage space, when the PML V intercepts the message emission for the first time. This interception can happen just after the message has been given to the network card for short messages, or just after the first bytes of the message have been given to the network card, and the network card cannot send more without blocking, for longer messages.

Conv.: When converting the user data to a serialized form usable by the network cards, the Open MPI data type engine can optimize the operation by establishing a pipeline. Instead of sending a single very large message, the convertor data type component splits the message in fragments, and it sends multiple messages of a predetermined maximal size on the network cards. Up to four messages can be given to the network card simultaneously, which will send one after the other. The data type engine tries to keep this pipeline as full as possible, to ensure that a network card has always some data ready to send. In the Conv. payload copy method, the PML V intercepts the convertor component's production of fragments, and introduces the message payload copy at this time. If the pipeline is enabled, each time a chunk of data is copied from the user data to the network card, the PML V copies the same amount of bytes from the user data to the backend storage. The size of the chunks in the pipeline is a parameter of this method.

Thread.: The last copying method is based on a thread. A copying thread is created during the initialization. This thread waits on a queue for copy orders. When this queue is not empty, the thread pops the first element of the queue, and copies the whole user memory onto the backend storage, using the memcpy libc call. When a message emission is intercepted by the PML V, if the message is short, it is copied as for the Pack method. If the message is long enough and could not be sent to the network in one go, a copy request is created and pushed at the end of the request queue. When the application returns from the MPI call, it synchronizes with the copy thread, and waits, to guarantee message integrity, that the related messages have been entirely logged before returning from the MPI call. To ensure a fair comparison, at constant hardware resources, this thread is pinned on the same core as the MPI process that produces the message.

3.5.4 Sender-Based Copy: Experimental Evaluation

The full description of the experimental conditions can be found in [7].

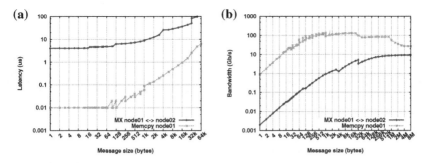

Fig. 3.9 Reference MPI MX NetPIPE performance between two dancer nodes compared to memcopy. **a** Latency. **b** Bandwidth

Figure 3.9 presents the reference latency (Fig. 3.9a) and bandwidth (Fig. 3.9b) of Open MPI on the specified network, and of the memory bus of the machines used. These figures are presented here as a absolute reference of peak performance achievable without message logging. A first observation is that the high-memory bandwidth and low latency compared to the High-Speed network card should enable a logging in memory with little performance impact for messages of less than 1 MB. For larger messages, the bandwidth of the memory bus will become a bottleneck for the logging, and unless the time taken to transfer the message on the network can be recovered by the logging mechanism, overheads are to be expected.

A few characteristics of the underlying network and the Open MPI implementation can moreover be observed from these two figures: one can clearly see the gaps in performance for messages of 4 KB (default size of the MX frame), and 32 KB (change of communication protocol from eager to rendezvous in the Open MPI library). In the rest of the paper, all other measurement will be presented relative to the bandwidth performance of the high-speed network card, to highlight the overheads due to message logging.

Each of the first figures grouped under Figure 3.10 consider a specific storage medium, and compare for a given medium the overheads of the different logging methods as function of the message size.

First, we consider Fig. 3.10a that uses as a storage medium the "Ideal" Storage. As described in Sect. 3.5.1, the Ideal storage uses a single memory area to log all the messages (thus overriding existing log with new messages). The goal of this experiment is to demonstrate the overheads due to the copy itself (and when it happens) without other effects, like page faults, etc. One can see that the logging method has no significant impact up to (and excluding) messages of 4 KB. At 4 KB, the Thread method suffers a huge overhead that decreases the performance by 80%, while the other methods suffer a lower overhead.

A single MX frame is of 4 KB (on this platform). Thus, for messages of 4 KB of payload, or more, multiple MX frames are necessary to send the message (this

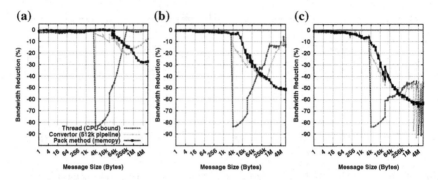

Fig. 3.10 NetPIPE MX bandwidth between two dancer nodes, according to the storage method. **a** Ideal storage memory. **b** Anonymous memory. **c** File map

is true for messages of 4 KB of payload too, since the message header must also be sent). When the message fits in a single frame, the logging thread can be scheduled while the message circulates on the network and is handled by the receiving peer. When the message does not fit in a single MX frame, the Open MPI engine requires scheduling to ensure the lowest possible latency. Since both threads are bound on the same core, they compete for the core, and the relative performances decrease.

On one hand, when the number of frames needed for a single message is low, the MPI thread and the logging thread must alternate with a high frequency on the core (since the MPI call exits only when the message has been sent and logged). On the other hand, when the number of frames needed is high, the thread that is scheduled on the CPU can either log the whole message in one quantum, or use all available frames in the MX NIC to send as much data as possible in one go. Thus, when the number of frames increases, the relative overhead due to the logging thread decreases.

The pack method decreases almost linearly with the message size, since all copies are made sequentially after the send. Because the network is eventually saturated, the relative overhead reaches a plateau. The Convertor method uses a pipeline of 512 KB. Thus, until messages are 512 KB long, it behaves similarly as the Pack method. The difference is due to a slightly better cache reuse from the Pack method that send the message, then logs it, instead of first logging it during the pack operation, then sending it on the network. When the messages size is larger than the pipeline threshold, the Convertor method introduces some parallelism (although not as much as the Thread method), that is used to recover the communication time with logging time.

The other two figures Fig. 3.10b, c demonstrate a similar behavior before 4 K, although the overheads begin to be notifiable a little sooner for all methods when logging on a File. This is due to file system overheads (inodes and free blocks accounting), and memory management (TLB misses) when more pages are needed

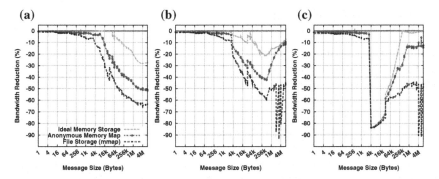

Fig. 3.11 NetPIPE MX bandwidth between two dancer nodes, according to the copy method. **a** Pack method. **b** Convertor method. **c** Thread method

to log the messages. When the file system is effectively used (messages of 4 and 8 MB end up consuming all available buffer caches of the file system), a high variability in the relative overhead becomes observable (Fig. 3.10c).

These phenomenon are more observable on the second group of three figures under Fig. 3.11. These figures consider each a specific logging method, and expose the impact of the medium on the overheads due to a logging method, as function of the message size. As can be seen, using a mmaped file as a storage space introduces the highest overhead, significantly higher than the overhead due to in-memory storage, even when the kernel buffers of the file system are large enough to hold this amount of data. This is due to accounting in the file system (free blocks lists, inodes status), forced synchronization of the journaling information, and a conservative policy for the copy of the data to the file system.

The difference of overhead between an anonymous memory map (in the heap of the process virtual memory), and the Ideal storage space is mainly due to TLB misses introducing additional page reclaims. This cost is unavoidable to effectively log the messages, but it is small for small messages, and amortized for very large messages. As a consequence, logging should happen in memory as long as the log can be kept small enough to fit there, and the system should resort to mmaped files only when necessary.

Figure 3.10b, c lead us to the conclusion that an hybrid approach, with different thresholds depending on the storage medium, and on the message size, should be taken: up to messages of 2 KB, the method has little influence, however after this, the Pack or the convertor methods should be preferred up to messages of 128 KB. For messages higher than 128 KB, the use of an asynchronous thread, even if it must share the core with the application thread, is the preferred method of logging.

3.5.5 Concluding Remarks

In this section, we studied three techniques to log the payload of messages in a sender-based approach, in the Open MPI PML V framework that implement message logging fault tolerance. Because the copying of the message payload must be achieved before the corresponding MPI emission is complete (either when the blocking send function exits, or when the corresponding wait operation exits), copying this payload is a critical efficiency bottleneck of any message logging approach.

One of the techniques proposed is to use an additional thread to process the copying asynchronously with the communication; a second uses the pipeline installed by the Open MPI communication engine to interlace transmissions toward the network, and copies in memory; the third simply copy the payload after it has been sent, and before the completion of the communication at the application level.

We also demonstrated that the medium used to store the payload has a significant impact on the performances of the payload logging process. We concluded that depending on the medium for storage, and the message size, different strategies should be chosen, advocating for a hybrid approach that will have to be tuned specifically for each hardware. Even with advanced payload logging methods, the experimental results indicate that devising a strategy to decrease the amount, or totally eliminate payload logging for some messages would be highly beneficial. We are presenting a technique that achieves such a goal in the next section.

3.6 Correlated Sets Coordination to Decrease Message Logging

Recent advances in message logging have decreased the cost of event logging [11]. As a consequence, more than the logging scheme adopted (a thorough survey of possible approaches is given in [30]), the prominent source of overhead in message logging is the copy of message payload caused by *in-transit* messages [10]. While attempts at decreasing the cost of payload copy have been successful to some extent [7], these optimizations are hopeless at improving shared-memory communication speed. Although the low mean time to failure of Exascale machines calls for preferring an uncoordinated checkpoint approach, the overhead on communication of message logging is bound to increase with the advent of many-core nodes. Uncoordinated checkpointing has been designed with the idea that failures are mostly independent, which is not the case in many-core systems, where multiple cores crash when the whole node is struck by a failure. Not only do simultaneous failures negate the advantage of uncoordinated recovery, but the logging of messages between cores is also a major performance issue. All interactions between two uncoordinated processes have to be logged, and a copy of the transaction must be kept for future replay. Since making a copy has the same cost as doing the transaction itself (as the processes are on the same node we consider the cost of communications equal to the cost of memory copies), the overhead is unacceptable. It is disconcerting that the most resilient fault-tolerant automatic method is also the most bound to suffer, in terms of performance, on expected future systems.

In this section, we consider the case of *correlated failures*: we say that two processes are correlated or codependent if they are likely to be subject to a simultaneous failure. We propose a hybrid approach between coordinated and noncoordinated checkpointing, that prevents the overhead of keeping message copies for communications between correlated processes, but retains the more scalable uncoordinated recovery of message logging for processes whose failure probability is independent. This is a special case of a hierarchical checkpointing protocol presented in Sect. 1.2.4 and analyzed in Sect. 1.3.3, and we detail here how this case can be handled and is evaluated. The coordination protocol we present is a split protocol, which takes into account the fragmentation of messages, to avoid long waiting cycles, while still implementing a transactional semantic for whole messages. Additionally, we demonstrate that application's communication pattern are likely to adopt a topology which is beneficial to the correlated set approach we propose, and leads to a drastic reduction of log volume.

3.6.1 Background

In this section we present our approach, designed to reduce the performance penalty due to message logging, suffered by distributed applications on many-core systems. On such systems the communication subsystem moving data between processes on the same physical node is usually implemented on top of a shared-memory substrate. Taking advantage of this geographical proximity of processes on a many-core system, our message logging protocol significantly reduces the amount of payload to be logged, by emphasizing characteristics linked to the process location.

3.6.1.1 Shared Memory and Message Logging

In uncoordinated checkpoint schemes, the ordering between checkpoint and message events is arbitrary. As a consequence, every message is potentially *in-transit*, and must be copied. Although the cost of the sender-based mechanism involved to perform this necessary copy is not negligible, the cost of a memory copy is often one order of magnitude lower than the cost of the network transfer. Furthermore, the copy and the network operation can overlap. As a result, proper optimization greatly mitigates the performance penalty suffered by network communications (typically to less than 10 %, [7, 11]). One can hope that future engineering advances will further reduce this overhead.

Unlike a network communication, a shared-memory communication is a strongly memory-bound operation. In the worst case, memory copy induced by message logging doubles the volume of memory transfers. Because it competes for the same scarce resource—memory bandwidth—the cost of this extra copy cannot be over-lapped, hence the time to send a message is irremediably doubled.

A message is *in-transit* (and needs to be copied) if it crosses the recovery line from the past to the future. The emission and reception dates of messages are beyond the control of the fault-tolerant algorithm: one could delay the emission or reception dates to match some arbitrary ordering with checkpoint events, but these delays would obviously defeat the goal of improving communication performance. The only events that the fault-tolerant algorithm can alter, to enforce an ordering between message events and checkpoint events, are checkpoint dates. Said otherwise, the only way to suppress *in-transit* messages is to synchronize checkpoints.

3.6.1.2 Correlated Failures

Fortunately, although many-core machines put a strain on message logging perfor-mance, a new opportunity opens, thanks to the side effect that failures do not have an independent probability on such an environment. All the processes hosted by a single many-core node are prone to fail simultaneously: they are located on the same piece of silicon, share the same memory bus, network interface, cooling fans, power supplies, operating system, and are subject to the same physical interferences (rays, heat, vibrations, …). One of the motivating properties of message logging is that it tolerates a large number of independent failures very well. If failures are correlated, the fault-tolerant algorithm can be more synchronous without decreasing its effective efficiency.

The leading idea of our approach is to propose a partially coordinated fault-tolerant algorithm, that retains message logging between sets of processes experiencing inde-pendent failure probability, but synchronize the checkpoints of processes that have a strong probability of simultaneous failures, what we call a *correlated set*. It leverages the correlated failures property to avoid message copies that have a high chance of being useless.

3.6.1.3 Related Works

Group coordinated checkpoint has been proposed in MVAPICH2 [37] to solve I/O storming issues in coordinated checkpointing. In that context, the group coordination refers to a particular scheduling of the checkpoint traffic, intended to avoid over-whelming the I/O network. Unlike our approach, which is partially uncoordinated, this algorithm builds a completely coordinated recovery set.

In [47], Ho et al. propose a group-based approach that combines coordinated and uncoordinated checkpointing, similar to the technique we use in this paper, to reduce the cost of message logging in uncoordinated checkpointing. Their work, however, focuses on communication patterns of the application, to reduce the amount of mes-sage logging. Similarly, in the context of AMPI [61], Meneses, Mendes, and Kalé have proposed in [58] a team-based approach to reduce the overhead of message logging. The Charm++ model advocates a high level of oversubscription, with a ratio of *Chares* threads per hardware thread much larger than one. In their work, teams are of fixed, predetermined sizes. The paper does not explicitly explain how teams are built, but an emphasis on communication patterns seems preferred. In a similar manner, in [70] the authors propose to study the communication pattern of the application in order to devise a static process grouping. In contrast, our work takes advantage of hardware properties of the computing resources, proposing to build correlated groups based on likeliness of failures, and relative efficiency of the communication medium. Our approach effectively circumvents the inherent limitation of competing for the same memory bandwidth on the shared-memory transport by completely eliminating the need for copies inside many-core processors.

3.6.2 Correlated Set Coordinated Message Logging

Whenever a process of a correlated set needs to take a checkpoint, it forces a synchronization with all other processes of the set. If a failure hits a process, all processes of that set have to roll back to their last checkpoint (see the recovery line in example execution depicted in Fig. 3.12). Considering a particular correlated set (as an example S_1), every message can be categorized as either *ingoing* (m_1, m_2), *outgoing* (m_5), or *internal* (m_3, m_4). Between sets, no coordination is enforced. A process failing in another correlated set does not trigger a rollback, but messages between sets have no guaranteed properties with respect to the recovery line, and can still be *orphan* or *in-transit*. Therefore, regular message logging, including payload copy and event logging must continue for outgoing and ingoing messages.

As checkpoints are coordinated, all *orphan* and *in-transit* messages are eliminated between processes of the correlated set. However, as the total recovery set does contain *in-transit* and *orphan* messages, the consistency proof of coordinated checkpoint does not hold for the recovery set formed by the union of the coordinated sets. In an uncoordinated protocol, a recovery set is consistent if all *in-transit* messages are

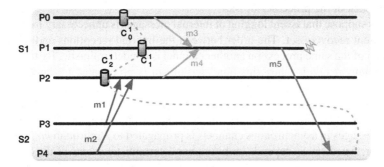

Fig. 3.12 An execution of the correlated set coordinated message logging algorithm

available, and no *orphan* message depends on the outcome of a nondeterministic event. In the next paragraphs, we demonstrate that payload copy can be disabled for internal messages, but that event logging must apply to all types of messages.

3.6.2.1 Intra-Set Payload Copy

Theorem 3.1 *There is no recovery set containing an* in-transit *message between two processes of the same correlated set.*

If two processes are part of the same correlated set, they rollback together to the recovery line containing their last checkpoint. By the direct application of the coordination algorithm, no message is *in-transit* between any pair of synchronized processes at the time of checkpoint (in the case of the Chandy–Lamport algorithm, occasional *in-transit* messages are integrated inside the receiver's checkpoint, hence they are considered as already delivered).

Because an internal message cannot be *in-transit*, it is never sent before the recovery line and received after. Therefore, the payload copy mechanism, used to recover past sent messages during the recovery phase, is unnecessary for internal messages.

3.6.2.2 Intra-Set Event Logging

Theorem 3.2 *In a fault-tolerant protocol creating recovery sets with at least two distinct correlated sets, if the nondeterministic outcome of any internal messages preceding an outgoing message is omitted from the recovery set, there exists an execution that reaches an inconsistent state.*

Outgoing messages are crossing a noncoordinated portion of the recovery line, hence the execution follows an arbitrary ordering between checkpoint events and message events. Therefore, for any outgoing message there is an execution in which it is *orphan*. Consider the case of the execution depicted in Fig. 3.12. In this execution, the message m_5, between the sets S_1 and S_2 is *orphan* in the recovery line produced by a rollback of the processes of S_1.

Let us suppose that Event logging of internal messages is unnecessary for building a consistent recovery set. The order between the internal receptions and any other reception of the same process on another channel is nondeterministic. By transitivity of the Lamport relationship, this nondeterminism is propagated to the dependent outgoing message. Because an execution in which this outgoing message is *orphan* exists, the recovery line in this execution is inconsistent. The receptions of messages m_3, m_4 are an example: the nondeterministic outcome created by the unknown ordering of messages in asynchronous channels is propagated to P_4 through m_5. The state of the correlated set S_2 depends on future nondeterministic events of the correlated set S_1, therefore the recovery set is inconsistent. One can also remark that the same proof holds for ingoing messages (as illustrated by m_1 and m_2).

As a consequence of this theorem, it is necessary to log all message reception events, even if the emitter is located in the same correlated set as the receiver. Only the payload of this message can be spared.

3.6.2.3 Implementation

We have implemented the correlated set coordinated message logging algorithm inside the pessimistic Vprotocol in Open MPI [11]. In order to evaluate the performance of our new approach, we have extended this fault-tolerant component with the capabilities listed below.

Construction of the Correlated Set, Based on Hardware Proximity: Open MPI enables the end user to select a very precise mapping of his application on the physical resources, up to pinning a particular MPI rank to a particular core. As a consequence, the Open MPI's runtime instantiates a process map detailing node hierarchies and ranks allocations. The detection of correlated sets parses this map and extracts the groups of processes hosted on the same node.

Internal Messages Detection: In Open MPI, the couple formed by the rank and the communicator is translated into a list of endpoints, each one representing a channel to the destination (eth0, ib0, shared memory, ...). During the construction of the correlated set, all endpoints pertaining to a correlated process are marked, so that set membership can be resolved directly. When the fault-tolerant protocol considers making a sender-based copy, the endpoint's mark is simply checked to determine if the message payload has to be copied.

Checkpoint Coordination in a Correlated Set: The general idea of a network-silence based coordination is simple: processes send a marker in their communication channels to notify other processes that no other message will be sent before the end of the phase. When all output channels and input channels have been notified, the network is silenced, and the processes can start communicating again. However, MPI communications do not exactly match the theoretical model, which assumes message emissions or receptions are atomic events. In practice, an MPI message is split into several distinct events. The most important include the emission of the first fragment (also called eager fragment), the matching of an incoming fragment with a receive request, and the delivery of the last fragment. Most of those events are unordered, in particular, a fragment can overtake another fragment, even from the same message (especially with channel bonding). Fortunately, because the MPI matching has to be FIFO, in Open MPI, eager fragments are FIFO, an advantageous property that our algorithm leverages. Our coordination algorithm has three phases: it silences eager fragments so that all posted sends are matched; it completes any matched receives; it checkpoints processes in the correlated set.

Eager silence: When a process enters the checkpoint synchronization, it sends a token to all correlated opened endpoints. Any send targeting a correlated endpoint, if posted afterwards, is stalled upon completion of the algorithm. When a process not yet synchronizing receives a token, it enters the synchronization immediately. The eager silence phase is complete for a process when it has received a token from every opened endpoint. Because no new message can inject an eager fragment after the token, and eager fragments are FIFO, at the end of this phase, all posted sends of processes in the correlated set have been matched.

Rendezvous Silence: Unlike eager fragments, the remainder fragments of a message can come in any order. Instead of a complex non-FIFO token algorithm, the property that any fragment left in the channel belongs to an already matched message can be leveraged to drain remaining fragments. In the rendezvous silence phase, every receive request is considered in turn. If a request has matched an eager fragment from a process of the correlated set, the progress engine of Open MPI is called repeatedly until it is detected that this particular request completed. When all such requests have completed, all fragments of internal messages to this process have been drained.

Checkpoint phase: When a process has locally silenced its internal inbound channels, it enters a local barrier. After the barrier, all channels are guaranteed to be empty. Each process then takes a checkpoint. A second barrier denotes that all processes finished checkpointing and that subsequent sends can be resumed.

3.6.3 Experimental Evaluation

In this section we assess the performance benefit of the correlated set coordination approach on a variety of platforms. First, we investigate the behavior of coordinated message logging on large multicore nodes. Second application performance on a cluster of multicore nodes is presented. Last we measure the log volume for several widely used collective communication patterns.

The Pluto platform features 48 cores, and is our main testbed for large shared-memory performance evaluations. The Dancer cluster is an 8 nodes (64 cores total) Infiniband DDR cluster. Vanilla Open MPI means that no fault-tolerant protocol is enabled, regular message logging means that the pessimistic algorithm is used, and correlated set message logging denotes that the pessimistic algorithm is used but cores of the same node undergo coordinated checkpoint. A full description of the experimental conditions is available in [12].

3.6.3.1 Shared-Memory Performance

Coordination Cost: The cost of coordinating a growing number of cores is presented in the Fig. 3.13. The first token exchange is a complete all-to-all, that cannot rely on a spanning tree algorithm. Although, all other synchronizations are simple barriers, the token exchange dominates the execution time, which grows quadratically with the number of processes. Note, however, that this synchronization happens only during a checkpoint, and that its average cost is comparable to sending a 10 KB message.

Fig. 3.13 Time to synchronize a correlated set (Pluto platform, log/log scale)

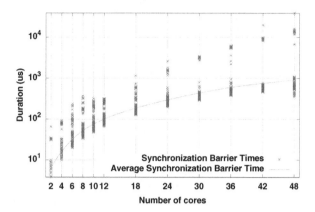

Fig. 3.14 Ping-pong performance (dancer node, shared memory, log/log scale)

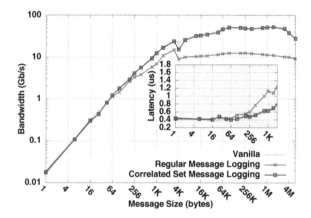

Clearly, the cost of transmitting a checkpoint to the I/O nodes overshadows the cost of this synchronization.

Ping-Pong: Figure 3.14 presents the results of the NetPIPE benchmark on shared memory with a logarithmic scale. Processes are pinned to two cores sharing an L2 cache, a worst case scenario for regular message logging. The maximum bandwidth reaches 53 Gb/s, because communication cost is mostly related to accessing the L2 cache. The sender-based algorithm decreases the bandwidth to 11 Gb/s, because it copies data to a buffer that is never in the cache. When the communication happens between processes of the same correlated set, the the sender-based mechanism is inactive and only event logging remains, which enables correlated set message logging to obtain the same bandwidth as the non-fault-tolerant execution.

NAS Benchmarks: Figure 3.15 presents the performance of the NAS benchmarks on the shared-memory Pluto platform. BT and SP run on 36 cores, all others run on 32. The results presented are the best run out of 10 for each benchmark protocol combination. One can see that avoiding payload copy enables the correlated set

Fig. 3.15 NAS performance (pluto platform, shared memory, 32/36 cores)

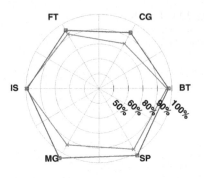

Regular Message Logging / Vanilla ─×─
Correlated Set Message Logging / Vanilla ─□─

Fig. 3.16 HPL cluster performance (dancer cluster, IB20G, 8 nodes, 64 cores)

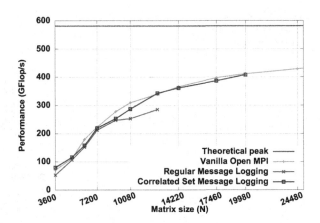

message logging algorithm to experience at most a 7% slowdown, and often no overhead, while the regular message logging suffers from up to 17% slowdown.

Cluster of Multicore Performance: Figure 3.16 presents the performance of the HPL benchmark on the Dancer cluster, with a one process per core deployment. For small matrix sizes, the behavior is similar between the three MPI versions. However, for slightly larger matrix sizes, the performance of regular message logging suffers. Conversely, the correlated set message logging algorithm performs better, and only slightly slower than the non fault-tolerant MPI, regardless of the problem size.

On the Dancer cluster, the available 500 MB of memory per core is a strong limitation. In this memory envelope, the maximum computable problem size on this cluster is $N = 28260$. The extra memory consumed by payload copy limits the maximum problem size to only $N = 12420$ for regular message logging, while the reduction on the amount of logged messages enables the correlated set message logging approach to compute problems as large as $N = 19980$. Not only does partial coordination of the message logging algorithm increase communication performance, it also decreases memory consumption.

3.6.3.2 Collective Communications and Log Volume

In the experiments presented in Fig. 3.17, the benefit of the correlated set message logging approach is compared with the legacy sender-based approach in terms of logged message volume, for a variety of collective operations. We consider the hierarchical collective communication implementations provided by Open MPI's "Hierarch" module. These operations have the particularity of being implemented in a topology-aware, hierarchical, manner; this is interesting to demonstrate the symbiotic relationship between the desired property of the application—reducing internode communication volume—and the desired property of the message logging scheme—reducing log volume. For a given collective, the colormap presents the ratio between the remaining log volume of the correlated set approach over the total log volume incurred by legacy message logging. The brighter the color, the less log is incurred by correlated set message logging, compared to legacy message logging. Each line represents the volume ratio on a particular rank (i.e., a core), the horizontal grid boundaries, every 8 cores, denote ranks allocated on the same node. The columns, as separated by the grid, pertain to a particular message size.

The first figure presents the log volume of the Hierarchical Broadcast algorithm. For small messages, between 4 bytes to 1k bytes, half of the processes are sending only messages crossing node boundaries that consequently incur logging in any cases. The other half of the processes are leaves in the topology and therefore do not send any message (denoted by the white area). This illustrates the typical behavior of a non-topology-aware algorithm; for small messages, it is more beneficial to favor operation parallelism over locality. The choice of this latency optimizing binary tree algorithm has dire consequences on message logging, as it incurs the logging of the complete volume, even with correlated sets. Furthermore, it generates log volume imbalance across nodes. However, one should notice that latency avoiding strategies are, by definition, beneficial only for small messages, where logging has no performance penalty on remote link bandwidth and generate small absolute log volume. In contrast, for large messages, which are obviously generating a much larger communication volume, the broadcast algorithm favors a hierarchical, topology-aware pipeline algorithm. This algorithm features the optimal volume of cross-node communications and maximizes bandwidth (the operation extract parallelism between progression of fragments in the pipeline, hence does not require to extract node parallelism with an aggressive dissemination strategy, as is the case for short messages). As a consequence, on large messages, the correlated set approach reaches the optimal log volume per node: only one core per node logs messages, all other messages payloads, having local destinations, are ignored. The overall log volume of the operation is hence divided by the number of cores per node. For intermediate messages, both bandwidth and latency are important, hence the broadcast algorithm undergoes an hybrid approach that starts by parallelizing the operation as much as possible, first sending fragments along a binary tree, and then finishes the broadcast by transmitting large fragments through a pipeline chain. The resulting log volume for the correlated set approach, in this case, reflects the dual nature of the algorithm, which is still imbalanced, but does not require to log all messages.

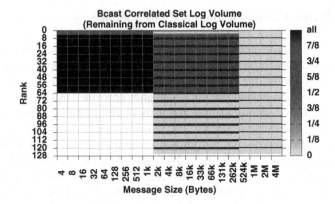

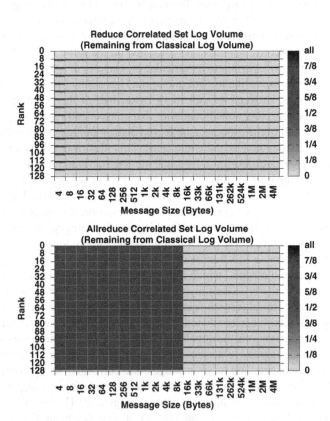

Fig. 3.17 Remaining log volume of correlated set message logging (as a ratio over regular sender-based message logging, lower is better)

The Reduce algorithm is completely hierarchical, no matter the size of the message. Thanks to the topology-aware nature of that algorithm, it can divide the log volume per node by the optimal factor: the number of cores per nodes.

The Allreduce operation presents the behavior of a many-to-many algorithm. For small and intermediate messages, up to 8 k bytes, the collective operation is implemented by an algorithm that enables large instant parallelism, at the expense of internode communication volume. Still, because the many-to-many nature of the algorithm incurs a significant stress on the network cross-section bandwidth, the implementors have taken great care of balancing the communication volume per node. They also favor to some extent intra-node communications, as is illustrated by the overall 3/8 log volume reduction in correlated set message logging. For large messages, similarly to the one-to-all communications, the main concern is to maximize cross-section bandwidth, which results in favoring the hierarchical pipeline chain algorithm, which is also optimal for correlated set message logging.

Overall, as soon as the collective operation incurs a significant message volume, the implementation of the collective systematically favors an algorithm that (1) minimizes the internode communication volume, as this is the crucial performance impacting factor, and (2) balances communication volume per node. As a consequence, the correlated set message logging approach spares the optimal amount of log volume ratio for operations involving the largest absolute communication volume. We advocate that inherently, scalable collective operation implementations aim at reducing the internode communication volume. Hence they induce a symbiotic relation between the communication pattern and the correlated set message logging approach.

3.6.4 Discussion on Process Grouping

In this section, we compare the interest of dynamically discovering the groups of processes based on their communication patterns, as is often proposed in the literature, versus defining the groups from the physical hierarchy of the machine, as proposed in this work. We proved in Theorem 3.2 that all nondeterministic events must be logged in order to maintain the recovery line consistency; independently of the shape or size of the process groups. Therefore, establishing checkpoint synchronization groups is only beneficial by reducing the payload logging mechanism.

Logging the payload of intra-node communications introduces an overhead that is of the same order of magnitude as the communications themselves. On the opposite, logging the payload of internode communications is orders of magnitude faster than the communication themselves. Hence, as long as storage memory is available, the logging of internode communications does not introduce a significant slowdown of the application. As a result, grouping coordinated sets according exclusively to communication patterns, without taking into account the relative overhead of the logging operation, can lead to suboptimal performance.

For applications using a large amount of internode communications, memory consumption might become a dominating problem. When the storage space for message payload is exhausted, forced checkpoints must be regularly taken, or payload logging must be transferred to a larger, and usually slower, storage. In that respect, grouping according to discovered communication patterns could yield better results than according to hardware process mapping, as it is specifically designed to decrease internode communication volume. However, we argue that on practical applications, the difference in volume of logged communications between the two approaches tend to be minimal. Indeed, as seen in the performance evaluation (Sect. 3.6.3.2), collective communications in MPI tend to be implemented in a hierarchical way, naturally grouping processes per levels of hierarchy in the underlying system. As a result, processes physically located on the same machine log only external communications, automatically realizing the same gain as a dynamic group discovery algorithm. Even for applications that rely on point-to-point to express their communication patterns, an application that would intensely communicate between nodes would present scalability and performance issues on a machine with a deep hierarchy. Hence application programmers and users have a strong incentive to map the communication pattern of the application according to the hardware topology. This results in a symbiotic mapping between the communication patterns and the correlated set message logging, enabling, in practice, close to optimal message logging volume reductions.

3.6.5 Concluding Remarks

In this section, we proposed a novel approach combining the most advantageous features of coordinated and uncoordinated checkpointing. The resulting fault-tolerant protocol, belonging to the event logging protocol family, spares the payload logging for messages belonging to a correlated set, but retains uncoordinated recovery scalability. We demonstrate formally, on one hand, that any pessimistic logging protocol must log all nondeterministic event outcomes, regardless of the type of communication generating them, to fulfill the piecewise deterministic assumption. On the other hand, we prove that payload logging of messages within the group can be safely avoided.

The benefit on shared-memory point-to-point performance is significant, translating directly into an observable performance improvement for many types of applications. Even though internode communications are not modified by this approach, the shared-memory speedup translates into a reduced overhead on cluster of multicore type platforms. Moreover, the memory required to hold message payload is greatly reduced; our algorithm provides a flexible control of the tradeoff between synchronization and memory consumption. Our discussion emphasizes that the hardware-conscious mapping of the correlated sets not only accounts for failure probability, but also tends toward minimizing the volume of payload logging per node. Overall, this work greatly improves the applicability of message logging in the context of distributed systems based on a large number of many-core nodes.

3.7 Supporting User-Level Recovery with Standard MPI

The primary form of fault tolerance today is rollback recovery with periodical checkpoints to disk. While this method is effective in allowing applications to recover from failures using a previously saved state, it causes serious scalability concerns [16]. Traditional approaches based on periodic checkpointing and rollback recovery, incurs a steep overhead, as much as 25 % [74], on failure-free operations. Moreover, periodic checkpointing requires precise heuristics for fault frequency to minimize the number of superfluous, expensive protective actions [17, 22, 39, 65, 92]. Even the advanced checkpoint/restart techniques we have discussed in this chapter can suffer from significant overhead given an adverse failure frequency, or a communication pattern that generates too many sender-based copies.

In contrast, *forward recovery* leverages algorithms' properties to complete operations despite failures. In naturally fault-tolerant applications, the algorithm can compute the solution while totally ignoring the contributions of failed processes. In Algorithm Based Fault Tolerance (ABFT) applications, a recovery phase is necessary, but failure damaged data can be reconstructed only by applying mathematical operations on the remaining dataset [48]. A recoverable dataset is usually created by initially computing redundant data, dispatched so as to avoid unrecoverable loss of information from failures. At each iteration, the algorithm applies the necessary mathematical transformations to update the redundant data (at the expense of more communication and computation). Despite great scalability and low overhead [13, 21, 57], the adoption of such algorithms has been hindered by the requirement that the support environment must continue to consistently deliver communications, even after being crippled by failures. This demand from the MPI library largely exceeds the specifications of the current MPI Standard [88] and has proven to be an unrealistic requirement, considering that only a handful of MPI implementations provide it.

The current MPI Standard (MPI-3.0, [88]) does not provide significant help to deal with the required type of behavior. Section 2.8 states in the first paragraph: *MPI does not provide mechanisms for dealing with failures in the communication system. [...] Whenever possible, such failures will be reflected as errors in the relevant communication call. Similarly, MPI itself provides no mechanisms for handling processor failures.* Failures, be they due to a broken link or a dead process, are considered resource errors. Later, in the same section: *This document does not specify the state of a computation after an erroneous MPI call has occurred. The desired behavior is that a relevant error code be returned, and the effect of the error be localized to the greatest possible extent.* So, for the current standard, process or communication failures are to be handled as errors, and the behavior of the MPI application after an error has been returned is left unspecified by the standard. However, the standard does not prevent implementations from going beyond its requirements, and on the contrary, encourages high-quality implementations to *return* errors once a failure is detected. Unfortunately, most of the implementations of the MPI Standard have taken the path of considering process failures as unrecoverable errors, and the processes of the application are most often killed by the runtime system when a failure hits any of them, leaving no opportunity for the user to mitigate the impact of failures.

Several proposals have emerged during the efforts of the MPI forum toward the MPI-3 standard.[2] However, these proposals are still in their infancy and it is expected that several years will pass before they are blessed by the forum in a future revision and become generally deployed and available.

The current MPI-3 standard leaves open an optional behavior regarding failures to qualify as a "high-quality implementation." According to this specification, when using the MPI_ERRORS_RETURN error handler, the MPI library should return control to the user when it detects a failure. In this paper, we propose the idea of Checkpoint-on-Failure (Checkpoint-on-Failure (CoF)) as a minimal impact feature to enable MPI libraries to support forward recovery strategies, while relying exclusively on the features of a high quality implementation, as defined by the current MPI Standard, and thereby enable advanced forward recovery techniques despite the default undefined state of MPI which does not permit continued communication in case of a failure. We demonstrate that an implementation that enables CoF is simple and yet effectively supports ABFT recovery strategies that completely avoid costly periodic checkpointing. The CoF protocol undergoes checkpoint *after* a failure has struck, thereby creating an optimal number of checkpoints (exactly one per actual failure). The MPI application is then restarted in order to restore a fresh, functional MPI library. The dataset is reloaded from checkpoints where possible, otherwise it is restored through a scalable, application-specific forward recovery protection scheme. It is notable that in this scheme, checkpoint actions are taken only *after* a failure is detected; hence the checkpoint interval is optimal by definition, as there will be one checkpoint interval per effective fault. We then extend the analysis to the broader case of general applications where only part of the computations are handled by MPI routines. In Sect. 3.7.4, we explain how such applications, for which periodic checkpoint–restart is generally not practical, can still efficiently integrate the subset of their MPI operations with the CoF approach. Additionally, this type of deployment also eliminates the checkpoint overhead: the non-MPI part of the application can remain dormant during the redeployment of MPI, so that the dataset remains resident in memory without paying the cost of checkpoint I/O.

3.7.1 The Checkpoint-on-Failure Protocol

In this section, we advocate that an extremely efficient form of fault tolerance can be implemented, strictly based on the MPI Standard, for applications capable of taking advantage of forward recovery. ABFT methods are a family of forward recovery algo-

---

[2]http://meetings.mpi-forum.org/mpi3.0_ft.php.

1. MPI returns an error at surviving processes	**ABFT**
2. Surviving processes checkpoint	**Recovery**
3. Surviving processes exit	
4. A new MPI application is started	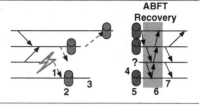
5. Processes load from checkpoint (if any)	
6. Processes enter ABFT dataset recovery	
7. Application resumes	

Fig. 3.18 The checkpoint-on-failure protocol

rithms, capable of restoring missing data from redundant information located on other processes. In the remainder of this text, we will consider the case of ABFT without loss of generality: any other forward recovery technique could be substituted. Forward recovery requires that communication between processes can resume, and we acknowledge that, in light of the current standard, requiring the MPI implementation to maintain service after failures is too demanding. However, a high-quality MPI library should at least allow the application to regain control following a process failure. We note that this control gives the application the opportunity to save its state and exit gracefully, rather than the usual behavior of being aborted by the MPI implementation.

Based on these observations, we propose a new approach for supporting application based forward recovery, called Checkpoint-on-Failure (CoF). The algorithm in Fig. 3.18 presents the steps involved in the CoF method. In the associated explanatory figure, horizontal lines represent the execution of processes in two successive MPI applications. When a failure eliminates a process, other processes are notified and regain control from ongoing MPI calls (1). Surviving processes assume the MPI library is dysfunctional and do not call further MPI operations (in particular, they do not yet undergo ABFT recovery). Instead, they checkpoint their current state independently and abort (2, 3). When all processes exited, the job is usually terminated, but the user (or a managing script, batch scheduler, runtime support system, etc.) can launch a new MPI application (4), which reloads the dataset from the checkpoint (5). In the new application, the MPI library is functional and communications possible; the ABFT recovery procedure is called to restore the data of the process(es) that could not be restarted from checkpoint (6). When the global state has been repaired by the ABFT procedure, the application is ready to resume normal execution.

Compared to periodic checkpointing, in CoF, a process pays the cost of creating a checkpoint only when a failure, or multiple simultaneous failures have happened, hence an optimal number of checkpoints during the run (and no checkpoint overhead on failure-free executions). Moreover, in periodic checkpointing, a process is protected only when its checkpoint is stored on safe, remote storage, while in CoF, local checkpoints are sufficient: the forward recovery algorithm reconstructs datasets of processes which cannot restart from checkpoint. Of course, CoF also exhibits the same overhead as the recovery technique employed by the application. In an ABFT approach, the application might need to do extra computation, even in the absence of failures, to maintain internal redundancy (whose degree varies with the maximum number of simultaneous failures) used to recover data damaged by failures. However, ABFT techniques often demonstrate excellent scalability; for example, the overhead on failure-free execution of the ABFT-QR operation (used as an example in Sect. 3.7.3) is inversely proportional to the number of processes [13].

3.7.1.1 MPI Requirements for Checkpoint-on-Failure

Returning Control over Failures: In most MPI implementations, the default (and often, only functional) error handler is MPI_ERRORS_ABORTS. However, the MPI Standard also defines the MPI_ERRORS_RETURN handler. To support CoF, the MPI library should never deadlock because of failures, but invoke the error handler, at least on processes doing direct communications with the failed process, and returns control to the application.

Preparing for Checkpoint: If the MPI implementation intends to support a system-based checkpoint module (such as MTCP [68]), before invoking the error handler, the MPI library must dispose of its internal state, so that it is not restored by the checkpoint library upon restart. This cleanup consists in releasing all acquired resources and freeing internal buffers and structures, which is safe as the MPI library is not supposed to be functional anymore. If the MPI implementation intends to support only user-directed checkpoint, this effort can be spared.

Termination After Checkpoint: A process that detects a failure ceases to use MPI. It only checkpoints on some storage and exits without calling MPI_FINALIZE. Exiting without calling MPI_Finalize is an error from the MPI perspective, hence the failure cascades following the communication pattern of the application, and MPI eventually returns with a failure notification on every process, which triggers their own checkpoint procedure and termination. Only processes that do not communicate may reach MPI_Finalize without detecting a failure, adding a Barrier before MPI_Finalize will result in the expected error in that case.

3.7.2 Implementation Issues

Open MPI is an MPI 3.0 implementation architected such that it contains two main levels, the runtime (ORTE) and the MPI library (OMPI). As with most MPI library implementations, the default behavior of Open MPI is to abort after a process failure. This policy was implemented in the runtime system, preventing any kind of decision from the MPI layer or the user level. The major change requested by the CoF protocol was to make the runtime system resilient, and leave the decision in case of failure to the MPI library policy, and ultimately to the user application.

3.7.2.1 Failure Resilient Runtime

For full support of the CoF protocol, it is sufficient for the runtime to delay the cleanup of the failed MPI application until it terminates itself. However, a persistent runtime that remains available for spawning replacement MPI processes confers a number of advantages compared to this simple design. Indeed, it eliminates the downtime resulting from the complete redeployment of the parallel job infrastructure and the supplementary wait time from losing the batch scheduler reservation. In addition, it can serve as a local storage for checkpoints.

The ORTE runtime layer depends on an out-of-band communication mechanism (OOB); therefore, node failures not only impact the MPI communications, but also disrupt the OOB overlay routing topology. Fortunately, restoring TCP-based OOB communications is easier than it is to repair MPI. The default routing policy in the Open MPI runtime has been amended to allow for self-healing behaviors. The OOB overlay topology now automatically routes around failed processes. In some routing topologies, such as a star, this is a trivial operation and only requires excluding the failed process from the routing tables. For more elaborate topologies, such as a binomial tree, the healing operation involves computing the closest neighbors in the direction of the failed process and reconnecting the topology through them. The repaired topology is not rebalanced, resulting in degraded performance but complete functionality after failures. Although in-flight messages that were currently "hopping" through the failed processes are lost, newer in-flight messages are safely routed on the repaired topology. Thanks to self-healing topologies, the runtime remains responsive and can continue to support the replacement MPI application.

3.7.2.2 Failure Notification

Although not strictly necessary to support CoF, rapid dissemination of failure notifications has a significant influence on the delay before the recovery can start. The runtime has therefore been augmented with a failure detection and dissemination service. To track the status of the failures, an incarnation number has been included in the process names. Following a failure, the name of the failed process (including the incarnation number) is broadcasted over the OOB topology. By including this incarnation number, we can identify transient process failures, prevent duplicate detections, and track message status. ORTE processes monitor the health of their neighbors in the OOB routing topology. Detection of other processes rely on a failure resilient broadcast that overlays on the OOB topology. This broadcast algorithm has a low probability of creating a bipartition of the routing topology, hence ensuring a high accuracy of the failure detector. We will show in the experiments that the underlying OOB routing algorithm has a significant influence on the propagation time. Finally, on each node, the ORTE runtime layer forwards failure notifications to the MPI layer, which has been modified to invoke the appropriate MPI error handler.

3.7.3 Example: The QR Factorization

In this section, we propose to illustrate the applicability of CoF by considering a representative routine of a widely used class of algorithms: dense linear factorizations. The QR factorization is a cornerstone building block in many applications, including solving $Ax = b$ when matrices are ill-conditioned, computing eigenvalues, least square problems, or solving sparse systems through the GMRES iterative method. For an $M \times N$ matrix A, the QR factorization produces Q and R, such that $A = QR$ and Q is an $M \times M$ orthogonal matrix and R is an $M \times N$ upper triangular matrix. The most commonly used implementation of the QR algorithm on a distributed memory machine comes from the ScaLAPACK linear algebra library [26], based on the block QR algorithm. It uses a 2D block-cyclic distribution for load balance, and is rich in level 3 BLAS operations, thereby achieving high performance.

3.7.3.1 ABFT-QR Factorization

In the context of FT-MPI, the ScaLAPACK-QR algorithm has been rendered fault tolerant through an ABFT method in previous work [13]. This ABFT algorithm protects both the left (Q) and right (R) factors from fail-stop failures at any time during the execution, and is similar to the ABFT LU algorithm presented in Sect. 1.5.2. At the time of failure, every surviving process is notified by FT-MPI. FT-MPI then spawns a replacement process that takes the same grid coordinates in the $P \times Q$ block-cyclic distribution. Missing checksums are recovered from duplicates, a reduction collective communication recovers missing data blocks in the right factor from checksums. The left factor is protected by the Q-parallel panel checksum, it is either directly recovered from checksum, or by recomputing the panels in the current Q-wide section (see [13]). Although this algorithm is fault tolerant, it requires continued service from the MPI library after failures—which is a stringent requirement that can be waived with CoF.

3.7.3.2 Checkpoint-on-Failure QR

Checkpoint Procedure: In our current implementation of CoF, system-level checkpointing is not supported and would result in restoring the state of the broken MPI library upon restart. Instead, the application provides a custom MPI error handler, which invokes an algorithm specific checkpoint procedure to dump the matrices and the value of important loop indices into a file.

State Restoration: In the theoretical version of the ABFT algorithm, regardless of when the failure is detected, the current iteration is completed before entering the recovery procedure, so that all updates are applied to the checksums. In the case of the CoF protocol, failures interrupt the algorithm immediately, the current iteration cannot be completed due to lack of communication capabilities. A ScaLAPACK program has a deep call stack, layering functions from multiple software packages, such as PBLAS, BLACS, LAPACK and BLAS. Because failure notification happens only in MPI, lower level, local procedures (BLAS, LAPACK) are never interrupted. However, PBLAS operations may be incomplete, and therefore checksums only partially updated.

To resolve this issue, the call stack must be restored on every process. The current indices in the loop nest of the QR algorithm, down to the PBLAS level, are adjunct to the checkpoint. When restarted from a checkpoint, a process undergoes a "dry run" phase that mimics the already completed loop nests, without actually applying modifications to or exchanging data. When the same loop indices as before the failure are reached, the matrix content is loaded from the checkpoint; the state is then identical to that immediately preceding the failure. The regular ABFT recovery procedure can then be applied: the current iteration of the factorization is completed to update all checksums and the dataset is finally rebuilt using the ABFT reduction.

3.7.4 *In-Memory* CoF *Protocol*

The CoF protocol circumvents one of the major limitations of current MPI implementations: the lack of confidence that the MPI library is capable of successfully completing communications once a failure happened. As illustrated above, forward recovery strategies are capable of taking advantage of this technique and provide efficient fault-tolerance support that does not require periodic checkpointing. However, when a failure strike, the CoF protocol still incurs checkpoint I/O overhead. In this section, we explain how the CoF protocol can be efficiently integrated with already resilient, non-MPI applications to completely eliminate all checkpointing activity. We will illustrate the approach with a fault-tolerant database management system, SciDB.

Fault-tolerant database management exposes a set of requirements that is best addressed today using replication and transactional operations. SciDB [89] combines database operations and many scientific specific operations (including linear algebra routines) to create a highly expressive request query language suitable for scientists to solve their data analysis problems. The SciDB system is not implemented on top of MPI, mainly because of the lack of fault-tolerance capabilities from the MPI Standard. It makes use, however, of the MPI based distributed linear algebra operations in ScaLAPACK, to provide, among other things, various factorization routines. Because most MPI implementations are not usable after a process failure, and high availability is a necessity in database management systems, the SciDB implementation cannot integrate the MPI library in its main process. As a result, the linear algebra operations are called from separate processes: a query coordinator orders the distributed database managers to locate the data on which the factorization operation must be applied and to expose this data in the expected ScaLAPACK layout using one shared memory segment per node; it then launches a ScaLAPACK/MPI application that attaches to this memory segment and applies the operation on it. If a failure hits a node, the MPI application aborts, and the *mpirun* child process

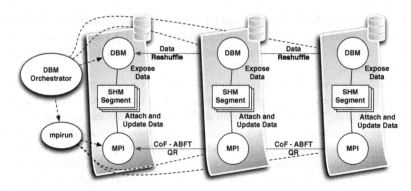

Fig. 3.19 SciDB/CoF ABFT-ScaLAPACK integration

reports the error to the data query coordinator. The original data is recovered from the database management system (using database-specific fault-tolerant techniques), and the linear algebra operation is relaunched from scratch on the original data.

This approach can be improved using the CoF protocol and an ABFT implementation of the factorization operation. The idea is depicted in Fig. 3.19. SciDB and ScaLAPACK are coupled in a similar way; however, the DB managers compute the initial checksum of the original data, and expose both the data and checksum to the ABFT-ScaLAPACK process. The ABFT operation is applied, and if no failure happens, the result of the factorization is accessible in the shared-memory segments (the checksum data can then be discarded by the DB managers). If a failure occurs, the MPI process updates the shared-memory segments with the meta information of the checkpoint (values of the loop counters, etc.); the content of the shared-memory segments is analog to the checkpoints performed in the normal CoF protocol. Then, the MPI processes quit and the *mpirun* child process reports the error to the database coordinator. Instead of fixing the data issue at the DB level, the coordinator immediately relaunches a new ABFT-ScaLAPACK operation on the same set of nodes plus a spare node with an empty shared-memory segment, the ABFT algorithm recovers the data and the original operation resumes. Upon successful completion, the *mpirun* child process reports to the database coordinator that the result is in the shared memory segments.

This approach is advantageous compared to both the original design and the checkpoint-based CoF approach. Instead of restarting from scratch after each failure, the factorization incurs only the small recovery overhead of ABFT, ensuring a faster time-to-solution for the linear algebra operation. In exchange, a small overhead, for creating and maintaining the checksum data during the operation, is imposed on the failure-free case. Second, this approach removes the cost of writing the checkpoint to a file: the shared-memory segment that survives the exit of the MPI processes where the node was not subject to a failure and the checksum information maintained by the ABFT algorithm are sufficient to recover the missing data. The segment of memory on which the operation is computed is made remanent, creating the bulk of the checkpoint data and reducing to an insignificant value the cost of checkpointing when a failure occurs. This will be demonstrated in the experimental section, below.

3.7.5 Performance Discussion

In this section, we use our Open MPI and ABFT -QR implementations to evaluate the performance of the CoF protocol. We use two test platforms. The first machine, "Dancer", is a 16-node cluster. All nodes are equipped with two 2.27 GHz quad-core Intel E5520 CPUs with a 20 GB/s Infiniband interconnect. Solid State Drive (SSD) disks are used as the checkpoint storage media. The second system is the "Kraken" supercomputer. Kraken is a Cray XT5 machine with 9,408 compute nodes. Each node has two Istanbul 2.6 GHz six-core AMD Opteron processors, 16 GB of memory, and is connected to other nodes through the SeaStar2+ interconnect. The scalable cluster file system "Lustre" is used to store checkpoints.

3.7.5.1 MPI Library Overhead

One of the concerns when evaluating the performance of fault tolerance techniques is the amount of overhead introduced by the fault-tolerance management additions. Our implementation of fault detection and notification is mostly implemented in the noncritical ORTE runtime. Typical HPC systems feature a separated service network (usually Ethernet-based) and a performance interconnect, hence health monitoring traffic, which happens on the OOB service network, is physically separated from the MPI communications, leaving no opportunity for network jitter. Changes to MPI functions are minimal: the same condition that used to trigger unconditional abort has been repurposed to trigger error handlers. As expected, no impact on MPI bandwidth or latency was measured. The memory usage of the MPI library is slightly increased, as the incarnation number doubles the size of process names; however, this is negligible in typical deployments.

3.7.5.2 Failure Detection

According to the requirement specified in Sect. 3.7.1.1, only in-band failure detection is required to enable CoF. Processes detecting a failure checkpoint then exit, cascading the failure to processes communicating with them. However, no recovery action (in particular checkpointing) can take place before a failure has been notified. Thanks to asynchronous failure propagation in the runtime, responsiveness can be greatly improved, with a high probability for the next MPI call to detect the failures, regardless of communication pattern or checkpoint duration.

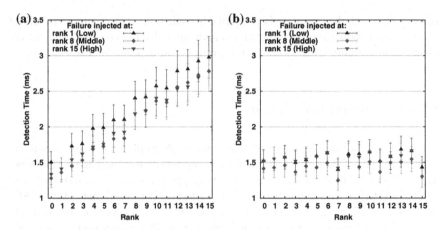

Fig. 3.20 Failure detection time, sorted by process rank, depending on the OOB overlay network used for failure propagation. **a** Linear OOB routing. **b** Binomial OOB routing

We designed a microbenchmark to measure failure detection time as experienced by MPI processes. The benchmark code synchronizes with an MPI_BARRIER, stores the reference date, injects a failure at a specific rank, and enters a ring algorithm until the MPI error handler stores the detection date. The OOB routing topology used by the ORTE runtime introduces a nonuniform distance to the failed process, hence failure detection time experienced by a process may vary depending on both the used OOB overlay topology and the position of the failed process in the topology. Figure 3.20a, b present the case of the linear and binomial OOB topologies, respectively. The curves "Low, Middle, High" present the behavior for failures happening at different positions in the OOB topology. On the horizontal axis is the rank of the detecting process, on the vertical axis is the detection time it experienced. The experiment uses 16 nodes, with one process per node, MPI over Infiniband, OOB over Ethernet, an average of 20 runs, and the MPI barrier latency is four orders of magnitude lower than measured values.

In the linear topology (Fig. 3.20a) every runtime process is connected to the *mpirun* process. For a higher rank, failure detection time increases linearly because it is notified by the *mpirun* process only after the notification has been sent to all lower ranks. This issue is bound to increase with scale. The binomial tree topology (Fig. 3.20b) exhibits a similar best failure detection time. However, this more scalable topology has a low output degree and eliminates most contentions on outgoing messages, resulting in a more stable, lower average detection time, regardless of the failure position. Overall, failure detection time is on the order of milliseconds, a much smaller figure than typical checkpoint time.

3.7.5.3 Checkpoint-on-Failure QR Performance

Supercomputer Performance: Figure 3.21 presents the performance on the Kraken supercomputer. The process grid is 24 × 24 and the block size is 100. ABFT-QR (no failure) presents the performance of the CoF-QR implementation, in a fault-free execution; it is noteworthy, that when there are no failures, the performance is exactly identical to the performance of the unmodified ABFT-QR implementation. The ABFT-QR (with CoF recovery, latter called CoF-QR for brevity) curves present the performance when a failure is injected after the first step of the PDLARFB kernel. The performance of the non-fault-tolerant ScaLAPACK-QR is also presented for reference.

Without failures, the performance overhead compared to the regular ScaLA-PACK is caused by the extra computation to maintain the checksums inherent to the ABFT algorithm [13]; this extra computation is unchanged when applying the CoF method to the ABFT-QR. Only on runs where failures occur does the CoF protocol undergoes the supplementary overhead of storing and reloading checkpoints. However, the performance of CoF-QR remains very close to the no-failure case. For instance, at matrix size $N = 100{,}000$, CoF-QR still achieves 2.86 Tflop/s after recovering from a failure, which is 90 % of the performance of the non-fault-tolerant ScaLAPACK-QR. This demonstrates that the CoF protocol enables efficient, practical recovery schemes on supercomputers.

Impact of Local Checkpoint Storage: Figure 3.22a presents the performance of the CoF-QR implementation on the Dancer cluster with a 8 × 16 process grid. Although a smaller test platform, the Dancer cluster features local storage on nodes and a variety of performance analysis tools unavailable on Kraken. As expected (see [13]), the ABFT method has a higher relative cost on this smaller machine (with a smaller number of processors and a smaller problem size, the cost in supplementary operations to update checksums is relatively larger). Compared to the Kraken platform, the

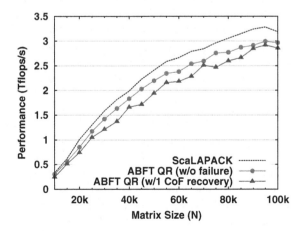

Fig. 3.21 ABFT-QR and one CoF recovery on Kraken (Lustre)

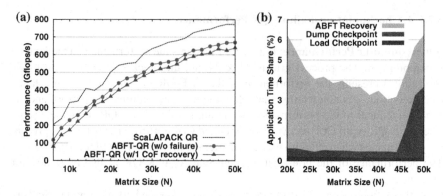

Fig. 3.22 ABFT-QR and one CoF recovery on dancer (local SSD). **a** Performance. **b** Time breakdown of one COF recovery

relative cost of CoF failure recovery is smaller on Dancer. The CoF protocol incurs disk accesses to store and load checkpoints when a failure hits, hence the recovery overhead depends on I/O performance. By breaking down the relative cost of each recovery step in CoF, Fig. 3.22b shows that checkpoint saving and loading only takes a small percentage of the total runtime, thanks to the availability of solid state disks on every node. Since checkpoint reloading immediately follows checkpointing, the OS cache satisfies most disk accesses, resulting in high I/O performance. For matrices larger than N = 44,000, the memory usage on each node is high and decrease the available space for disk cache, explaining the decline in I/O performance and the higher cost of checkpoint management. Overall, the presence of fast local storage can be leveraged by the CoF protocol to speedup recovery (unlike periodic checkpointing, which depends on remote storage by construction). Nonetheless, as demonstrated by the efficiency on Kraken, while this is a valuable optimization, it is not a mandatory requirement for satisfactory performance.

In-Memory Checkpoint-on-Failure: An interesting optimization to CoF is to avoid the checkpointing cost by using the SM-CoF approach described in Sect. 3.7.4. In this paragraph, we present the performance of the QR factorization, when applied by a fragile helper MPI application, onto a dataset exported through a shared-memory segment from a resilient, non-MPI application. Figure 3.23 compares the overhead incurred by introducing a failure with checkpoint-based CoF recovery versus a shared-memory-CoF recovery where a master application maintains the dataset resident in memory.

The cost of the ABFT recovery is unchanged by the use of SM-CoF; the obvious consequence is that, for very small matrix sizes, when the relative cost of ABFT checksum inversion represents a large portion of the overall compute time, the difference between the shared-memory optimization and the checkpoint-based CoF is small. A similar result is observed for very large matrices: for a matrix of size N, checkpointing time is $O(N^2)$ while compute time is $O(N^3)$, thus the cost of storing and reloading checkpoints is dwarfed by the total execution time of the appli-

Fig. 3.23 ABFT -QR and one recovery on Kraken: comparing CoF and SM-CoF overheads

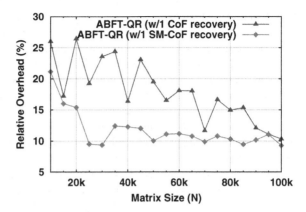

cation and achieve similar asymptotic performance. For intermediate matrix sizes, however, the cost of checkpointing represents a significant share of the overhead experienced by the application during the recovery procedure. In that case, which is the most relevant in production deployments, the SM-CoF optimization successfully suppresses the checkpoint overhead and performs similarly to ABFT -QR on a fully fault-tolerant MPI implementation, although at the expense of more complexity in the application code.

3.7.6 Concluding Remarks

In this section, we presented an original scheme to enable forward recovery using only features of the current MPI Standard. Rollback recovery, which relies on periodic checkpointing, has a variety of issues. The ideal period between checkpoints, a critical parameter, is particularly hard to assess. Too short a period wastes time and resources on unnecessary Input/Output. Overestimating the period results in dramatically increasing the lost computation when returning to the distant last successful checkpoint. Although Checkpoint-on-Failure involves checkpointing, it takes checkpoint images at optimal times by design: only after a failure has been detected. This small modification enables the deployment of ABFT techniques, without requiring a complex, unlikely to be available MPI implementation that itself survives failures. The MPI library needs only to provide the feature set of a high-quality implementation of the MPI Standard: the MPI communications may be dysfunctional after a failure, but the library must return control to the application instead of aborting brutally.

We demonstrated, by providing such an implementation in Open MPI, that this feature set can be easily integrated without noticeable impact on communication performance. We then converted an existing ABFT -QR algorithm to the CoF protocol. Beyond this example, the CoF protocol is applicable on a large range of applications that already feature an ABFT version, for example the dense direct solvers LLT, LU [24] and the dense iterative solver CG [21]. Similarly, ABFT algorithms exist for sparse computation [77]. Beside ABFT, many master–slave and iterative methods enjoy an extremely inexpensive forward recovery strategy where the damaged domains are simply discarded, and therefore can also benefit from the CoF protocol.

The performance on the Kraken supercomputer reaches 90 % of the non fault-tolerant algorithm, even when including the cost of recovering from a failure (a figure similar to regular, noncompliant MPI ABFT). In addition, on a platform featuring node local storage, the CoF protocol can leverage low overhead checkpoints (unlike rollback recovery that requires remote storage). To the extreme, the cost of checkpointing can be completely avoided when the application uses a master process to actively retain the dataset in memory during the MPI restart.

3.8 User-Level Fault Tolerance with Extended MPI

Although the CoF protocol permits the deployment of ABFT forward recovery techniques, it does not spare the application the cost of a full restart of the MPI runtime, including the restart of processes originally not impacted by failures. In order to avoid the full redeployment of the application, the MPI library itself must be able to restore its capability to communicate. Considering a context in which the application is fully aware of the volatile nature of compute resources, and is ready to take protective actions of its own (like forward recovery), an important question is the degree of control the application should have over the repair of the MPI internal state, and what gains can be expected from providing a precise interface to monitor only relevant failures and repair only necessary objects.

Basic coordinated checkpoint can be implemented without meaningful support for MPI fault tolerance, but the literature is rich in alternative recovery strategies permitting better performance in a volatile, high failure rate environment. The variety of techniques employed is very wide, and notably include checkpoint–restart variations based on uncoordinated rollback recovery [12], replication [33], algorithm-based fault tolerance where mathematical properties are leveraged to avoid checkpoints [13, 24], etc. A common feature required by most of these advanced failure recovery strategies is that, unlike historical rollback recovery where the entire application is interrupted and later restarted from a checkpoint, the application is expected to continue operating despite processor failures, thereby reducing the incurred I/O, downtime and computation loss overheads. However, the MPI Standard doesn't define a precise behavior for MPI implementations when disrupted by failures. As a consequence, the deployment of advanced fault-tolerance techniques is challenging, taking a strain on software development productivity in many applied science communities, and fault-tolerant applications suffer from the lack of portability of ad hoc solutions.

Several issues prevented the standardization of recovery behavior by the MPI Standard. Most prominently, the diversity of the available recovery strategies is, in itself, problematic: there does not appear to be a single best practice, but a complex ecosystem of techniques that apply best to their niche of applications. The second issue is that, without a careful, conservative design, fault tolerance additions generally take an excruciating toll on bare communication performance. Many MPI implementors, system vendors and users are unwilling to accept this overhead, an attitude further reinforced by the aforementioned diversity of fault-tolerance techniques which results in costly additions being best suited for *somebody else's* problem.

In this section, we describe a set of extended MPI routines and definitions that were introduced in Sect. 1.5.1 of the overview Chapter. This extension is called User-Level Failure Mitigation (User-Level Failure Mitigation (ULFM)), and it permits MPI applications to continue communicating across failures, while avoiding the two issues described above.

3.8.1 Communication Substrate Recovery Background

Many communication layers embed fault tolerance, or, at least, a graceful failure handling as a major feature of the specification. In the socket interface, given the appropriate initiation parameters, error codes are returned when a remote peer is disconnected (as the result of a network or crash error). However, the stream semantic of sockets comes with copy overheads, and maintaining an open connection with every peer is not a scalable solution. Closer to the HPC field, MapReduce [25] and PVM [38] both feature strong resilience capabilities. In the case of MapReduce, the programming model is essentially master–worker based, which makes fault tolerance easier to manage; workers that are discovered dead are simply eliminated from the pool and masters are protected with checkpoints. Since there is little shared global state, it is simple to restore the application and the communication context. It is striking that such a simple use case is hard to deploy with MPI. In PVM, the feature set is even more advanced, applications can register a callback to be triggered when failures (or other erroneous conditions) happen. The callback offers the flexibility to actively monitor tasks and processes, but as the monitoring has to inserted with explicit calls in the application code, it can be cumbersome to develop, in practice. Once a failure is detected, the application can use one of the control operations of the virtual machine, to add supplementary hosts to the resource pool. This rich feature set partially plays a role in the observation that none of these approaches can rival in terms of raw bandwidth, latency and optimized collective communication patterns with MPI, which is spared the effort of maintaining a workable state when failures disrupt ongoing communication [42].

In [43], Gropp and Lusk describe methods using the then current version of the MPI Standard to perform fault tolerance. They described methods including checkpointing, MPI_ERRHANDLERs, and using intercommunicators to provide some form of fault tolerance. They outline the goals of providing fault tolerance without requiring changes to the MPI Standard. However, at the time of writing, fault tolerance was still an emerging topic of research with few solutions beyond checkpointing and simple ABFT in the form of master–worker applications. As fault tolerance has evolved to include those paradigms mentioned in Sect. 3.8.4, the requirements on the MPI implementation have also grown, and the limited functionality emphasized are insufficient for general communication purposes.

Another notable effort was FT-MPI [31]. The overreaching goal was to support ABFT techniques, it thereby provide three failure modes adapted to this type of recovery techniques, but difficult to use in other contexts. In the *Blank* mode, failed processes were automatically replaced by MPI_PROC_NULL; messages to and from them were silently discarded and collective communications had to be significantly modified to cope with the presence of MPI_PROC_NULL processes in the communicator. In the *Replace* mode, faulty processes were replaced with new processes. In the *Shrink* mode, the communicator would be changed to remove failed processes (and ranks reordered accordingly). In all cases, only MPI_COMM_WORLD would be repaired and the application was in charge of rebuilding any other communicators. No standardization effort was pursued, and it was mostly used as a playground for understanding the fundamental concepts. A major distinction with the ULFM design is that when FT-MPI detects a failure, it repairs the state of MPI internally according to the selected recovery mode, and then only triggers the coordinated user recovery handle at all nodes. Library composition is rendered difficult by the fact that recovery preempts the normal flow of execution and returns to the highest level of the software stack without alerting intermediate layers that a failure happened.

A more recent effort to introduce failure handling mechanisms was the Run-Through Stabilization (RTS) proposal [51]. This proposal introduced many new constructs for MPI including the ability to "validate" communicators as a way of marking failure as recognized and allowing the application to continue using the communicator. It included other new ideas such as Failure Handlers for uniform failure notification. Because of the implementation complexity imposed by resuming operations on failed communicators, this proposal was eventually unsuccessful in its introduction to the MPI Standard.

From a higher perspective, it can be noted that the reasons why these fault-tolerant communication libraries did not enjoy wide adoptions are varied. In some cases, the feature set is too rich and the consistency guarantees after a failure are too strong, which incurs too large an impact on performance (PVM, RTS). In some instances, the feature set is too limited, and although performance is satisfactory, the lack of features renders the implementation of any fault-tolerant strategy excruciatingly difficult (MPI with connect/accept/disconnect). In other instances, the feature set permits effective deployment of fault-tolerant applications with adequate performance, but is too specific to a particular recovery model, and lacking generality cannot support the full ecosystem of HPC applications (Hadoop, FT-MPI).

3.8.2 Establishing a Flexible Feature Set

After evaluating the strengths and weaknesses of the previous efforts toward fault tolerance both within MPI and with other models, we converged on four overarching goals for ULFM. In this work, we consider the effect of fail-stop failures (that is, when a processor crashes and stops responding completely). Network failures are equally important to tolerate, but are generally handled at the link protocol level, thereby relieving MPI programs from experiencing their effect. Silent errors that damage the dataset of the application (memory corruption) without hindering the capacity to deliver messages (or resulting in a crash), are the sole responsibility of the application to correct. The motivation behind the design choices are weighted against alternatives, a task that requires simultaneously considering MPI from the viewpoint of both the implementor (in Sect. 3.8.3) and the user (in Sect. 3.8.4). The overreaching design goals we have identified are the following:

Flexibility in fault response is paramount: not all applications have identical requirements. In the simple case of a Monte Carlo master–worker application that can continue computations despite failures, the application should not have to pay for the cost of any recovery actions; on the contrary, consistency restoration interfaces must be available for applications that need to restore a global context (a typical case for applications with collective communications). As a consequence, and in sharp contrast with previous approaches (see Sect. 3.8.1), we believe that MPI should not attempt to define the failure recovery model or to repair applications. It should inform applications of specific conditions that prevent the successful delivery of messages, and provide constructs and definitions that permit applications to restore MPI objects and communication functionalities. Such constructs must be sufficient to express advanced high-level abstractions (without replacing them), such as transactional fault tolerance, uncoordinated checkpoint/restart, and programming language extensions. The failure recovery strategies can then be featured by independent portable packages that provide tailored, problem specific recovery techniques and drive the recovery of MPI on behalf of the applications.

Resiliency refers to the ability of the MPI application not only to survive failures, but also to recover into a consistent state from which the execution can be resumed. One of the most strenuous challenges is to ensure that no MPI operation stalls as a consequences of a failure, for fault tolerance is impossible if the application cannot regain full control of the execution. An error must be raised when a failure prevents a communication from completing. However, we propose that such a notice indicates only the local status of the operation, and does not permit inferring whether the associated failure has impacted MPI operations at other ranks. This design choice avoids expensive consensus synchronizations from obtruding into MPI routines, but leaves open the danger of some processes proceeding unaware of the failure. Therefore, supplementary constructs must be sparingly employed in the application code to let processes which have received an error resolve their divergences. but only when necessary (see MPI_COMM_REVOKE in Sect. 3.8.5).

Productivity and the ability to handle the large number of legacy codes already deployed in production is another key feature. Backward compatibility (i.e., supporting unchanged non-fault-tolerant applications) and incremental migration are necessary. A fault tolerant API should be easy to understand and use in common scenarios, as complex tools have a steep learning curve and a slow adoption rate by the targeted communities. To this end, the number of newly proposed constructs must be small, and have clear and well-defined semantics that are familiar to users.

Performance impact outside of recovery periods should be minimal. Indeed, minimalism is a design principle that we believe must be embraced, for it limits the number and extent of modifications to the MPI implementation. Failure protection actions within the implementation must be outside the performance critical path, and recovery actions triggered by the application only when necessary. As most functions are left unmodified (as an example, the implementation of collective operations), they continue to deliver the extraordinary performance resulting from years of careful optimization. Overheads are tolerated only as a consequence of actual failures.

3.8.3 The Implementor's Perspective

In this section we discuss the rationale behind the proposed design by taking the view of MPI implementors in analyzing the challenges and performance implications that result from possible implementations, and explain why sometime counterintuitive designs are superior. One of the most important design goasl is performance, which explain why we take this stance first; our claims that the proposed design does indeed achieve excellent performance is supported by an implementation, presented in [5].

3.8.3.1 Failure Detection

Failure detection has proven to be a complex but crucial area of fault-tolerance research. Although in the most adverse hypothesis of a completely asynchronous system, failures are intractable in theory [34], the existence of an appropriate failure detector permits resolving most of the theoretical impossibilities [19]. One of the crucial goals of ULFM is to prevent deadlocks from arising, which indeed requires the use of some failure detection mechanism (in order to discriminate between arbitrarily long message delays and failures). However, because the practicality of implementing a particular type of failure detector strongly depends on the hardware features, the specification is intentionally vague and refrains from forcing a particular failure detection strategy. Instead, it leaves open to the implementations choices that better match the target system. On some systems, hardware introspection may be available and provide total awareness of failures (typically, an IPMI capable batch scheduler). However, on many systems, a process may detect a failure only if it has an active open connection with the failed resource, or if it is actively monitoring its status with heartbeat messages. In the latter situation requiring complete awareness of failures of every process by every process would generate an immense amount of system noise (from heartbeat messages injected into the network and the according treatments on the computing resources to respond to them), and it is known that MPI communication performance is very sensitive to system noise [63]. Furthermore, processes that are not trying to communicate with the dead process do not need to be aware of its failure, as their operations are with alive processors and therefore deadlock-free. As a consequence, to conserve generality and avoid extensive generation of system noise, failure detection in ULFM requires only to detect failures of processes that are active partners in a communication operation, so that this operation eventually returns an appropriate error. In the ideal case, the implementation should be able to turn on failure monitoring only for the processes it is expecting events from (like the source or destination in a point-to-point operation). Some cases (like wildcard receives from any source) may require a wider scoped failure detection scheme, as any processor is a potential sender. However, the triggering of active failure detection can be delayed according to implementation internal timers, so that latency critical operations don't have to suffer a performance penalty.

3.8.3.2 Communication Objects Status

A natural conception is to consider that detection of failures results in MPI automatically altering the state of all communication objects (i.e., communicators, windows, etc.) in which the associated process appears. In such a model, it is understood that the failure "damages" the communication object and renders it inappropriate for further communications. However, a complication is hidden in such an approach: the state of MPI communication objects is the aggregate state of individual views by each process of distributed system. As failure awareness is not expected to be global, the implementation would then require internal and asynchronous propagation of failure detection, a process prone to introduce jitter. Furthermore, MPI messages would be able to cross the toggling of the communication object into an invalid state, resulting in a confuse semantic where operations issued on a valid communication object would still fail, diluting the meaning of a valid and invalid state of communication objects.

We decided to take the opposite stance on the issue, failures never automatically modify the state of communication objects. Even if it contains failed processes, a communicator remains a valid communication object. Instead, error reporting is not intended to indicate that a process failed, but to indicate that an operation cannot complete. As long as no failures happen, the normal semantic of MPI must be respected. When a failure has happened, but the MPI operation can proceed without disruption, it completes normally. Obviously, when the failed process is supposed to participate to the result of the operation, it is impossible for the operation to succeed, and an appropriate error is returned. Posting more operations that involve the dead processes is allowed, but is expected to result in similar errors.

There are multiple advantages to this approach. First, the consistency of MPI objects is always guaranteed, as their state remains unchanged as long as users don't explicitly change it with one of the recovery constructs. Second, there is no need to introduce background propagation of failure detections to update the consistent state of MPI objects, because operations that need to report an error do actively require the dead process' participation, thereby active failure detection is forced only at the appropriate time and place.

3.8.3.3 Local or Uniform Error Reporting

In the ULFM design, errors notify the application that an operation couldn't satisfy its MPI specification. However, most MPI operations are collective, or have a matching call at some other process. Should the same error be returned *uniformly* at all ranks that participated in the communication? Although such a feature is desirable for some users, as it permits easily tracking the global progress of the application (and then infer a consistent synchronized recovery point), the consequences on performance are dire. This would require that each communication conclude with a global agreement operation to determine the success or failure of the previous communication as viewed by each process. Such an operation cannot be possibly achieved in less than the cost of an Allreduce, even without accounting for the cost of actually tolerating failures in the operation, and would thus impose an enormous overhead on communication. In regard to the goal of maintaining unchanged level of performance, it is clearly unacceptable to double, at best, the cost of all communication operations, even when no failure happened.

As a consequence, in ULFM, the reporting of errors has a local semantic: the local completion status (in error, or successfully) cannot be used to assume if the operation has failed or succeeded at other ranks. In many applications, this uncertainty is manageable, because the communication pattern is simple enough. When the communication pattern does not allow such flexibility, the application is required to resolve this uncertainty itself (by explicitly changing the state of the communication object to *Revoked* with one of our proposed additional API). Indeed, it is extremely difficult for MPI to assess if a particular communication pattern is still consistent (it would require computing global snapshots after any communication), while the user can know through algorithm invariants when it is the case. Thanks to that flexibility, the cost associated with consistency in error reporting is paid only after an actual failure has happened, and applications that do not need consistency can enjoy better recovery performance.

3.8.3.4 Restoring Consistency and Communication Capabilities

Revoking a communication object results in a definitive alteration of the state of the object, that is consistent across all processes. This alteration is not to be seen as the (direct) consequence of a failure, but as the consequence of the user calling a specific operation on the communication object. In a sense, Revoking a communication object explicitly achieves the propagation of failure knowledge that has intentionally not been required, but is provided when the user deems necessary. Another important feature of that change of state is that it is definitive. After a communication object has been revoked, it can never be repaired. The rationale is to avoid the matching to have to check for stale messages from past incarnations of a repaired communication object. Because the object is discarded definitively, any stale message matches the revoked object and is appropriately ignored without modifications in the matching logic. In order to restore communication capacity, the repair function derive new, fresh communication objects, that do not risk intermixing pre-failure operations.

3.8.4 The Users' Perspective

Since the focus of this work is to design an extension to the MPI runtime to enable effective deployment of advanced fault-tolerance techniques, it is critical to understand how the proposed design goals interact with the specificities, issues, common features and opportunities offered by this wide range of application recovery techniques. We classify application recovery in four crude categories, presented in Fig. 3.24, depending on the type of features that are required from the MPI layer to enable their effective production deployment, or simply to improve their performance or portability.

3.8.4.1 Ignoring Failures

Master/Worker is the simplest case in terms of application recovery strategy. Despite its simplicity, it is a model that is currently unsupported by MPI, and has been poorly supported in many past attempts at providing meaningful fault semantics to MPI. It is a model that commands important applications, like Monte Carlo simulation. The ULFM design has indeed been used in significant multilevel MC simulations to estimate partial differential equations [62]. In this work, the application can continue

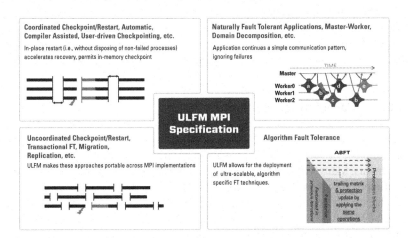

Fig. 3.24 ULFM is a set of MPI interfaces that enables varied recovery strategies, that match a wide range of applications

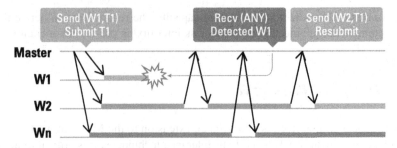

Fig. 3.25 A master/worker application illustrates a communication pattern in which MPI layer repair is a useless cost

without the contribution of the failed processes, either producing a degraded, less accurate result, or computing supplementary iterations before convergence [62].

Figure 3.25 presents the mockup communication pattern of a master/worker application with a very simple communication pattern. A master process submits tasks to a set of workers. In order to monitor the completion of the submitted tasks, the master posts a single MPI_ANY_SOURCE receive in order to receive completion notices from the workers. When a workers reports such a successful job completion, the master submits a new work unit. In this example, the worker W_1 fails. The only things required for this application to reach a successful completion is that the master is informed of the failure, and can thereafter resume posting point-to-point send and receives with non-failed workers, in order to resubmit the failed work package to another worker. In any case, rebuilding a full communication context in which collective operations can be employed, eliminating completely failed ranks from the communicator, or restoring a full consistent global view of the entire system is utterly unnecessary in this application. For maximum performance, the MPI layer must refrain from taking any of these corrective actions automatically, and simply maintain a communication capability on the preexisting communicator context.

3.8.4.2 Full Capability Restoration

As fault tolerance continues to evolve, checkpoint/restart will continue to be a popular design for legacy codes and therefore should be supported by any new fault-tolerant environments. Many MPI libraries provide coordinated checkpointing automatically, without application knowledge or involvement [15, 49]. Because of the use of system-based checkpoint routines [28], these libraries have to be internally modified to remove the MPI state from the checkpoints. However, these modifications do not alter the interface presented to users and the performance hit on communication routines is usually insignificant. More generally, coordinated rollback recovery has been widely deployed by message passing applications without any specific requirements from the MPI implementation. The program code flow is designed so that checkpoints are

taken at points when no messages have been injected into MPI (hence the network
is empty and the checkpoint set consistent) [78].

A recent development in checkpointing techniques is the increasing importance of
decreasing the cost of checkpointing, and the reliance of ultrafast, micro-checkpoints.
Advanced checkpointing libraries like Scalable Checkpoint–Restart (SCR) [60] and
Fault Tolerance Interface (FTI) [4] automate the deployment of multilevel, hierarchical checkpoint storage, in local memory, in nonvolatile memory, in the memory
of peers (possibly employing diskless checkpointing [56]), and finally, if no failure
has disrupted the application in the meantime, on the global shared file system. Even
the most customary checkpoint/restart strategy can benefit immensely from being
deployed on a fault-tolerant MPI framework: if the application processes are not
wiped out by failure, they can serve as hot caches for checkpoint storage, and the
high-performance network can be employed to save, and redistribute to the replacement hot swap processes the checkpoint data. In order to deploy these optimizations,
the MPI layer must provide the capability to report failures, synchronize the state
by aborting ongoing operations, and recreate a communication environment that
respect the former process mapping. Because ULFM targets a wider audience than
just checkpoint/restart, the proposed interface proposes a *shrink* recovery model.
The *shrink* model is not very popular with users of checkpoint/restart (it requires to
rebalance the application and shuffles the ranks), but, as is illustrated in Fig. 3.26, it
can be employed in two different ways as the basic building block to support check-
point/restart, or any application that needs to restore an identical world with the same
rank mapping after a failure. In this mockup application, processes participate to some
collective operations. When a failure strikes, processes undergo a Shrink operation
(through an additional API in ULFM). In order to restore the process/rank mapping,
two strategies can be employed. The first one is to overallocate spare processes
for the application. The application initially splits the MPI_COMM_WORLD in two
separate communicators, one contains the active processes partaking in the computation, while the other are spare processes kept available to replace dead processes
(these processes can also serve as checkpoint storage cache, etc.). When a failure
strikes, the MPI_COMM_WORLD communicator is shrunk, and, given that enough
spare processes were still available, a replacement for the active processes commu-

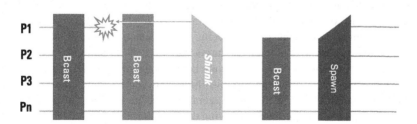

Fig. 3.26 Some applications can continue on a reduced world, in which failed processes are
excluded, and the communication context is sanitized. Many more need to maintain the process
topology grid a and rank mapping, and thereby require to replace failed processes

nicator is recreated. The second approach is to use a sequential combination of the shrink operation and an MPI-2 spawn operation. After the shrink operation has completed, any collective MPI operation is valid, in particular, the MPI_COMM_SPAWN function can be employed to spawn spares on demand, rather than relying on over-allocations. Again, from the intercommunicator created from spawn, a replacement communicators with the same process/rank mapping can be derived. Indeed, frameworks that harness these capabilities of ULFM to present a simple to use, but powerful, user-directed checkpointing interface are already emerging [36, 86].

Transactional based computation (sometimes referred to as containment domain fault tolerance) takes advantage of micro-checkpoints to integrate them in the programming model itself. It can be seen as a form of speculative progress with lightweight checkpoints. The basic idea is that the algorithm is divided into blocks of code. Each block is concluded with a construct that decides the status of all communication operations which occurred within the block, as opposed to checking the status of each communication operation independently [46]. If the block completion construct determines that the block is completed, the necessary data is saved in memory (a form of micro-checkpoint) and the algorithm executes the next block with the optimistic assumption that no failures are happening. If at the end of the block, a process failure had occurred, the application to return to the status before the beginning of the block, giving it the opportunity to execute the block again (after replacing the failed process). One of the major blocks to perform transactional fault tolerance therefore a resilient collective operation that can be employed to validate when a transaction has been successful, or to trigger the recovery if a failure has disturbed the current algorithmic phase. Depending on the application, the recovery procedure may redo the transaction only on the surviving processes, or may require the replacement of the failed processes and an in-place restart.

3.8.4.3 Portability in Automatic Methods

In Checkpoint/Restart with partial rollback recovery, processes that have not been damaged by a failure are kept alive and can continue computing as long as they don't depend on a message from a failed process. Although the interface presented to users is unchanged, major modifications have to be integrated deep within the MPI library to enable continued MPI communications. This in turn makes deployment of Message Logging dependent upon a specific implementation of MPI that may not be best suited or even available on the target hardware. With an MPI specification providing clear semantics for its post-failure behavior, uncoordinated checkpointing approaches could be expressed as portable libraries that could be deployed on top of any MPI implementation. The message logging interposition layer can be inserted in the original MPI code, either bye employing the PMPI profiling interface [75], or by performing automated code injection in the original source code. Then, when a failure happens, it needs to be able to direct the cleanup of the damaged communicators, and silently swap them with replacements obtained from the ULFM communicator recreation functions [55, 86].

In replication [33] an application executes multiple concurrent copies of itself simultaneously. In most variations, the replicates need to remain strongly synchronized, and messages' delivery are effectively atomic commits to multiple targets. Process migration [18, 91] is a form of fault tolerance which combines advanced, proactive failure detectors with some other form of fault tolerance, often checkpoint/restart. To reduce the increasing overhead of other forms of fault tolerance at scale, process migration detects that a failure is likely to occur at a particular process and moves it (or replicates it) to a node in the system less likely to fail. Migration requires accurate failure predictors to be useful, but when successful, it can reduce the overhead of other fault-tolerance mechanisms significantly. For both Replication and Migration, the PMPI hooks are usually already leveraged in order inject the appropriate code to redirect messages to the appropriate target, or to integrate the consistent delivery protocol to multiple replicates. For example, the library might redirect the communication to point to the currently active target depending on which replica is being used or if a process has migrated. When a failure happens, the ULFM specification can be employed to hot swap the communication object in order to restore optimal collective communication performance (an crucial feature when point-to-point message have to be converted to broadcasts), ensure the safe completion of a transaction with a failure resistant agreement operation, and spawn new clones of failed processes on the fly, to insure continued protection from further failures.

3.8.4.4 Forward Recovery with Complex Patterns

Algorithm-Based Fault Tolerance (ABFT) [6, 13, 21, 24] is a family of recovery techniques based on algorithmic properties of the application. In some *Naturally Fault-Tolerant* applications, when a failure occurs, the application can simply continue while ignoring the lost processes. In other cases, the application uses intricate knowledge of the structure of the computation to maintain supplementary, redundant data, that is updated algorithmically and forms a recovery dataset that does not rely on checkpoints. Although generally exhibiting excellent performance and resiliency, ABFT requires that the algorithm is innately able to incorporate fault tolerance and therefore might be a less generalist approach. In ABFT applications that require the restoration of a full set of processes [1], the recovery procedure for the MPI layer actually has strikingly similar requirement to the deployment of coordinated checkpoint with in-place restart. Unlike checkpoint/restart, this is not only an optimization, but a hard requirement. Nonetheless, many forward recovery strategies prove more malleable and can cope with a reduced number of processes. However, if the communication pattern is complex, the occurrence of failures has the potential to deeply disturb the application and prevent an effective recovery from being implemented. Consider the example in Fig. 3.27; four processes are communicating in a point-to-point pattern. process P3 is waiting to receive a message from P2, which is itself waiting to receive a message from P1. In the meantime, P1 has failed, but, as only P2 communicates with P1 directly, other processes do not detect this condition, and

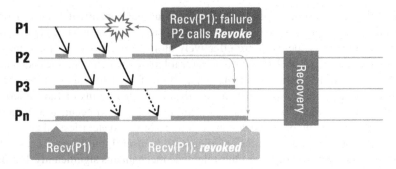

Fig. 3.27 A complex communication pattern that requires to revoke the existing communication context when switching to the recovery phase, in order to avoid deadlocks resulting from transitive dependencies between processes

only P2 is informed of the failure of P1. Process 2 is faced with a dilemma: it knows that P1 has failed, and that the application should branch into its recovery procedure. However, if it were to branch abruptly to the recovery procedure, it would cease matching the receives P3 is waiting on. At this point, without an additional MPI construct, the application would reach a deadlock: the messages that P3 to Pn are waiting for will never arrive. To resolve this scenario, before switching to the recovery procedure, P2 calls MPI_COMM_REVOKE, a new API which notifies all other processes in the communicator that a condition requiring recovery actions has been reached. When receiving this notification, any communication on the communicator (ongoing or future) is interrupted and a special error code returned. Then, all surviving processes can safely enter the recovery procedure of the application, knowing that no alive process belonging to that communicator can deadlock.

3.8.5 The User-Level Failure Mitigation API

ULFM was proposed as an extension to the MPI Forum[3] to introduce fault-tolerance constructs in the MPI standard. It is designed according to the criterion identified in the previous section: to be the minimal interface necessary to restore the complete MPI capability to transport messages after failures. As requested by our flexibility goal, it does not attempt to define a specific application recovery strategy. Instead, it defines the set of functions that can be used by applications (or libraries and languages that provide high-level fault-tolerance abstractions) to repair the state of MPI.

---

[3]The interested reader may refer to Chap. 17 of the complete draft, available from http://fault-tolerance.org/ulfm/ulfm-specification.

3.8.5.1 Failure Reporting

The failure reporting mechanism of ULFM piggybacks on existing features of MPI. In the current MPI standard, erroneous conditions arising from within MPI functions are reported through exceptions, which can be captured by error handlers. The default error handler aborts the application, but it can be replaced with an user provided handler, or a predefined handler that returns the error code as an output of the MPI function. Even when the handler doesn't abort the application, it is considered that the state of MPI is corrupted when any exception is raised. To this mechanism, we add supplementary exception classes to denote a process failure. When a process failure exception is raised, the state of the MPI library remains well-defined.

Failures are reported on a per-operation basis, and indicate essentially that the operation could not be carried out successfully because a failure occurred on one of the processes involved in the operation. For performance reasons, not all failures need to be propagated, in particular, processes that do not communicate with the failed process are not expected to detect its demise. Similarly, during a collective communication, some processes may detect the failure, while some other may consider that the operation was successful; a particularity that we name nonuniform error reporting (see Sect. 3.8.3.3). Let's imagine a broadcast communication using a tree-based topology. The processes that are high in the tree topology, close to the root, complete the broadcast earlier than the leaves. Consequently, these processes may report the successful completion of the broadcast, before the failure disrupts the communication, or even before the failure happens, while processes below a failed process cannot deliver the message and have to report an error.

Once a failure condition has been reported, users can employ the recovery functions summarized in Fig. 3.28 to inspect process local, known information about failed processes, to propagate failure notifications, and last to restore a sane communication context.

- MPI_Comm_failure_ack(comm)
  - Resumes matching for MPI_ANY_SOURCE
- MPI_Comm_failure_get_acked(comm, &group)
  - Returns to the user the group of processes acknowledged to have failed

- MPI_Comm_revoke(comm)
  - Non-collective, interrupts all operations on comm (future or active, at all ranks) by raising MPI_ERR_REVOKED

- MPI_Comm_shrink(comm, &newcomm)
  - Collective, creates a new communicator without failed processes (identical at all ranks)
- MPI_Comm_agree(comm, &mask)
  - Agree on the AND value on binary mask, ignoring failed processes (reliable AllReduce)

Fig. 3.28 Summary of the ULFM recovery functions. Functions can be classified into three groups, the introspection of locally known failure, the propagation of failure knowledge, and the recovery of a stable context

3.8.5.2 Failure Introspection

The next two functions, `MPI_COMM_FAILURE_ACK` and `MPI_COMM_`
`FAILURE_GET_ACKED` are introduced as a lightweight mechanism to continue
using point-to-point operations on a communicator that contains failed processes.
Using these functions, the application can determine which processes are known to
have failed, and inform the MPI library that it acknowledges that no future receive
operation can match sends from any of the reported dead processes. `MPI_COMM_`
`FAILURE_GET_ACKED` returns the group containing all processes which were
locally known to have failed at the time the last `MPI_COMM_FAILURE_ACK` was
called. These functions can be used on any type of communicator, be it revoked
or not.

The operation of retrieving the group of failed processes is split into two functions
for two reasons. First, it permits multiple threads to synchronize on the acknowledge,
to prevent situations were multiple thread read a different group of failed processes.
Second, the acknowledge acts as a mechanism for alerting the MPI library that the
application has been notified of a process failure, permitting to relax error reporting rules for "wildcard" `MPI_ANY_SOURCE` receives. Without an acknowledgment
function, the MPI library would not be able to determine if the failed process is a
potential matching sender, and would have to take the safe course of systematically
returning an error, thereby preventing any use of wildcard receives after the first
failure. Once the application has called `MPI_COMM_FAILURE_ACK`, it becomes
its responsibility to check that no posted "wildcard" receive should be matched by
a send at a reported dead process, as MPI stops reporting errors for such processes.
However, it will continue to raise errors for named point-to-point operations with the
failed process as well as collective communications.

3.8.5.3 Failure Knowledge Propagation

`MPI_COMM_REVOKE`, is a key additional construct: and is intended to resolve the
issues resulting from nonuniform error reporting. As seen above, if nonuniform error
reporting is possible, the view of processes, and accordingly the actions that they
will undergo in the future, may diverge. Processes that have detected the failure may
need to initiate a recovery procedure, but they have the conflicting need to match
pending operations that have been initiated by processes that have proceeded unaware
of the failure, as otherwise these may deadlock while waiting for their operation
to complete. When such a situation is possible, according to the communication
pattern of the application, processes that have detected that recovery action is needed
and intend to interrupt following the normal flow of communication operations can
release other processes by explicitly calling the `MPI_COMM_REVOKE` function on
the communication object. Like many other MPI constructs `MPI_COMM_REVOKE` is
a collective operation over the associated communicator. However, unlike any other
collective MPI constructs it does not require a symmetric call on all processes, a single
processes in the communicator calling the revoke operation ensure the communicator

will be eventually revoked. In other words, it has a behavior similar to `MPI_ABORT` with the exception that it does not abort processes, instead it terminate all ongoing operations on the communicator and mark the communicator as improper for future communications.

3.8.5.4 Rebuilding Communicators

The next construct provides a recovery mechanism: `MPI_COMM_SHRINK`. Although the state of a communicator is Although the state of a communicator is left unchanged by process failures, and point-to-point operations between non-failed processes are still functional, it is to be expected that most collective communication will always raise an error, as they involve all processes in the communicator. Therefore, to restore full communication capacity, MPI communicators objects must be repaired. The `MPI_MPI_COMM_SHRINK` function create a new functional communicator based on an existing, revoked communicator containing failed processes. It does this by creating a duplicate communicator (in the sense of `MPI_COMM_DUP`) but omitting any processes which are agreed to have failed by all remaining processes in the shrinking communicator. If there are new process failures which are discovered during the shrink operation, these failures are absorbed as part of the operation.

3.8.5.5 Ensuring a Consistent State

The last function permits deciding on the completion of an algorithmic section: `MPI_COMM_AGREE`. This function, which is intrinsically costly, is designed to be used sparingly, for example when a consistent view of the status of a communicator is necessary, such as during algorithm completion, or at the end of an application's transaction. This operation performs an agreement algorithm, computing the conjunction of boolean values provided by all alive processes in a communicator. It is important to note that this function will continue successfully even if a communicator has known failures (or if failures happen during the operation progress).

3.8.5.6 Beyond Communicators

While communicator operations are the historic core of MPI, the standard has been extended over the years to support other type of communication contexts, namely shared-memory windows (with explicit put/get operations) and collective file I/O. The same principles described in this paper are extended to these MPI objects in the complete proposal; in particular, windows and files have a similar Revoke function. A notable difference though, is that file and window object don't have repair functions. These objects are initially derived from a communicator object, and the expected recovery strategy is to create a repaired copy of this communicator, before using it to create a new instance of the window or file object. While windows also have the failure introspection function MPI_WIN_GET_FAILED, which is useful for continuing active target operations on the window when failed processors can be ignored (similarly to point-to-point operations on a communicator), all file operations are collective, hence this function is not provided, as the only meaningful continuation of a failure impacting a file object is to revoke the file object. It is to be noted that in the case of file objects, only failures of MPI processes (that may disrupt collective operations on the file) are addressed. Failures of the file backend itself are already defined in MPI-2.

3.8.6 Performance Assessment

3.8.6.1 Impact on Failure-Free Operations

Memory: Because a communicator cannot be repaired, tracking the state of failed processes imposes a minimal memory overhead. From a practical perspective each node needs a global list of detected failures, shared by all communicators; its size grows linearly with the number of failures, and it is empty as long as no failures occur. Within each communicator, the supplementary state is limited to two values: whether the communicator is revoked or not, and an index in the global list of failures denoting the last acknowledged failure (with MPI_COMM_FAILURE_ACK). For efficiency reasons, an implementation may decide to cache the fact that some failures have happened in the communicator so that collective operations and MPI_ANY_SOURCE receptions can bail out quickly. Overall, the supplementary memory consumption from fault-tolerant constructs is small, independent of the total number of nodes, and unlikely to affect the cache and TLB hit rates.

Conditionals: Another concern is the number of supplementary conditions on the latency critical path. Indeed, most completion operations require a supplementary conditional statement to handle the case where the underlying communication context has been revoked. However, the prediction branching logic of the processor can be hinted to favor the failure-free outcome, resulting in a single load of a cached value and a single, mostly well-predicted, branching instruction, unlikely to affect the instruction pipeline. It is notable that non-blocking operations raise errors related to process failure only during the completion step, and thus do not need to check for revocation before the latency critical section.

Matching logic: MPI_COMM_REVOKE does not have a matching call on other processes on which it has an effect. As such, it might add detrimental complexity to the matching logic. However, any MPI implementation needs to handle unexpected messages. The order of revocation message delivery is loose enough that the handling of revocation notices can be integrated within the existing unexpected message matching logic. In our implementation in Open MPI, we leverage the active message low level transport layer to introduce revocation as a new active message tag, without a single change to the matching logic.

Collective operations: A typical MPI implementation supports a large number of collective algorithms, which are dynamically selected depending on criteria such as communicator or message size and hardware topology. The loose requirements of the proposal concerning error reporting of process failures in collective operations limits the impact it has on collective operations. Typically, the collective communication algorithms and selection logic are left unchanged. The only new requirement is that failures happening at any rank of the communicator cause all processes to exit the collective (successfully for some, with an error for others). Due to the underlying loosely connected topologies used by some algorithms, a point-to-point based implementation of a collective communication is unlikely to detect all process failures. Fortunately, a practical implementation exists that does not require modifying any of the collective operations: when a rank raises an error because of a process failure, it can revoke an internal, temporary communication context associated with the collective operation. As the revocation notice propagates on the internal communicator, it interrupts the point-to-point operations of the collective. An error code is returned to the high-level MPI wrapper, which in turn raises the appropriate error on the user's communicator.

3.8.6.2 Recovery Routines

Some of the recovery routines described in Sect. 3.8.5 are unique in their ability to deliver a valid result despite the occurrence of failures. This specification of correct behavior across failures calls for resilient, more complex algorithms. In most cases, these functions are intended to be called sparingly by users, only after actual failures have happened, as a means of recovering a consistent state across all processes. The remainder of this section describes the algorithms that can be used to deliver this specification and their cost.

Agreement: The agreement can be conceptualized as a failure resilient reduction on a boolean value. Many agreement algorithms have been proposed in the literature; the log-scaling two-phase consensus algorithm used by the ULFM prototype is one of many possible implementations of MPI_COMM_AGREE operation based upon prior work in the field. Specifically, this algorithm is a variation of the multilevel two-phase commit algorithms [59]. The algorithm first performs a reduction of the input values to an elected coordinator in the communicator. The coordinator then makes a decision on the output value and broadcasts that value back to all of the alive processes in the communicator. The complexity of the agreement algorithm appears when adapting to an emerging process failure of the coordinator and/or participants. A more extensive discussion of the algorithmic complexity has been published by Hursey, et.al. [50]. The algorithmic complexity of this implementation is $O(log(n))$ for the failure-free case, matching that of an MPI_ALLREDUCE operation over the alive processes in the communicator.

Revoke: Although the revoke operation is not collective, the revocation notification needs to be propagated to all alive processes in the specified communicator, even when new failures happen during the revoke propagation. These requirements are not without recalling those from the *reliable broadcast* [45]. Among the four defining qualities of a reliable broadcast (*Termination, Validity, Integrity, Agreement*), the termination and integrity criteria can be relaxed in the context of the revoke algorithm. If a failure during the Revoke algorithm kills the initiator as well as all the already notified processes, the Revoke notification is indeed lost, but the observed behavior, from the view of the application, is indiscernible from a failure at the initiator before the propagation started. As the algorithm still ensures agreement, there are no opportunities for inconsistent views.

In the ULFM implementation, the initiator marks the communicator as revoked and sends a Revoke message to every processes in the groups (local and remote) of the communicator. Upon reception of a revoke message, if the communicator is not already revoked, it is revoked and the process acts as a new initiator.

Shrink: The Shrink operation is, algorithmically, an agreement on which the consensus is done on the group of failed processes. Hence, the two operations have the same algorithmic complexity. Indeed, in the prototype implementation, `MPI_COMM_AGREE` and `MPI_COMM_SHRINK` share the same internal implementation of the agreement.

3.8.6.3 Experimental Evaluation

The following analysis uses a prototype of the ULFM proposal based on the development trunk of Open MPI [35] (r26237). The test results presented were gathered from the Smoky system at Oak Ridge National Laboratory. Each node contains four quad-core 2.0 GHz AMD Opteron processors with 2 GB of memory per compute core. Compute nodes are connected with gigabit Ethernet and InfiniBand. Some shared-memory benchmarks were conducted on Romulus, a 6×8-core AMD Opteron 6180 SE with 256 GB of memory (32 GB per socket) at the University of Tennessee.

The NetPIPE-3.7 benchmark [82] was used to assess the 1-byte latency and bandwidth impact of the modifications necessary for the ULFM support in Open MPI. We compare the vanilla version of Open MPI (r26237) with the ULFM enabled version on Smoky. Table 3.2 highlights the fact that the differences in performance are well below the noise limit, and that the standard deviation is negligible proving the performance stability and lack of impact.

The impact on shared-memory systems, which are sensitive even to small modifications of the MPI library, has been further assessed on the Romulus machine—a large shared-memory machine—using the IMB benchmark suite (v3.2.3). As shown in Fig. 3.29, the duration difference of all the benchmarks (point-to-point and collective) remains below 5 %, thus within the standard deviation of the implementation on that machine.

Table 3.2 NetPIPE results on Smoky

Interconnect	Vanilla	Std. Dev.	Enabled	Std. Dev.	Difference
1-byte Latency (microseconds) (cache hot)					
Shared memory	0.8008	0.0093	0.8016	0.0161	0.0008
TCP	10.2564	0.0946	10.2776	0.1065	0.0212
OpenIB	4.9637	0.0018	4.9650	0.0022	0.0013
Bandwidth (Mbps) (cache hot)					
Shared memory	10,625.92	23.46	10,602.68	30.73	−23.24
TCP	6,311.38	14.42	6,302.75	10.72	−8.63
OpenIB	9,688.85	3.29	9,689.13	3.77	0.28

Fig. 3.29 The intel MPI benchmarks: relative difference between ULFM and the vanilla Open MPI on shared memory (Romulus). Standard deviation ≈5 % on 1,000 runs

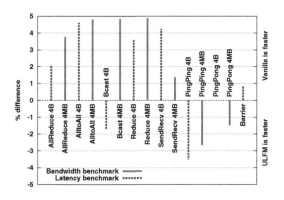

To measure the impact of the prototype on a real application, we used the Sequoia AMG benchmark.[4] This MPI intensive benchmark is an Algebraic Multigrid (AMG) linear system solver for unstructured mesh physics. A weak scaling study was conducted up to 512 processes following the problem *Set 5*. In Fig. 3.30, we compare the time slicing of three main phases (Solve, Setup, and SStruct) of the benchmark, with, side by side, the vanilla version of the Open MPI implementation, and the ULFM enabled one. The application itself is not fault tolerant and does not use the features proposed in ULFM. The goal of this benchmark is to demonstrate that a careful implementation of the proposed semantic does not impact the performance of the MPI implementation, and ultimately leaves the behavior and performance of legacy applications unchanged. The results show that the performance difference is negligible.

To assess the overheads of recovery constructs, we developed a synthetic benchmark that mimics the behavior of a typical fixed-size tightly coupled fault-tolerant

---

[4]https://asc.llnl.gov/sequoia/benchmarks/#amg.

Fig. 3.30 Comparison of the
vanilla and ULFM versions of
Open MPI running Sequoia
AMG at different scales
(smoky)

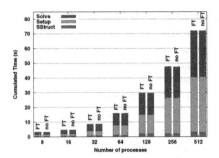

application. Unlike a normal application it performs an infinite loop, where each iteration contains a failure and the corresponding recovery procedure. Each iteration consists of 5 phases: in the first phase (*Detection*), all processes but a designated victim enter a Barrier on the intracommunicator. The victim dies, and the failure detection mechanism makes all surviving processes exit the Barrier, some with an error code. In Phase 2 (*Revoke*), the surviving processes that detected a process failure-related error during the previous phase invoke the new construct MPI_COMM_REVOKE. Then they proceed to Phase 3 (*Shrink*) where the intracommunicator is shrunk using MPI_COMM_SHRINK. The two other phases serve to repair a full-size intracommunicator using spawn and intercommunicator merge operations to allow the benchmark to proceed to the next round.

In Fig. 3.31, we present the timing of each phase, averaged upon 50 iterations of the benchmark loop, for a varying number of processes on the Smoky machine. We focus on the three points related to ULFM: failure detection, revoke and shrink. The failure detection is mildly impacted by the scale. In the prototype implementation, the detection happens at two levels, either in the runtime system or in the MPI library (when it occurs on an active link). Between the two detectors, all ranks get notified within 30 ms of the failure (this compares to the 1 s timeout at the link level). Although the revoke call will inject a linear number of messages (at each rank) in the network to implement the level of reliability required for this operation, the duration of this call itself is under 50 μs and is not visible in the figure. The network is disturbed

Fig. 3.31 Evaluation of the
fault injection benchmark
with full recovery at different
scales (smoky)

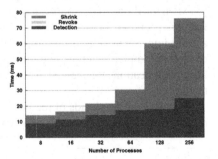

for a longer period, due to the processing of the messages, but this disturbance will appear in the network only after a failure occurred. The last call shown in the figure is the shrink operation. Although its duration increases linearly with the number of processes (the figure has a logarithmic scale on the x-axis), this cost must only be paid after a failure, in order to continue using collective operations. In its current implementation, shrink requires an agreement, the allocation of a new communicator identifier, and the creation of the communicator (with `MPI_COMM_SPLIT`). Most of the time spent in the shrink operation is not in the agreement (which scales logarithmically), but in the underlying implementation of the communicator creation.

3.8.7 Concluding Remarks

Simple communication interfaces, such as sockets or streams, have been featuring robust fault tolerance for decades. It may come as a surprise that specifying the behavior of MPI when fail-stop failures strike is so challenging. In this chapter we have identified the contentious issues, rooted in the fact that the state of MPI objects is implicitly distributed and that specifying the behavior of collective operations and communication routines requires a careful, precise investigation of unexpected consequences on the concepts as well as on the performance. We first took a review of the field of fault tolerance and recovery methods; most require that MPI can restore the full set of communication functionalities after a failure happened. Then, we proposed the ULFM interface, which responds to that demand, and took the critical viewpoint of the implementor unwilling to compromise performance, on a number of hidden, but crucial issues regarding the state of MPI objects when failure happen. Lastly, we took the viewpoint of MPI users, and depicted how the ULFM specification can be used to support high level recovery strategies.

The landscape for fault-tolerant applications is varied, with differing needs and opportunities. Some applications can afford to pay the engineering cost of a full application rewrite, taking into account inherent properties of the algorithms in order to decrease the cost of the fault protection strategy. Some legacy applications are not ready to take that leap, yet, or do not have algorithmic features that can be easily leveraged to significantly improve above checkpoint/restart. In this chapter, we have presented techniques that support both these usage models: the MPI standard is active, and getting ready to support the effective, portable deployment of advanced fault-tolerance techniques.

Acknowledgments The author would like to thank the coworkers who participated in the creation of the software or participated in the elaboration of some source material in this chapter, in particular, Christine Morin and Thomas Ropars, for leading the effort with the optimistic message logging protocol; Wesley Bland and Joshua Hursey which helped with the implementation of ULFM; and George Bosilca and Thomas Herault. This work has received support from the NSF under award Adapt #1339763, and from the CREST project of the Japan Science and Technology Agency (JST).

References

1. Ali MM, Southern J, Strazdins PE, Harding B (2014) Application level fault recovery: using fault-tolerant open MPI in a PDE solver. In: 2014 IEEE international parallel and distributed processing symposium workshops, Phoenix, AZ, USA, 19–23 May 2014, pp 1169–1178. IEEE
2. Alvisi L, Marzullo K (1995) Message logging: pessimistic, optimistic, and causal. In: Proceedings of the 15th international conference on distributed computing systems (ICDCS 1995). IEEE CS Press, pp 229–236
3. Alvisi L, Elnozahy E, Rao S, Husain SA, Mel AD (1999) An analysis of communication induced checkpointing. In: 29th symposium on fault-tolerant computing (FTCS'99). IEEE CS Press
4. Bautista-Gomez L, Tsuboi S, Komatitsch D, Cappello F, Maruyama N, Matsuoka S (2011) FTI: high performance fault tolerance interface for hybrid systems. In: International conference high performance computing, networking, storage and analysis SC'11
5. Bland W, Bouteiller A, Herault T, Hursey J, Bosilca G, Dongarra JJ (2013) An evaluation of user-level failure mitigation support in MPI. Computing 95(12):1171–1184
6. Bosilca G, Delmas R, Dongarra J, Langou J (2009) Algorithm-based fault tolerance applied to high performance computing. J Parallel Distrib Comput 69(4):410–416
7. Bosilca G, Bouteiller A, Herault T, Lemarinier P, Dongarra JJ (2010) Dodging the cost of unavoidable memory copies in message logging protocols. In: Keller R, Gabriel E, Resch MM, Dongarra J (eds) EuroMPI, Lecture Notes in Computer Science, Vol 6305. pp 189–197
8. Bouteiller A, Collin B, Herault T, Lemarinier P, Cappello F (2005) Impact of event logger on causal message logging protocols for fault tolerant MPI. In: IPDPS'05: proceedings of the 19th IEEE international parallel and distributed processing symposium (IPDPS'05)—papers. IEEE Computer Society, Washington, DC, USA, p 97
9. Bouteiller A, Herault T, Krawezik G, Lemarinier P, Cappello F (2006) MPICH-V project: a multiprotocol automatic fault tolerant MPI. Int J High Perform Comput Appl 20:319–333
10. Bouteiller A, Ropars T, Bosilca G, Morin C, Dongarra J (2009) Reasons to be pessimist or optimist for failure recovery in high performance clusters. In: IEEE, editor, proceedings of the 2009 IEEE cluster conference
11. Bouteiller A, Bosilca G, Dongarra J (2010) Redesigning the message logging model for high performance. Concurr Comput: Pract Exp 22(16):2196–2211
12. Bouteiller A, Herault T, Bosilca G, Dongarra JJ (2013) Correlated set coordination in fault tolerant message logging protocols for many-core clusters. Concurr Comput: Pract Exp 25(4):572–585
13. Bouteiller A, Herault T, Bosilca G, Du P, Dongarra J (2015) Algorithm-based fault tolerance for dense matrix factorizations, multiple failures and accuracy. ACM Trans Parallel Comput 1(2):10:1–10:28
14. Bronevetsky G (2007) Portable checkpointing for parallel applications. Ph.D. thesis, Cornell University, Department of Computer Science
15. Buntinas D, Coti C, Herault T, Lemarinier P, Pilard L, Rezmerita A, Rodriguez E, Cappello F (2008) Blocking versus non-blocking coordinated checkpointing for large-scale fault tolerant MPI protocols. Future Gener Comput Syst 24(1):73–84
16. Cappello F, Geist A, Gropp B, Kalé LV, Kramer B, Snir M (2009) Toward exascale resilience. Int J High Perform Comput Appl 23(4):374–388
17. Cappello F, Casanova H, Robert Y (2011) Preventive migration versus preventive checkpointing for extreme scale supercomputers. Parallel Process Lett, pp 111–132
18. Chakravorty S, Mendes CL, Kalé LV (2006) Proactive fault tolerance in MPI applications via task migration. In: HiPC 2006, the IEEE high performance computing conference. IEEE Computer Society Press, pp 485–496
19. Chandra TD, Toueg S (1996) Unreliable failure detectors for reliable distributed systems. J ACM (JACM) 43(2):225–267
20. Chandy KM, Lamport L (1985) Distributed snapshots: determining global states of distributed systems. Trans Comput Syst 3(1):63–75. ACM

21. Chen Z, Fagg GE, Gabriel E, Langou J, Angskun T, Bosilca G, Dongarra J (2005) Fault tolerant high performance computing by a coding approach. In: Proceedings of the tenth ACM SIGPLAN symposium on principles and practice of parallel programming, PPoPP'05. ACM, New York, USA, pp 213–223

22. Daly JT (2006) A higher order estimate of the optimum checkpoint interval for restart dumps. Future Gener Comput Syst 22:303–312

23. Damani OP, Wang Y-M, Garg VK (2003) Distributed recovery with K-optimistic logging. J Parallel Distrib Comput 63:1193–1218

24. Davies T, Karlsson C, Liu H, Ding C, Chen Z (2011) High performance Linpack benchmark: a fault tolerant implementation without checkpointing. In: Proceedings of the 25th ACM international conference on supercomputing (ICS 2011). ACM

25. Dean J, Ghemawat S (2004) MapReduce: simplified data processing on large clusters. In: Proceedings of the 6th conference on symposium on operating systems design and implementation—vol 6, OSDI'04. USENIX Association, Berkeley, CA, USA, pp 10–10

26. Dongarra J, Blackford L, Choi J et al (1997) ScaLAPACK user's guide. Society for Industrial and Applied Mathematics, Philadelphia

27. Dongarra J, Beckman P, Moore T, Aerts P, Aloisio G, Andre J-C, Barkai D, Berthou J-Y, Boku T, Braunschweig B, Cappello F, Chapman B, Chi X, Choudhary A, Dosanjh S, Dunning T, Fiore S, Geist A, Gropp B, Harrison R, Hereld M, Heroux M, Hoisie A, Hotta K, Jin Z, Ishikawa Y, Johnson F, Kale S, Kenway R, Keyes D, Kramer B, Labarta J, Lichnewsky A, Lippert T, Lucas B, Maccabe B, Matsuoka S, Messina P, Michielse P, Mohr B, Mueller MS, Nagel WE, Nakashima H, Papka ME, Reed D, Sato M, Seidel E, Shalf J, Skinner D, Snir M, Sterling T, Stevens R, Streitz F, Sugar B, Sumimoto S, Tang W, Taylor J, Thakur R, Trefethen A, Valero M, Van Der Steen A, Vetter J, Williams P, Wisniewski R, Yelick K (2011) The international exascale software project roadmap. Int J High Perform Comput Appl 25(1):3–60

28. Duell J (2002) The design and implementation of Berkeley lab's Linux checkpoint/restart. Technical report LBNL-54941, Livermore-Berkeley National Laboratory

29. Elnozahy E, Zwaenepoel W (1992) Manetho: transparent rollback-recovery with low overhead, limited rollback and fast output. IEEE Trans Comput 41(5):526–531

30. Elnozahy M, Alvisi L, Wang YM, Johnson DB (2002) A survey of rollback-recovery protocols in message-passing systems. ACM Comput Surv (CSUR) 34(3):375–408

31. Fagg G, Dongarra J (2000) FT-MPI: fault tolerant MPI, supporting dynamic applications in a dynamic world. In: 7th Euro PVM/MPI user's group meeting 2000, vol 1908/2000, Balaton-füred, " Hungary". Springer, Heidelberg

32. Fagg GE, Bukovsky A, Dongarra JJ (2001) HARNESS and fault tolerant MPI. Parallel Comput 27(11):1479–1495

33. Ferreira K, Stearley J, Laros J, Oldfield R, Pedretti K, Brightwell R, Riesen R, Bridges P, Arnold D (2011) Evaluating the viability of process replication reliability for exascale systems. In: International conference for high performance computing, networking, storage and analysis (SC), 2011. ACM Request Permissions, pp 1–12

34. Fischer M, Lynch N, Paterson M (1985) Impossibility of distributed consensus with one faulty process. J ACM 32:374–382

35. Gabriel E, Fagg GE, Bosilca G, Angskun T, Dongarra JJ, Squyres JM, Sahay V, Kambadur P, Barrett B, Lumsdaine A, Castain RH, Daniel DJ, Graham RL, Woodall TS (2004) Open MPI: goals, concept, and design of a next generation MPI implementation. In: Proceedings of the 11th European PVM/MPI users' group meeting, Budapest, Hungary, pp 97–104

36. Gamell M, Katz DS, Kolla H, Chen J, Klasky S, Parashar M (2014) Exploring automatic, online failure recovery for scientific applications at extreme scales. In: Proceedings of the international conference for high performance computing, networking, storage and analysis, SC'14. IEEE Press, Piscataway, NJ, pp 895–906

37. Gao Q, Huang W, Koop MJ, Panda DK (2007) Group-based coordinated checkpointing for MPI: a case study on infiniband. In: International conference on parallel processing, 2007. ICPP 2007

38. Geist GA, Kohl JA, Papadopoulos PM (1996) PVM and MPI: a comparison of features. Calc Paralleles 8:137–150
39. Gelenbe E (1979) On the optimum checkpoint interval. J ACM 26:259–270
40. Geoffray P (2002) OPIOM: off-processor i/o with myrinet. Future Gener Comput Syst 18(4):491–499
41. Goglin B (2008) Improving message passing over ethernet with i/oat copy offload in open-mx. In: Proceedings of the 2008 IEEE international conference on cluster computing. IEEE, pp 223–231
42. Gropp W, Lusk E (1997) Why are PVM and MPI so different? In: Bubak M, Dongarra J, Waśniewski J (eds) Recent advances in parallel virtual machine and message passing interface, Lecture Notes in Computer Science, vol 1332. Springer, Berlin, pp 1–10
43. Gropp W, Lusk E (2004) Fault tolerance in message passing interface programs. Int J High Perform Comput Appl 18:363–372
44. Guermouche A, Ropars T, Snir M, Cappello F (2012) HydEE: failure containment without event logging for large scale send-deterministic MPI applications. In: IEEE 26th international parallel distributed processing symposium (IPDPS), 2012, pp 1216–1227
45. Hadzilacos V, Toueg S (1993) Fault-tolerant broadcasts and related problems. In: Mullender S (ed) Distributed systems, 2nd edn. ACM/Addison-Wesley, Boston, pp 97–145 (chapter 5)
46. Hassani A, Skjellum A, Brightwell R (2014) Design and evaluation of FA-MPI, a transactional resilience scheme for non-blocking MPI. In: International conference on dependable systems and networks (DSN), 2014 44th annual IEEE/IFIP, pp 750–755
47. Ho JCY, Wang C-L, Lau FCM (2008) Scalable group-based checkpoint/restart for large-scale message-passing systems. In: Proceedings of the 22nd IEEE international symposium on parallel and distributed processing (IPDPS). IEEE, pp 1–12
48. Huang K, Abraham J (1984) Algorithm-based fault tolerance for matrix operations. IEEE Trans Comput 100(6):518–528
49. Hursey J, Squyres J, Mattox T, Lumsdaine A (2007) The design and implementation of checkpoint/restart process fault tolerance for open MPI. In: IEEE international parallel and distributed processing symposium, 2007. IPDPS 2007, pp 1–8
50. Hursey J, Naughton T, Vallee G, Graham RL (2011) A log-scaling fault tolerant agreement algorithm for a fault tolerant MPI. In: EuroMPI 2011: proceedings of the 18th EuroMPI conference, Santorini, Greece
51. Hursey J, Graham RL, Bronevetsky G, Buntinas D, Pritchard H, Solt DG (2011) Run-through stabilization: an MPI proposal for process fault tolerance. In: EuroMPI 2011: proceedings of the 18th EuroMPI conference, Santorini, Greece
52. Hélary J-M, Mostefaoui A, Raynal M (1999) Communication-induced determination of consistent snapshots. IEEE Trans Parallel Distrib Syst 10(9):865–877
53. Lamport L (1978) Time, clocks, and the ordering of events in a distributed system. Commun ACM 21(7):558–565
54. Lemarinier P, Bouteiller A, Herault T, Krawezik G, Cappello F (2004) Improved message logging versus improved coordinated checkpointing for fault tolerant MPI. In: IEEE international conference on cluster computing. IEEE CS Press
55. Liu X, Xu X, Ren X, Tang Y, Dai Z (2013) A message logging protocol based on user level failure mitigation. In: Kołodziej J, Di Martino B, Talia D, Xiong K (eds) Algorithms and architectures for parallel processing, Lecture Notes in Computer Science, vol 8285. Springer International Publishing, Switzerland, pp 312–323
56. Lu C-D, Reed DA (2005) Scalable diskless checkpointing for large parallel systems. Ph.D. thesis, University of Illinois at Urbana-Champain
57. Luk F, Park H (1988) An analysis of algorithm-based fault tolerance techniques. J Parallel Distrib Comput 5(2):172–184
58. Meneses E, Mendes CL, Kalé LV (2010) Team-based message logging: preliminary results. In: Proceedings of the 2010 10th IEEE/ACM international conference on cluster, cloud and grid computing, CCGRID'10. IEEE Computer Society, Washington, pp 697–702

59. Mohan C, Lindsay B (1985) Efficient commit protocols for the tree of processes model of distributed transactions. SIGOPS OSR, vol 19. ACM, New York, pp 40–52
60. Moody A, Bronevetsky G, Mohror K, de Supinski BR (2010) Design, modeling, and evaluation of a scalable multi-level checkpointing system. In: Proceedings of the 2010 ACM/IEEE international conference for high performance computing. Networking, storage and analysis, pp 1–11
61. Negara S, Zheng G, Pan K-C, Negara N, Johnson RE, Kalé LV, Ricker PM (2011) Automatic MPI to AMPI program transformation using photran. In: Proceedings of the 2010 conference on parallel processing, Euro-Par 2010. Springer, Berlin, pp 531–539
62. Pauli S, Kohler M, Arbenz P (2013) A fault tolerant implementation of multi-level monte carlo methods. In: Bader M, Bode A, Bungartz H, Gerndt M, Joubert GR, Peters FJ (eds) Parallel computing: accelerating computational science and engineering (CSE), proceedings of the international conference on parallel computing, ParCo 2013, 10–13 September 2013, Garching (near Munich), Germany, advances in parallel computing, vol 25. IOS Press, pp 471–480
63. Petrini F, Frachtenberg E, Hoisie A, Coll S (2003) Performance evaluation of the quadrics interconnection network. Clust Comput 6(2):125–142
64. Plank JS (1993) Efficient checkpointing on MIMD architectures. Ph.D. thesis, Princeton University
65. Plank JS, Thomason MG (2001) Processor allocation and checkpoint interval selection in cluster computing systems. J Parallel Distrib Comput 61:1590
66. Rao S, Alvisi L, Vin HM (1998) The cost of recovery in message logging protocols. In: 17th symposium on reliable distributed systems (SRDS). IEEE CS Press, pp 10–18
67. Rao S, Alvisi L, Vin HM (1999) Egida: an extensible toolkit for low-overhead fault-tolerance. In : 29th symposium on fault-tolerant computing (FTCS'99). IEEE CS Press, pp 48–55
68. Rieker M, Ansel J, Cooperman G (2006) Transparent user-level checkpointing for the native posix thread library for linux. In: The 2006 international conference on parallel and distributed processing techniques and applications, Las Vegas, NV
69. Ropars T, Morin C (2011) Active optimistic and distributed message logging for message-passing applications. Concurr. Comput. : Pract. Exper. 23(17):2167–2178
70. Ropars T, Guermouche A, Uçar B, Meneses E, Kalé LV, Cappello F (2011) On the use of cluster-based partial message logging to improve fault tolerance for MPI HPC applications. In: Proceedings of the 17th international conference on parallel processing—volume Part I, Euro-Par'11. Springer, Berlin, pp 567–578
71. Roy-Chowdhury A, Banerjee P (1996) Algorithm-based fault location and recovery for matrix computations on multiprocessor systems. IEEE Trans Comput 45(11):1239–1247
72. Ruscio JF, Heffner MA, Varadarajan S (2006) DejaVu: transparent user-level checkpointing, migration and recovery for distributed systems. In: SC'06: proceedings of the 2006 ACM/IEEE conference on supercomputing. ACM Press, New York, USA, pp 158
73. Sankaran S, Squyres JM, Barrett B, Lumsdaine A, Duell J, Hargrove P, Roman E (2003) The LAM/MPI checkpoint/restart framework: system-initiated checkpointing. In: Proceedings, LACSI symposium, Sante Fe, New Mexico, USA
74. Schroeder B, Gibson G (2007) Understanding failures in petascale computers. In: Journal of physics: conference series, vol 78. IOP Publishing, pp 12–22
75. Schulz M, de Supinski B (2006) A flexible and dynamic infrastructure for MPI tool interoperability. In: International conference on parallel processing, 2006. ICPP 2006, pp 193–202
76. Schulz M, Bronevetsky G, Supinski BR (2008) On the performance of transparent MPI piggyback messages. In: Proceedings of the 15th European PVM/MPI users' group meeting on recent advances in parallel virtual machine and message passing interface. Springer, Berlin, pp 194–201
77. Shantharam M, Srinivasmurthy S, Raghavan P (2012) Fault tolerant preconditioned conjugate gradient for sparse linear system solution. In: Proceedings of the 26th ACM international conference on supercomputing, ICS'12. ACM, New York, pp 69–78

78. Silva LM, Silva JG (1998) System-level versus user-defined checkpointing. In: Proceedings of the the 17th IEEE symposium on reliable distributed systems, SRDS'98. IEEE Computer Society, Washington, DC, p 68

79. Singhal M, Kshemkalyani A (1992) An efficient implementation of vector clocks. Inf Process Lett 43(1):47–52

80. Sistla AP, Welch JL (1989) Efficient distributed recovery using message logging. In: PODC '89: proceedings of the eighth annual ACM symposium on principles of distributed computing. ACM Press, New York, pp 223–238

81. Smith SW, Johnson DB, Tygar JD (1995) Completely asynchronous optimistic recovery with minimal rollbacks. In: FTCS-25: 25th international symposium on fault tolerant computing digest of papers. Pasadena, California, pp 361–371

82. Snell QO, Mikler AR, Gustafson JL (1996) NetPIPE: a network protocol independent performance evaluator. In: IASTED international conference on intelligent information management and systems

83. Stellner G (1996) CoCheck: checkpointing and process migration for MPI. In: Proceedings of the 10th international parallel processing symposium (IPPS'96), Honolulu, Hawaii. IEEE CS Press

84. Stricker T, Gross T (1995) Optimizing memory system performance for communication in parallel computers. In: ISCA'95: proceedings of the 22nd annual international symposium on computer architecture. ACM, New York, pp 308–319

85. Strom R, Yemini S (1985) Optimistic recovery in distributed systems. ACM Trans Comput Syst 3(3):204–226

86. Teranishi K, Heroux MA (2014) Toward local failure local recovery resilience model using MPI-ULFM. In: Proceedings of the 21st European MPI users' group meeting, EuroMPI/ASIA'14. ACM, New York, pp 51:51–51:56

87. The MPI Forum (1993) MPI: a message passing interface. In: Supercomputing'93: proceedings of the 1993 ACM/IEEE conference on supercomputing. ACM Press, New York, pp 878–883

88. The MPI Forum (2012) MPI: a message-passing interface standard, version 3.0. The Universtity of Tennessee, Knoxville

89. The SciDB Development Team (2010) Overview of SciDB: large scale array storage, processing and analysis. In: Proceedings of the 2010 ACM SIGMOD international conference on management of data, SIGMOD'10. ACM, New York, pp 963–968

90. Vaidyanathan K, Chai L, Huang W, Panda DK (2007) Efficient asynchronous memory copy operations on multi-core systems and I/OAT. In: CLUSTER'07: proceedings of the 2007 IEEE international conference on cluster computing. IEEE Computer Society, Washington, DC, pp 159–168

91. Wang C, Mueller F, Engelmann C, Scott SL (2008) Proactive process-level live migration in hpc environments. In: SC'08: proceedings of the 2008 ACM/IEEE conference on supercomputing. IEEE Press, Piscataway, NJ, pp 1–12

92. Young JW (1974) A first order approximation to the optimum checkpoint interval. Commun ACM 17:530–531

Chapter 4
Using Replication for Resilience on Exascale Systems

Henri Casanova, Frédéric Vivien and Dounia Zaidouni

Abstract High-performance computing applications must be resilient to faults. The traditional fault tolerance solution is checkpoint–recovery, by which application state is saved to and recovered from secondary storage throughout execution. It has been shown that, even when using an optimal checkpointing strategy, the checkpointing overhead precludes high parallel efficiency at large-scale. Additional fault tolerance mechanisms must thus be used. Such a mechanism is replication, which can be used in addition to checkpoint–recovery. Using replication, multiple processors perform the same computation so that a processor failure does not necessarily mean application failure. While at first glance replication may seem wasteful, it may be significantly more efficient than using solely checkpoint–recovery at large scale. In this work we investigate two approaches for replication. In the first approach, entire application instances are replicated. In the second approach, each process in a single application instance is (transparently) replicated. We provide a theoretical study of these two approaches, comparing them to the pure checkpoint–recovery approach in terms of expected application execution times.

4.1 Introduction

As plans are made for deploying post-petascale high-performance computing (HPC) systems [8, 27], solutions need to be developed to ensure that applications on such systems are resilient to faults. Resilience is particularly critical for applications that enroll large numbers of processors. For such applications, processor failures

H. Casanova
University of Hawai'i, Manoa, USA
e-mail: henric@hawaii.edu

F. Vivien (✉) · D. Zaidouni
INRIA & Ecole Normale Supérieure de Lyon, Lyon, France
e-mail: frederic.vivien@inria.fr

D. Zaidouni
e-mail: dounia.zaidouni@inria.fr

© Springer International Publishing Switzerland 2015   229
T. Herault and Y. Robert (eds.), *Fault-Tolerance Techniques*
for High-Performance Computing, Computer Communications and Networks,
DOI 10.1007/978-3-319-20943-2_4

are projected to be common occurrences [10, 23, 30]. For instance, the 45,208-processor Jaguar platform is reported to have experienced on the order of 1 failure per day [37], and its scale was modest compared to upcoming platforms. Failures occur because not all faults are automatically detected and corrected in current production hardware. To tolerate failures the standard approach is to use rollback and recovery for resuming application execution from a previously saved fault-free execution state, or *checkpoint*. Checkpoints are saved to resilient storage throughout execution, usually periodically. More frequent checkpointing leads to higher overhead during fault-free execution, but less frequent checkpointing leads to a larger loss when a failure occurs. A *checkpointing strategy* specifies when checkpoints should be taken. A large literature is devoted to identifying good checkpointing strategies, including both theoretical and practical efforts. The former typically rely on assumptions regarding the probability distributions of inter-failure times of the processors (e.g., Exponential, Weibull), while the latter rely on simulations driven by failure datasets obtained on real-world platforms.

In spite of these efforts, the necessary checkpoint frequency for tolerating failures in large-scale platforms can become so large that processors spend more time checkpointing than computing. Consider an ideal moldable parallel application that can be executed on an arbitrary number of processors and that is perfectly parallel. The makespan with p processors is the sequential makespan divided by p. In a failure-free execution, the larger p the faster the execution. But in the presence of failures, as p increases so does the frequency of processor failures, leading to (i) more time spent performing recoveries from these failures and (ii) more time spent saving more frequent checkpoints to avoid long re-executions after failure. Beyond some threshold values, increasing p actually increases the expected makespan when using checkpoint–recovery [10, 13, 23, 30]. This is because the MTBF (mean time between failures) of the platform becomes so small that the application performs too many recoveries and re-executions to make progress efficiently.

One possible solution to this problem is to increase the reliability of individual components, e.g., with more hardware redundancy. This increase comes at a higher cost. Since system acquisition costs are typically constrained when designing a parallel platform, vendors must instead use commercial off-the-shelf (COTS) components. The reliability of these COTS components is defined by the product lifetime, as driven by the market. HPC systems with COTS components will thus experience higher failure rates at higher scales [34], thereby limiting parallel efficiency if only checkpoint–recovery is used at these scales. Furthermore, even if the MTBF of an individual component is a high μ_{comp}, then the MTBF of a platform with p components is $\mu = \frac{\mu_{comp}}{p}$ (see Eq. 1.14 in Sect. 1.3.2.1). No matter how reliable the individual components, there is thus a value of p above which errors are so frequent that they can prevent any application progress with checkpoint–recovery.

In this work we focus on *replication*: several processors perform the same computation synchronously, so that a fault on one of these processors does not lead to an application failure. Replication is an age-old fault-tolerant technique, but it has gained traction in the HPC context only relatively recently [12, 28, 38]. While replication wastes compute resources in fault-free executions, it can alleviate the poor scalability of checkpoint–recovery.

We study two replication approaches. Consider a parallel application that is *moldable*, meaning that it can be executed on an arbitrary number of processors, which each processor running one application process. In the first approach, *group replication*, multiple application instances are executed. For example, 2 distinct *n*-process application instances could be executed on a 2*n*-processor platform. Each instance runs at a smaller scale, meaning that it has better parallel efficiency than a single 2*n*-process instance due to a smaller checkpointing frequency. Furthermore, once an instance saves a checkpoint, another instance can use this checkpoint immediately to "jump ahead" in its execution. Hence group replication is more efficient than the mere independent execution of several instances: each time one instance successfully completes a given "chunk of work," all the other instances immediately benefit from this success.

In the second approach, *process replication*, a single instance of an application is executed but each application process is (transparently) replicated. For the same example, one could execute the application with *n* processes so that there are two replicas of each process, each running on a distinct physical processor. This approach is sensible because the mean time to failure of a group of two replicas is larger than that of a single processor. The checkpointing frequency can thus be lowered and the parallel efficiency improved. In [13] Ferreira et al. have studied process replication, with a practical implementation and analytical results. Process replication has been introduced in Sect. 1.4.2.

Process replication largely outperforms group replication due to dramatically increased MTBF for each replica set. However, process replication may not always be a feasible option because it must be provided transparently as part of the runtime system. There are several popular programming models and runtimes (e.g., message passing, concurrent objects, distributed components, workflows, algorithmic skeletons). In some cases, e.g., for the Message Passing Interface (MPI) runtime, proof of concept implementations that provide process replication are available [13]. But in general, many existing and popular runtimes do not (yet) provide transparent process replication for the purpose of fault tolerance, and enhancing them with this capability may be nontrivial. A solution could be to implement process replication explicitly as part of the application, but this would be labor-intensive, especially for legacy applications. Group replication can be used whenever process replication is not available because it is agnostic to the parallel programming model, and thus views the application as an unmodified black box. The only requirement is that the application be moldable and that an instance be startable from a saved checkpoint file.

We note that (process or group) replication prevents the execution of an application that requires the aggregate memory of the full platform, and in this sense limits the scale of the application execution. However, such full-scale execution is likely impractical in the first place due to the need for a high checkpointing frequency. The processors would spend more time-saving state than computing state, thus leading to low parallel efficiency.

At first glance, it may seem paradoxical that better performance can be achieved by using (process or group) replication. After all in the above example, 50 % of the platform is "wasted" to perform redundant computation. As a result the application instance runs at a smaller scale. But, precisely because the scale is smaller, the application can use a lower checkpointing frequency, and can thus have better parallel efficiency when compared to an application instance running at full scale. The application makespan can then be comparable to or even shorter than that obtained when running a single application instance. In the end, the cost of wasting processor power for redundant computation can be offset by the benefit of the reduced checkpointing frequency.

In this chapter we study group and process replication from a theoretical perspective, with the following highlights:

- For group replication:

  - We propose a simple, yet effective algorithm for group replication.
  - For exponentially distributed failures, we derive a checkpointing period that minimizes a upper bound on application makespan.
  - For non-exponentially distributed failures we propose a Dynamic Programming approach that computes non-periodic checkpoint dates in a view to minimizing makespan.
  - For non-exponentially distributed failures we also propose a periodic checkpointing approach in which the period is computed based on a numerical search.
  - We perform simulation experiments assuming that failures follow Exponential or Weibull distributions, the latter being more representative of real-world failure behaviors [17, 18, 22, 29].

- For process replication:

  - We derive exact expressions for the *MNFTI* (Mean Number of Failures To Interruption) and the *MTTI* (Mean Time To Interruption) for arbitrary numbers of replicas assuming Exponential failures.
  - We extend these results to arbitrary failure distributions, notably obtaining closed-form solutions in the case of Weibull failures.
  - We perform simulation experiments and the results show that the choice of a good checkpointing period is no longer critical when process replication is used.

A broad and expected result for both approaches is that replication is beneficial at large scale. But more precisely, our results make it possible to determine in which conditions the use of replication is beneficial and to quantify the benefit in terms of expected application makespan. This chapter is organized as follows. Section 4.2 discusses related work. Section 4.3 defines our models and states our key assumptions. Section 4.4 presents our results for group replication. Section 4.5 presents our results for process replication. Finally, Sect. 4.6 provides concluding remarks and perspectives.

4.2 Related Work

Checkpointing policies have been widely studied in the literature. In [7], Daly studies periodic checkpointing for Exponential failures, generalizing the well-known bound obtained by Young [36]. Daly extended his work in [19] to study the impact of suboptimal checkpointing periods. In [32], the authors develop an "optimal" checkpointing policy, based on the popular assumption that optimal checkpointing must be periodic. In [4], Bouguerra et al. *prove* that the optimal checkpointing policy is periodic when checkpointing and recovery overheads are constant, for either Exponential or Weibull failures. But their results rely on the unstated assumption that all processors are rejuvenated after each failure and after each checkpoint, an assumption that is unreasonable for Weibull failures [3]. We have developed optimal solutions for Exponential failures and dynamic programming solutions for Weibull failures, demonstrating performance improvements over checkpointing approaches proposed in the literature in the case of Weibull failures [3]. The Weibull distribution is recognized as a reasonable approximation of failures in real-world systems [17, 29].

In spite of all the above advances in the areas of checkpointing policies, several studies have questioned the feasibility of pure checkpoint–recovery for large-scale systems (see [13] for a discussion of this issue and for references to such studies). This chapter studies the use of replication in addition to checkpoint–recovery, and is thus related to previous works on checkpointing policies. In particular, some of our results build on the algorithms and results developed in [3].

Replication has long been used as a fault tolerance mechanism in distributed systems [16]. The idea to use replication together with checkpoint–recovery has been studied in the context of grid computing [35]. One concern about replication in HPC is the induced resource waste. However, given the scalability limitations of pure checkpoint–recovery, replication has recently received more attention in the HPC literature [12, 30, 38].

In this chapter we study two replication techniques, group replication and process replication. While, to the best of our knowledge, no previous work has considered group replication, process replication has been studied by several authors. Process replication is advocated in [11] for HPC applications, and in [21] for grid computing with volatile nodes. The work by Ferreira et al. [13] studies the use of process replication for MPI (Message Passing Interface) applications, using 2 replicas per MPI process. They provide a theoretical analysis of parallel efficiency, an MPI implementation that supports transparent process replication (including failure detection, consistent message ordering among replicas, etc.), and a set of experimental and simulation results. Partial redundancy is studied in [9, 31] (in combination with coordinated checkpointing) to decrease the overhead associated to full replication. Adaptive redundancy is introduced in [15], where a subset of processes is dynamically selected for replication.

In Sect. 4.5 we provide a full-fledged theoretical analysis of the combination of process replication and checkpoint–recovery. While some theoretical results are provided in [13], they are based on an analogy between the process replication problem and the birthday problem. This analogy is appealing but, as seen in Sect. 4.5.1.1, does not make it possible to compute exact *MNFTI* (Mean Number of Failures To Interruption) and *MTTI* (Mean Time To Interruption) values. In addition, the authors use Daly's formula for the checkpointing period, even for Weibull or other distributions, simply using the mean of the distribution in the formula. This is a commonplace approach. However, a key observation is that using replication changes the optimal checkpointing period, even for Exponential distributions. This chapter provides the optimal value of the period for Exponential and Weibull distributions (either analytically or experimentally), taking into account the use of replication.

4.3 Models and Assumptions

We consider the execution of a tightly coupled parallel application, or *job*, on a large-scale platform composed of p processors. We use the term processor to indicate any individually scheduled compute resource (a core, a multi-core processor, a cluster node), so that our work is agnostic to the granularity of the platform. We assume that standard checkpoint–recovery is performed (with checkpointing either at the system level or at the application level, with some checkpointing overhead involved). At most one application process (replica) runs on one processor.

The job must complete $\mathcal{W}$ units of (divisible) work, which can be split arbitrarily into separate *chunks*. We define the work unit so that when the job is executed on a single processor one unit of work is performed in one unit of time. The job can be executed on any number $q \leq p$ processors. Defining $\mathcal{W}(q)$ as the time required for a failure-free execution on q processors, we consider three models:

- Perfectly parallel jobs: $\mathcal{W}(q) = \mathcal{W}/q$.
- Generic parallel jobs: $\mathcal{W}(q) = (1 - \gamma)\mathcal{W}/q + \gamma\mathcal{W}$. As in Amdahl's law [1], $\gamma < 1$ is the fraction of the work that is inherently sequential.
- Numerical kernels: $\mathcal{W}(q) = \mathcal{W}/q + \gamma\mathcal{W}^{2/3}/\sqrt{q}$, which is representative of a matrix product or a LU/QR factorization of size N on a 2D-processor grid, where $\mathcal{W} = O(N^3)$. In the algorithm in [2], $q = r^2$ and each processor receives $2r$ blocks of size N^2/r^2 during the execution; γ is the platform's communication-to-computation ratio.

Each participating processor is subject to *failures* that each cause a *downtime*. We do not distinguish between soft and hard failures, with the understanding that soft failures are handled via software rejuvenation (i.e., rebooting [5, 20]) and that hard failures are handled by processor sparing, a common approach in production systems. For simplicity we assume that a downtime lasts D time units, regardless of the failure type. After a downtime the processor is fault-free and begins a new lifetime. In the absence of replication, when a processor fails, the whole execution is stopped, and all processors must recover from the previous checkpointed state. The recovery lasts the time needed to restore the last checkpoint from persistent storage. We assume coordinated checkpointing [33] so that no message logging/replay is needed for recovery. We allow failures to happen during recovery or checkpointing, but not during downtime (otherwise, the downtime could be considered part of the recovery). We assume that processor failures are independent and identically distributed (i.i.d.). This assumption is commonplace in the literature because it makes analysis more tractable. In the real world, instead, failures are bound to be correlated. One source of failure correlation is the hierarchical structure of compute platforms (each rack comprises compute nodes, each compute node comprises processors, each processor comprises cores), which leads to simultaneous failures of groups of processors. Generalizing the theoretical results in this chapter to non-i.i.d. failures is an open question.

We let $C(q)$ denote the time needed to perform a checkpoint, and $R(q)$ the time needed to perform a recovery. Assuming that the memory footprint of an application checkpoint is V bytes, with each processor holding V/q bytes, we consider two scenarios:

- Proportional overhead: $C(q) = R(q) = \alpha V/q = C/q$ for some constant α. This is representative of cases where the bandwidth of the network card/link at each processor is the I/O bottleneck.
- Constant overhead: $C(q) = R(q) = \alpha V = C$, which is representative of cases where the bandwidth to/from the resilient storage system is the I/O bottleneck.

Since we consider tightly coupled parallel jobs, all q processors operate synchronously. These processors execute the same amount of work $\mathcal{W}(q)$ in parallel, chunk by chunk. The total time (on one processor) to execute a chunk of duration, or *size*, ω and then checkpoint it, is $\omega + C(q)$.

4.4 Group Replication

With group replication one executes multiple application instances on different processor groups. All groups compute the same chunk simultaneously, and do so until one of them succeeds, potentially after several failed trials. Then all other groups stop executing that chunk and recover from the checkpoint stored by the successful group. All groups then attempt to compute the next chunk. Group replication can be implemented easily with no modification to the application, provided that the recovery implementation allows a group to recover immediately from a checkpoint produced by another group. Hereafter we formalize group replication as an execution protocol we call ASAP (As Soon As Possible).

We consider g groups, where each group has q processors, with $g \times q \leq p$. A group is available for execution if and only if all its q processors are available. In case of a failure at a processor in a group the downtime of this group is a random variable $X_D(q) \geq D$. This random variable can take values strictly larger than D because while a processor in a group is experiencing a downtime another processor in that group can experience a failure, thus prolonging the groups' downtime beyond D seconds. If a group encounters a first processor failure at time t we say that the group is *down* between times t and $t + X_D(q)$.

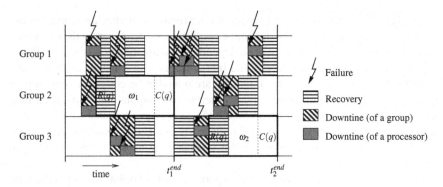

Fig. 4.1 Execution of chunks ω_1 and ω_2 (macro-steps 1 and 2) using the ASAP protocol. At time t_1^{end}, Group 1 is not ready, and Group 2 is the only one that does not need to recover

ASAP proceeds in k macro-steps, with a chunk of work processed during each macro-step. More formally, during macro-step j, $1 \leq j \leq k$, each group independently attempts to execute the jth chunk of size ω_j and to checkpoint it, restarting as soon as possible in case of a failure. As soon as one of the groups succeeds, say at time t_j^{end}, all the other groups are immediately stopped, macro-step j is over, and macro-step $(j + 1)$ starts (if $j < k$). The only two necessary inputs to the algorithm are (i) the number of chunks, k, and (ii) all chunk sizes, the ω_j's, chosen so that $\sum_{j=1}^{k} \omega_j = \mathscr{W}(q)$.

Before being able to start macro-step $(j + 1)$, a group that has been stopped must execute a recovery so that it can resume execution from the checkpoint saved by a successful group. Furthermore, this recovery may start later than time t_j^{end}, in the case where the group is down at time t_j^{end}. This is shown on an example execution in Fig. 4.1. At time t_1^{end}, Group 2 completes the computation and checkpointing of the chunk for macro-step 1. During that macro-step, Group 1 experiences two downtimes, each of duration D, while Group 3 experiences a single downtime of duration $> D$ due to a failure at a first processor followed by a failure at a second processor before the end of the first processor's downtime. At time t_1^{end}, Group 1 is down (experiencing a downtime caused by a sequence of three processor failures), so it cannot begin the recovery from the checkpoint saved by Group 2 immediately. Group 3, instead, can begin the recovery immediately a time t_1^{end}, but due to a failure it must reattempt the recovery. At time t_2^{end} it is Group 3 that completes the chunk for macro-step 2. The only groups that do not need to recover at the beginning of the next macro-step are the groups that were successful for the previous macro-step (except for the first macro-step for which all groups can start computing right away).

4.4.1 Exponential Failures

In this section we provide an analytical evaluation of ASAP assuming Exponential failures. More specifically, we are able to compute the optimal number of macro-steps

---

Algorithm 1: ASAP $(\omega_1, \ldots, \omega_k)$

---

for $j = 1$ *to* k **do**
    for *each group* **do in parallel**
        repeat
            Finish current downtime (if any)
            Try to perform a recovery, then a chunk of size ω_j, and finally to checkpoint
            if *execution successful* **then**
                Signal other groups to immediately stop their attempts
        until *one of the groups has a successful attempt*

---

k and the optimal values of the chunk sizes ω_j. Assume that individual processor failures are distributed following an Exponential distribution of parameter λ. For the sake of the theoretical analysis, we introduce a slightly modified version of the ASAP protocol in which all groups, including the successful ones, execute a recovery at the beginning of all macro-steps, including the first one. This version of ASAP is described in Algorithm 1. It is completely symmetric, which renders its analysis easier: for macro-step j to be successful, one of the groups must be up and running for a duration of $R(q) + \omega_j + C(q)$. Note however that all experiments reported in Sect. 4.4.4 use the original version of ASAP, without any superfluous recovery during execution (as depicted in Fig. 4.1).

Consider the jth macro-step, number the attempts of all groups by their start time, and let N_j be the index of the earliest started attempt that successfully computes chunk ω_j. Figure 4.2 zooms in on the execution of the second macro-step ($j = 2$). Each attempt is called Job_i in the order of its start time, and is followed by a downtime but for the last attempt, which is successful. In that example the successful computation of the chunk of size $R + \omega_2 + C$ is the fourth attempt, Job_4, executed by Group 3. Consequently, $N_2 = 4$, meaning that macro-step 2 requires 4 attempts. The duration of each attempt is the sum of a sample of two random variables X_i^j and Y_i^j, $1 \le i \le N_j$. X_i^j corresponds to the duration of the ith attempt at executing

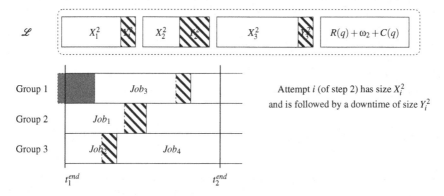

Fig. 4.2 Zoom on macro-step 2 of the execution depicted in Fig. 4.1, using the (X, Y) notation of Algorithm 2. Recall that Job_i has size $X_i^2 + Y_i^2$ for $1 \le i \le 3$, and Job_4 has size $R(q) + \omega_2 + C(q)$

Algorithm 2: Step j of ASAP $(\omega_1, \ldots, \omega_k)$

$i \leftarrow 1$ `/* number of attempts for the job */`

$\mathscr{L} \leftarrow \emptyset$ `/* list of attempts for the job */`

Sample X_i^j and Y_i^j using D_X and $D_{X_D(q)}$, respectively

while $X_i^j < R(q) + \omega_j + C(q)$ **do**

$\quad$ Add Job_i, with processing time $X_i^j + Y_i^j$, to $\mathscr{L}$

$\quad$ $i \leftarrow i + 1$

$\quad$ Sample X_i^j and Y_i^j using D_X and $D_{X_D(q)}$, respectively

$N_j \leftarrow i$

Add Job_{N_j}, with processing time $R(q) + \omega_j + C(q)$, to $\mathscr{L}$

$\quad$ `/* first successful job has size` $R(q)+\omega_j+C(q)$ `not` $X_{N_j}^j+Y_{N_j}^j$ `*/`

From time t_{j-1}^{end} on, execute a List Scheduling algorithm to distribute jobs in $\mathscr{L}$ to the different groups (recall that some groups may not be ready at time t_{j-1}^{end})

---

the chunk. Y_i^j corresponds to the duration of the ith downtime that follows the ith attempt (if $i \neq N_j$). Note that $X_i^j < R(q) + \omega_j + C(q)$ for $i < N_j$, and $X_{N_j}^j = R(q) + \omega_j + C(q)$. All the X_i^j's follow the same distribution D_X, an Exponential distribution of parameter $q\lambda$. And all the Y_i^j's follow the same distribution $D_{X_D}(q)$, that of the random variable $X_D(q)$ corresponding to the downtime of a group of q processors. The main idea is to view the N_j execution attempts as jobs, where the size of job i is $X_i^j + Y_i^j$, and to distribute them across the g groups using the classical online *list scheduling* algorithm for independent jobs [24, Sect. 5.6], as stated in the following proposition:

Proposition 4.1 *The jth ASAP macro-step can be simulated using Algorithm 2: the last job scheduled by Algorithm 2 ends exactly at time t_j^{end}.*

Proof The List Scheduling algorithm distributes the next job to the first available group. Because of the memoryless property of Exponential laws, it is equivalent (i) to generate the attempts a priori and greedily schedule them, or (ii) to generate them independently within each group.

Proposition 4.2 *Let $T_{truestart}^{(R(q)+\omega_j+C(q))}$ be the time elapsed between t_{j-1}^{end} and the beginning of Job_{N_j} (see Fig. 4.3). We have*

$$\mathbb{E}\left(T_{truestart}^{(R(q)+\omega_j+C(q))}\right) \leq \mathbb{E}(Y) + \frac{\mathbb{E}(N_j)\mathbb{E}(X) - \mathbb{E}(X_j^{N_j}) + (\mathbb{E}(N_j) - 1)\mathbb{E}(Y)}{g}$$

where X and Y are random variables corresponding to an attempt (sampled using D_X and $D_{X_D(q)}$ respectively). Moreover, we have $\mathbb{E}(N_j) = e^{\lambda q(R(q)+\omega_j+C(q))}$ and $\mathbb{E}(X_j^{N_j}) = \frac{1}{q\lambda} + R(q) + \omega_j + C(q)$.

Fig. 4.3 Notations used in
Proposition 4.2

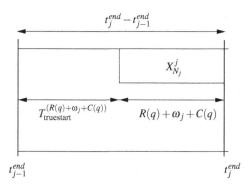

Proof For group x, $1 \le x \le g$, let $\tilde{Y}_x$ denote the time elapsed before it is ready for macro-step j. For example, in Fig. 4.2, we have $\tilde{Y}_1 > 0$ (group 1 is down at time t_{j-1}^{end}), while $\tilde{Y}_2 = \tilde{Y}_3 = 0$ (groups 2 and 3 are ready to compute at time t_{j-1}^{end}). Proposition 4.1 has shown that executing macro-step j can be simulated by executing a List Schedule on a job list $\mathscr{L}$ (see Algorithm 2). We now consider g "jobs" $\tilde{Job}_x$, $x = 1, \ldots, g$, so that $\tilde{Job}_x$ has duration $\tilde{Y}_x$. We now consider the augmented job list $\mathscr{L}' = \mathscr{L} \cup \bigcup_{x=1}^{g} \tilde{Job}_x$. Note that $\mathscr{L}'$ may contain more jobs than macro-step j: the jobs that start after the successful job Job_{N_j} are discarded from the list $\mathscr{L}'$. However, both schedules have the same makespan, and jobs common to both systems have the same start and completion dates. Thus, we have $T_{truestart}^{(R(q)+\omega_j+C(q))} \le \frac{\sum_{x=1}^{g}(\tilde{Y}_x)+\sum_{i=1}^{N_j-1}(X_i^j+Y_i^j)}{g}$: this key inequality is due to the property of list scheduling: the group which is assigned the last job is the least loaded when this assignment is decided, hence its load does not exceed the average load (which is the total load divided by the number of groups). Given that $\mathbb{E}(\tilde{Y}_x) \le \mathbb{E}(Y)$, we derive

$$\mathbb{E}\left(T_{truestart}^{(R(q)+\omega_j+C(q))}\right) \le \mathbb{E}(Y) + \frac{\mathbb{E}\left(\sum_{i=1}^{N_j-1} X_i^j\right) + \mathbb{E}\left(\sum_{i=1}^{N_j-1}(Y_i^j)\right)}{g}$$

But N_j is the stopping criterion of the (X_i^j) sequence; hence, using Wald's theorem [26], we have $\mathbb{E}(\sum_{i=1}^{N_j} X_i^j) = \mathbb{E}(N_j)\mathbb{E}(X)$ and $\mathbb{E}(\sum_{i=1}^{N_j-1} X_i^j) = \mathbb{E}(N_j) \mathbb{E}(X) - \mathbb{E}(X_j^{N_j})$. Moreover, as N_j and Y_i^j are independent variables, we have $\mathbb{E}(\sum_{i=1}^{N_j-1} Y_i^j) = (\mathbb{E}(N_j) - 1)\mathbb{E}(Y)$, and we get the desired bound for $\mathbb{E}(T_{truestart}^{(R(q)+\omega_j+C(q))})$. Finally, as the expected number of attempts when repeating independently until success an event of probability α is $\frac{1}{\alpha}$ (geometric law), we get $\mathbb{E}(N_j) = e^{\lambda q(R(q)+\omega_j+C(q))}$. The value of $\mathbb{E}(X_j^{N_j})$ can be directly computed from the definition, recalling that $X_j^{N_j} \ge R(q) + \omega_j + C(q)$ and each X_j^i follows an Exponential distribution of parameter $q\lambda$.

Theorem 4.1 *The expected makespan of ASAP has the following upper bound:*
$$\frac{g-1}{g}\mathscr{W}(q) + \frac{1}{g}\left(\frac{1}{q\lambda} + \mathbb{E}(Y)\right)e^{\lambda q(R(q)+C(q))}k^*e^{\lambda q\frac{\mathscr{W}(q)}{k^*}} + k^*\left(\frac{g-1}{g}(\mathbb{E}(Y)+R(q)+C(q)) - \frac{1}{g}\frac{1}{q\lambda}\right),$$
where Y is a random variable with distribution $D_{X_D(q)}$. This bound is obtained when using $k^* = \max(1, \lfloor k_0 \rfloor)$ or $k^* = \lceil k_0 \rceil$ same-size chunks, whichever leads to the smaller value, where

$$k_0 = \frac{\lambda q \mathscr{W}(q)}{1 + \mathbb{L}\left(\left(g - 1 + \frac{(g-1)q\lambda(R(q)+C(q))-g}{1+q\lambda\mathbb{E}(Y)}\right)e^{-(1+\lambda q(R(q)+C(q)))}\right)}.$$

$\mathbb{L}$, *the Lambert function, is defined as* $\mathbb{L}(z)e^{\mathbb{L}(z)} = z$.

Proof From Proposition 4.2, the expected execution time of ASAP has upper bound $T_{ASAP} = \sum_{j=1}^{k}\alpha_j$, where

$$\alpha_j = \mathbb{E}(Y) + \frac{\mathbb{E}(N_j)\mathbb{E}(X) - \mathbb{E}(X_j^{N_j}) + (\mathbb{E}(N_j) - 1)\mathbb{E}(Y)}{g} + (R(q) + \omega_j + C(q)).$$

Our objective now is to find the inputs to the ASAP algorithm, namely the number k of macro-steps together with the chunk sizes $(\omega_1, \ldots, \omega_k)$, that minimize this T_{ASAP} bound.

We first have to prove that any optimal (in expectation) policy uses only a finite number of chunks. Let α be the expectation of the ASAP makespan using a unique chunk of size $\mathscr{W}(q)$. According to Proposition 4.2,

$$\alpha = \mathbb{E}\left(T_{\text{truestart}}^{(R(q)+\mathscr{W}(q)+C(q))}\right) + C(q) + \mathscr{W}(q) + R(q),$$

and is finite. Thus, if an optimal policy uses k^* chunks, we must have $k^*C(q) \le \alpha$, and thus k^* is bounded.

In the proof of Theorem 1 in [3], we have shown that any deterministic strategy uses the same sequence of chunk sizes, whatever the failure scenario, thanks to the memoryless property of the Exponential distribution. We cannot prove such a result in the current context. For instance, the number of groups performing a downtime at time t_1^{end} depends on the scenario. There is thus no reason a priori for the size of the second chunk to be independent of the scenario. To overcome this difficulty, we restrict our analysis to strategies that use the same sequence of chunk sizes whatever the failure scenario. We optimize T_{ASAP} in that context, at the possible cost of finding a larger upper bound.

We thus suppose that we have a fixed number of chunks, k, and a sequence of chunk sizes $(\omega_1, \ldots, \omega_k)$, and we look for the values of $(\omega_1, \ldots, \omega_k)$ that minimize $T_{ASAP} = \sum_{j=1}^{k}\alpha_j$. Let us first compute one of the α_j term. Replacing $\mathbb{E}(N_j)$ and $\mathbb{E}(X_j^{N_j})$ by the values given in Proposition 4.2, and $\mathbb{E}(X)$ by $\frac{1}{q\lambda}$, we get

$$\alpha_j = \frac{g-1}{g}\omega_j + \frac{1}{g}e^{\lambda q(R(q)+\omega_j+C(q))}\left(\frac{1}{q\lambda} + \mathbb{E}(Y)\right)$$

$$+ \frac{g-1}{g}\left(\mathbb{E}(Y) + R(q) + C(q)\right) - \frac{1}{g}\frac{1}{q\lambda}$$

$$T_{ASAP} = \frac{g-1}{g}\mathcal{W} + \frac{1}{g}\left(\frac{1}{q\lambda} + \mathbb{E}(Y)\right)e^{\lambda q(R(q)+C(q))}\sum_{j=1}^{k}e^{\lambda q\omega_j}$$

$$+ k\left(\frac{g-1}{g}\left(\mathbb{E}(Y) + R(q) + C(q)\right) - \frac{1}{g}\frac{1}{q\lambda}\right)$$

By convexity, the expression $\sum_{j=1}^{k}e^{\lambda q\omega_j}$ is minimal when all ω_j's are equal (to $\mathcal{W}(q)/k$). Hence all the chunks should be equal for T_{ASAP} to be minimal. We obtain:

$$T_{ASAP} = \frac{g-1}{g}\mathcal{W} + \frac{1}{g}\left(\frac{1}{q\lambda} + \mathbb{E}(Y)\right)e^{\lambda q(R(q)+C(q))}ke^{\lambda q\frac{\mathcal{W}(q)}{k}}$$

$$+ k\left(\frac{g-1}{g}\left(\mathbb{E}(Y) + R(q) + C(q)\right) - \frac{1}{g}\frac{1}{q\lambda}\right).$$

Let $f(x) = \tau_1 x e^{\lambda q\frac{\mathcal{W}(q)}{x}} + \tau_2 x$, where

$$\tau_1 = \frac{1}{g}\left(\frac{1}{q\lambda} + \mathbb{E}(Y)\right)e^{\lambda q(R(q)+C(q))} \qquad \text{and}$$

$$\tau_2 = \left(\frac{g-1}{g}\left(\mathbb{E}(Y) + R(q) + C(q)\right) - \frac{1}{g}\frac{1}{q\lambda}\right).$$

A simple analysis using differentiation shows that f has a unique minimum, and solving $f'(x) = 0$ leads to $\tau_1 e^{\lambda q\frac{\mathcal{W}(q)}{k}}\left(1 - \frac{\lambda q\mathcal{W}(q)}{k}\right) + \tau_2 = 0$, and thus to $k = \frac{\lambda q\mathcal{W}(q)}{1+\mathbb{L}\left(\frac{\tau_2}{\tau_1\cdot e}\right)} = k^*$, which concludes the proof.

This theorem can in turn be used to compute numerically the number of chunks and an upper bound on the expected makespan, provided that $\mathbb{E}(Y) = \mathbb{E}(X_D(q))$ can be itself bounded. The following proposition provides such a bound:

Proposition 4.3 *Let $X_D(q)$ denote the downtime of a group of q processors. Then*

$$D \leq \mathbb{E}(X_D(q)) \leq \frac{e^{(q-1)\lambda D} - 1}{(q-1)\lambda}. \tag{4.1}$$

Proof In [3], we have shown that the optimal expectation of the makespan is computed as:

$$\mathbb{E}^*(q) = K^*(q) \left(\frac{1}{q\lambda} + \mathbb{E}(T_{rec}(q)) \right) \left(e^{\frac{q\lambda \mathcal{W}(q)}{K^*(q)} + q\lambda C(q)} - 1 \right) \qquad (4.2)$$

where $\mathbb{E}(T_{rec}(q))$ denotes the expectation of the recovery time, i.e., the time spent recovering from failure during the computation of a chunk. All chunks have the same recovery time because they all have the same size and because of the memoryless property of the Exponential distribution. It turns out that although we can compute the optimal number of chunks (and thus the chunk size), we cannot compute $\mathbb{E}^*(q)$ analytically because $\mathbb{E}(T_{rec}(q))$ is difficult to compute. We write the following recursion:

$$T_{rec}(q) = \begin{cases} X_D(q) + R(q) & \text{if no processor fails} \\ & \text{during } R(q) \text{ units of time,} \quad (4.3) \\ X_D(q) + T_{lost}(R(q)) + T_{rec}(q) & \text{otherwise.} \end{cases}$$

$X_D(q)$ is the downtime of a group of q processors, that is, the time between the first failure of one of the processors and the first time at which all of them are available (accounting for the fact a processor can fail while another one is down, thus prolonging the downtime). $T_{lost}(R(q))$ is the amount of time spent computing by these processors before a first failure, knowing that the next failure occurs within the next $R(q)$ units of time. In other terms, the compute time is wasted because checkpoint recovery was not completed. The time until the next failure of a group of q processors is the minimum of q iid Exponential random variables, and is thus Exponential with parameter $q\lambda$. We can compute $\mathbb{E}(T_{lost}(R(q))) = \frac{1}{q\lambda} - \frac{R(q)}{e^{q\lambda R(q)} - 1}$ (see [3] for details). Plugging this value into Eq. 4.3 leads to:

$$\mathbb{E}(T_{rec}(q)) = e^{-q\lambda R(q)} (\mathbb{E}(X_D(q)) + R(q))$$
$$+ (1 - e^{-q\lambda R(q)}) \left(\mathbb{E}(X_D(q)) + \frac{1}{q\lambda} - \frac{R(q)}{e^{q\lambda R(q)} - 1} + \mathbb{E}(T_{rec}(q)) \right)$$
$$(4.4)$$

Equation 4.4 reads as follows: after the downtime $X_D(q)$, either the recovery succeeds for everybody, or there is a failure during the recovery and another attempt must be made. Both events are weighted by their respective probabilities. Simplifying the above expression we get:

$$\mathbb{E}(T_{rec}(q)) = \mathbb{E}(X_D(q))e^{q\lambda R(q)} + \frac{1}{q\lambda}(e^{q\lambda R(q)} - 1) \qquad (4.5)$$

Plugging back this expression in Eq. 4.2, we obtain the Equation:

$$\mathbb{E}^*(q) = K^*(q) \left(\frac{1}{q\lambda} + \mathbb{E}(X_D(q)) \right) e^{q\lambda R(q)} \left(e^{\frac{q\lambda \mathcal{W}(q)}{K^*(q)} + q\lambda C(q)} - 1 \right) \qquad (4.6)$$

Now we establish the desired bounds on $\mathbb{E}(X_D(q))$ We always have $X_D(q) \geq X_D(1) \geq D$, hence the lower bound. For the upper bound, consider a date at which one of the q processors, say processor i_0, just had a failure and initiates its downtime period for D time units. Some other processors might be in the middle of their downtime period: for each processor i, $1 \leq i \leq q$, let t_i denote the remaining duration of the downtime of processor i. We have $0 \leq t_i \leq D$ for $1 \leq i \leq q$, $t_{i_0} = D$, and $t_i = 0$ means that processor i is up and running. Let $X_D^{t_1,\ldots,t_q}(q)$ be the *remaining* downtime of a group of q processors, knowing that processor i, $1 \leq i \leq q$, will still be down for a duration of t_i, and that a failure just happened (i.e., there exists i_0 such that $t_{i_0} = D$). Given the values of the t_i's, we have the following equation for the random variable $X_D^{t_1,\ldots,t_q}(q)$:

$$X_D^{t_1,\ldots,t_q}(q) = \begin{cases} D & \text{if none of the processors of the group fails during} \\ & \text{the next } D \text{ units of time} \\ T_{lost}^{t_1,\ldots,t_q}(D) + X_D^{t'_1,\ldots,t'_q}(q) & \text{otherwise.} \end{cases}$$

In the second case of the equation, consider the next D time units. Processor i can only fail in the *last* $D - t_i$ of these time units. Here the values of the t'_i's depend on the t_i's and on $T_{lost}^{t_1,\ldots,t_q}(D)$. Indeed, except for the last processor to fail, say i_1, for which $t'_{i_1} = D$, we have $t'_i = \max\{t'_i - T_{lost}^{t_1,\ldots,t_q}(D), 0\}$. More importantly, we always have $T_{lost}^{t_1,\ldots,t_q}(D) \leq T_{lost}^{D,0,\ldots,0}(D)$ and $X_D^{t_1,\ldots,t_q}(q) \leq X_D^{D,0,\ldots,0}(q)$ because the probability for a processor to fail during D time units is always larger than that to fail during $D - t_i$ time units. Thus, $\mathbb{E}(X_D^{t_1,\ldots,t_q}(q)) \leq \mathbb{E}(X_D^{D,0,\ldots,0}(q))$. Following the same line of reasoning, we derive an upper bound for $X_D^{D,0,\ldots,0}(q)$:

$$X_D^{D,0,\ldots,0}(q) \leq \begin{cases} D & \text{if none of the q-1 running processors of the group} \\ & \text{fails during the downtime } D \\ T_{lost}^{D,0,\ldots,0}(D) + X_D^{D,0,\ldots,0}(q) & \text{otherwise.} \end{cases}$$

Weighting both cases by their probability and taking expectations, we obtain

$$\mathbb{E}\left(X_D^{D,0,\ldots,0}(q)\right)$$
$$\leq e^{-(q-1)\lambda D}D + (1 - e^{-(q-1)\lambda D})\left(\mathbb{E}\left(T_{lost}^{D,0,\ldots,0}(D)\right) + \mathbb{E}\left(X_D^{D,0,\ldots,0}(q)\right)\right)$$

Hence, $\mathbb{E}\left(X_D^{D,0,\ldots,0}(q)\right) \leq D + (e^{(q-1)\lambda D} - 1)\mathbb{E}\left(T_{lost}^{D,0,\ldots,0}(D)\right)$, with

$$\mathbb{E}\left(T_{lost}^{D,0,\ldots,0}(D)\right) = \frac{1}{(q-1)\lambda} - \frac{D}{e^{(q-1)\lambda D} - 1}.$$

We derive

$$\mathbb{E}\left(X_D^{t_1,\dots,t_q}(q)\right) \le \mathbb{E}\left(X_D^{D,\dots,0}(q)\right) \le \frac{e^{(q-1)\lambda D} - 1}{(q-1)\lambda} .$$

which concludes the proof. As a sanity check, we observe that the upper bound is at least D, using the identity $e^x \ge 1 + x$ for $x \ge 0$.

4.4.2 General Failures

The analytical derivations in Sect. 4.4.1 hold only for Exponential failures. In the case of non-Exponential failures we propose two algorithms for determining an execution of ASAP that achieves good makespan in practice: a "brute-force" approach called BESTPERIOD and a Dynamic Programming approach called DPNEXTFAILURE.

4.4.2.1 Brute-Force Algorithm

The BESTPERIOD algorithm enforces a periodic execution of ASAP, meaning that all chunk sizes are identical. For a given number of groups, the period is computed via a numerical search among a set of candidate periods generated as follows. The work in [3] makes it possible to compute an optimal period, τ, for an application executed without replication on n processors subjected to Exponential failures. In our case, with g groups and p processors, we compute this period for $n = \lfloor p/g \rfloor$ processors. Besides τ, we then generate 360 candidates as $\tau(1 + 0.05 \times i)$ and $\tau/(1 + 0.05 \times i)$ for $i \in \{1, \dots, 180\}$, and 120 candidates as $\tau \times 1.1^j$ and $\tau/1.1^j$ for $j \in \{1, \dots, 60\}$, for a total of 481 candidate periods. We then evaluate each candidate period in simulation (see Sect. 4.4.3 for details on our simulation methodology) over 50 randomly generated experimental scenarios. We pick the candidate period that achieves the best average makespan over these 50 scenarios.

BESTPERIOD has two potential drawbacks. First, it enforces a periodic execution, even though there is no theoretical reason why the optimal should correspond to a periodic execution if failures are non-Exponential. Second, it requires running a large number of simulations ($50 \times 481 = 24,050$). With our current implementation each individual set of 481 simulations requires between 3 and 24 min on one core of a Quad-core AMD Opteron running at 2400 MHz. While this may indicate that BESTPERIOD is impractical, when compared to application makespans that can be several days the overhead of searching for the period may not be significant. Furthermore, the search for the period can be done in parallel since all simulations are independent. The search for the best period to execute an application on a large-scale platform can thus be done in a few seconds on that same large-scale platform.

4.4.2.2 Dynamic Programming Algorithm

As an alternative to the brute-force algorithm in the previous section, one can resort to Dynamic Programming (DP). We initially developed a DP algorithm to compute chunk sizes for each group at each step of the application execution. Even though this seems like a natural approach, it is only tractable (in terms of number of DP states) if the chunk sizes for each group are computed independently of those for the other groups. As a result, we found that the resulting algorithm does not achieve good results in practice.

Algorithm 3: DPNextCheckpoint(W, T, T_0, τ_1, ..., τ_{gq})

---

1 **if** $W = 0$ **then return** 0 $best\_work \leftarrow 0$;
2 $next\_chkpt \leftarrow T$;
3 $(W_1, ..., W_g) \leftarrow$ WorkAlreadyDone(T);
  ;   /* Work done since last recovery or checkpoint */
4 Sort groups by non-increasing of work done (W_1 is maximum);
5 **for** $t = T$ *to* $T + W - W_g$ *step* quantum; /* Loop on checkpointing date */
6 **do**
7 $\quad$ $cur\_work \leftarrow 0$;
8 $\quad$ **for** $x = 1$ *to* g; /* Loop on the first group to successfully work until $t + C(q)$ */
9 $\quad$ **do**
10 $\quad\quad$ $\delta \leftarrow (t + C(q)) - T_0$;   /* Total time elapsed until the checkpoint completion */
11 $\quad\quad$ $proba \leftarrow \left(\prod_{y=1}^{x-1} P_{fail}(\tau_{(y-1)q+1} + \delta, ..., \tau_{(y-1)q+q} + \delta \mid \tau_{(y-1)q+1}, ..., \tau_{(y-1)q+q}) \right)$;
$\quad\quad\quad\quad$ $\times P_{suc}(\tau_{(x-1)q+1} + \delta, ..., \tau_{(x-1)q+q} + \delta \mid \tau_{(x-1)q+1}, ..., \tau_{(x-1)q+q})$
12 $\quad\quad$ $\omega \leftarrow \min\{W - W_x, t - T\}$;
$\quad\quad$;   /* Work done between T and t by group x */
13 $\quad\quad$ $(rec\_\omega, rec\_t) \leftarrow$ DPNextCheckpoint($W - W_x - \omega$, $T + \omega + C(q) + R(q)$, T_0, τ_1, ..., τ_{gq});
14 $\quad\quad$ $cur\_work \leftarrow cur\_work + proba \times (W_x + \omega + rec\_\omega)$
15 $\quad$ **if** $cur\_work > best\_work$ **then**
16 $\quad\quad$ $best\_work \leftarrow cur\_work$;
17 $\quad\quad$ $next\_chkpt \leftarrow t$
18 **return** $(best\_work, next\_chkpt)$

When faced with an exponential number of DP states when using DP to minimize expected makespan, an alternate goal is to maximize the expected amount of completed work before the next failure [3]. We generalize this idea to the context of replication, doing away with the concept of chunk sizes altogether. More specifically, since the first failure only interrupts a single group, the objective is to maximize the expected amount of work completed before all groups have failed. This can be achieved with the DP algorithm presented hereafter. We make one simplifying assumption: we ignore that once a group has failed, it will eventually restart and resume computing. This is because keeping track of such restarts would again lead to an exponential number of DP states. The hope is that our approach will work well in spite of this simplifying assumption.

Our DP algorithm, DPNEXTCHECKPOINT, is shown in Algorithm 3. It does *not* define chunk sizes, i.e., amounts of work to be processed before a checkpoint is taken, but instead it defines *checkpoint dates*. The rationale is that one checkpoint date can correspond to different amounts of work for each group, depending on when the group has started to process its chunk, after either its last failure and recovery, or its last checkpoint, or its last recovery from another group's checkpoint. Input to the algorithm is the amount of work that remains to be done (W), the current time (T), the time at which the application started (T_0), and the times since the latest failure at each processor before time T_0 (the τ_i's). The output is the next checkpoint date and the expected amount of work completed before the next failure occurs.

Algorithm 4: DPNEXTFAILURE(W).

---

for *each group $x = 1$ to g* **do in parallel**
    while $W \neq 0$ **do**
        $(\tau_1, \ldots, \tau_{gq}) \leftarrow$ ALIVE($1, \ldots, gq$)
        $T_0 \leftarrow$ TIME()   `/* Current time */`
        $(work, date) \leftarrow$ DPNEXTCHECKPOINT(W, T_0, T_0, $\tau_1, \ldots, \tau_{gq}$)
        Signal all processors that the next checkpoint date is now *date*
        Try to work until *date* and then checkpoint
        if *successful work until date and checkpoint* **then**
            Let y be the longest running group without failure among the successful groups
            Let ω be the work performed by y since its last recovery or checkpoint
            $W \leftarrow W - \omega$
            if *group x's last recovery or checkpoint was strictly later than that of y* **then**
                Perform a recovery
        if *failure* **then** Complete downtime **if** *failure* or *signal* **then** Perform recovery from
        last successfully completed checkpoint

DPNEXTCHECKPOINT proceeds as follows. At Line 3 a function is called (WORK-ALREADYDONE) which returns, for each group, the time since it has started processing its current chunk (i.e., the amount of work it has done to date). The groups are sorted in decreasing order of work performed to date (Line 4). The algorithm then picks the next checkpoint date for all possible dates between the current time T and time $T + W - W_g$, i.e., the time at which the last group would finish computing if no failure were to occur (Line 5). At the checkpointing date, the amount of work completed is the maximum of the amount of work done by the different groups that successfully complete the checkpoint. Therefore, we consider all the different cases (Line 11), that is, which group x, among the successful groups, has done the most work. We compute the probability of each case (Line 11). All groups that started to work earlier than group x have failed (i.e., at least one processor in each of them has failed) but not group x (i.e., none of its processors have failed). We compute the expectation of the amount of work completed in each case (Lines 12 and 13). We then sum the contributions of all the cases (Line 14) and record the checkpointing date leading to the largest expectation (Line 15). Note that the probability computed at Line 11 explicitly states which groups have successfully completed the checkpoint, and which groups have not. We choose not to take this information into account when computing the expectation (recursive call at Line 13) so as to avoid keeping track of which groups have failed, thereby lowering the complexity of the dynamic program. This is why the conditions do not evolve in the conditional probability at Line 11.

Algorithm 4 shows the overall algorithm, DPNEXTFAILURE, which uses DP-NEXTCHECKPOINT (the ALIVE function returns, for a list of processors, the amount of time each has been up and running since its last downtime). Each time a group is affected by an event (a failure, a successful checkpoint by itself or by another group), it computes the next checkpoint date and broadcasts it to the g group leaders. Hence, a group may have computed the next checkpoint date to be t, and that date can be either unmodified, postponed, or advanced by events occurring at other groups and by their recomputation of the best next checkpoint date. In practice, as time is discretized, at each time quantum a group can check whether the current date is a checkpoint date or not.

Both Algorithms 3 and 4 have a complexity in $O\left(gq\left(\frac{W}{\text{quantum}}\right)^2\right)$. The gq factor comes from the computation of the probabilities at Line 11 in Algorithm 3. This complexity can be lowered using the methodology outlined in [3].

4.4.3 Simulation Methodology

In this section we detail our simulation methodology.

4.4.3.1 Evaluated Algorithms

Our simulator implements two versions of the ASAP protocol in the case of exponentially distributed failures. The first version, OPTEXP, simply uses for each group the optimal and periodic policy outlined in Sect. 1.3.1 (Proposition 1.1 and its discussion) and formally established in [3] for Exponential failure distributions and no replication. To use OPTEXP with g groups we use the period from [3] computed with $\lfloor p/g \rfloor$ processors. The second, OPTEXPGROUP, uses the periodic policy defined by Theorem 4.1. Both OPTEXP and OPTEXPGROUP compute the checkpointing period based solely on the MTBF, assuming that failures are exponentially distributed. We nevertheless include them in all our experiments, simply using the MTBF value even when failures are not exponentially distributed. The simulator also implements BESTPERIOD (Sect. 4.4.2.1) and DPNEXTFAILURE (Sect. 4.4.2.2). Based on the results in [3], we do not consider any additional checkpointing policy, such as those defined by Young [36] or Daly [7] for instance.

4.4.3.2 Failure Distributions

To choose failure distribution parameters that are representative of realistic systems, we use failure statistics from the Jaguar platform. Jaguar contained 45,208 processors and is said to have experienced on the order of 1 failure per day [37]. Assuming a 1-day platform MTBF leads to a processor MTBF equal to $\frac{45,208}{365} \approx 125$ years. We generate both Exponential and Weibull failures, the former serving as a best case yet unrealistic scenario and the latter being representative of failure behavior in production systems [17, 18, 22, 29]. For the Exponential distribution of failure inter-arrival times, we simply set $\lambda = \frac{1}{MTBF}$. For the Weibull distribution, which requires two parameters, a shape parameter k and a scale parameter λ, and has density $\frac{k}{\lambda}(\frac{x}{\lambda})^{k-1}e^{-(x/\lambda)^k}$ for $x \geq 0$, we have $\lambda = MTBF/\Gamma(1 + 1/k)$. Based on the results in [17, 18, 22, 29] we use $k = 0.5$ and $k = 0.7$. For small values of the shape parameter k, the Weibull distribution is far from an Exponential distribution, meaning that it is far from being memoryless.

4.4.3.3 Platform and Job Parameters

We consider platforms containing from 32,768 to 4,194,304 processors. We determine the job size $\mathcal{W}$ so that a job using the whole platform would use it for a significant amount of time in the absence of failures, namely ≈ 21 h on the largest platforms ($\mathcal{W} = 10,000$ years). In experiments we use $D = 60$ s, and $C = R = 60$ s, 600 s, and 6000 s, thus spanning the spectrum from relatively fast to relatively slow checkpointing/recovery. We also ran experiments with a very short $C = R = 6$ s, but the results are virtually identical to those obtained with $C = R = 60$ s and we do not present them. Finally, for all experiments we use $\gamma = 10^{-6}$ for generic parallel jobs, and $\gamma = 0.1$ for numerical kernels (see Sect. 4.3).

4.4.3.4 Generation of Failure Scenarios

Given a p-processor job, a failure trace is a set of failure dates for each processor over a fixed time horizon h (set to 2 years). The job start time is assumed to be 1 year. We use a nonzero start time to avoid side effects related to the synchronous initialization of all nodes/processors. Given the distribution of inter-arrival times at a processor, for each processor we generate a trace via independent sampling until the target time horizon is reached.

4.4.4 Simulation Results

In this section, we only present simulation results for perfectly parallel applications under the constant overhead model (see Sect. 4.3). All trends and conclusions are similar regardless of the application and overhead models. All results are averages over at least 50 instances, and all graphs show one-standard-deviation error bars.

4.4.4.1 Exponential Failures

Figure 4.4 shows the average makespan versus the number of processors for our algorithms, using $g = 1, 2$, or 3 groups, and assuming Exponential failures. A first observation is that many curves overlap each other: for a given g all algorithms lead to similar average makespan. For instance, for $C = R = 600$ s and $g = 2$, and taking OPTEXP as a reference, the relative difference between the average makespan of OPTEXP and that of the other three algorithms is at most 6.81% (and only 2.31% when averaged over all considered numbers of processors). In spite of such small differences, several trends emerge. OPTEXP almost always leads to higher average makespan than OPTEXPGROUP (note that for $g = 1$ the two algorithms are equivalent). Over the 8 numbers of processors considered, the 3 values for $R = C$, and the 3 values for g, i.e., 72 scenarios, OPTEXP leads to average makespans shorter than

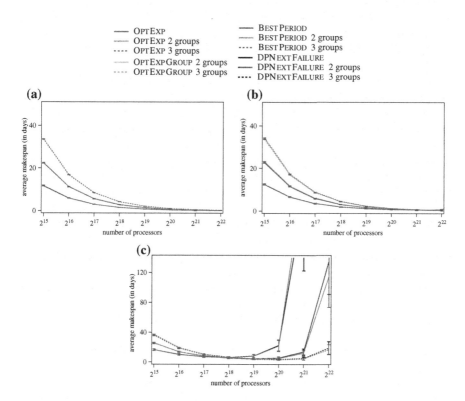

Fig. 4.4 Average makespan versus number of processors, Exponential failures, $MTBF = 125$ years. **a** $C = R = 60$ s. **b** $C = R = 600$ s. **c** $C = R = 6000$ s

that of OPTEXPGROUP only 4 times (for $R = C = 6000$ s, for 2^{18} to 2^{21} processors, and by at most 3.27 %). BESTPERIOD never leads to an average makespan higher than that of OPTEXP or OPTEXPGROUP, and outperforms them by up to several percent across all the $R = C$ and g values. DPNEXTFAILURE leads to mixed results, with equal or shorter average makespan than OPTEXPGROUP, resp. BESTPERIOD, for 31, resp. 24, of the 72 different scenarios.

A second observation is that the use of $g > 1$ (i.e., multiple groups) often does not help and can even lead to larger average makespans. For $R = C = 60$ s, increasing g from 1 to 2, or from 2 to 3, never leads to a lower average makespan for any of our algorithms. For $R = C = 600$ s, the only improvements are seen when going from 1 to 2 groups, for the OPTEXP, OPTEXPGROUP, and BESTPERIOD algorithms, and only with more than 2^{21} processors. The relative improvements are at most 7.75 % for 2^{21} processors, and between 25.40 and 41.09 % for 2^{22} processors. No improvements are achieved when going from 2 to 3 groups. More improvements are seen for $C = R = 6000$ s. When going from 1 to 2 groups, improvements are achieved starting at 2^{18} processors, with improvements up to between 93.64 and 95.17 % at large scale, for all four algorithms. When going from 2 to 3 groups, relative improvements are seen starting at 2^{19} processors, reaching up to between 85.09 and 85.78 % for all four algorithms.

For low and moderate checkpointing overheads, $C = R = 60$ s or 600 s, the average makespan decreases as the number of processors increases. Instead, for high checkpointing overheads, $C = R = 6000$ s, the average makespan initially decreases but starts increasing at large scale. This is particularly noticeable when using $g = 1$ group. For instance, the average makespan using OPTEXP goes from 21.83 s with 2^{20} processors to 249.39 s with 2^{21} processors, or an increase by a factor 11.42. The increase is similar with BESTPERIOD and marginally lower with DPNEXTFAILURE (a factor 9.72). The reason for this makespan increase is simply that with a high checkpointing overhead, the parallel efficiency is low as processors spend more time in checkpointing activities than in actual computation. This observation is precisely the motivation for using $g > 1$ (see Sect. 3.1). With $g = 2$, we still see increases in average makespans, but only by a factor between 2.46 and 2.53 when going from 2^{20} processors to 2^{21} processors for all algorithms. With $g = 3$, this factor is between 1.34 and 1.39 for all algorithms. Therefore, the use of group replication improves parallel efficiency and can lead to scalability improvements. For instance, with $g = 1$ or $g = 2$, regardless of the algorithm in use, it is not advisable to use 2^{20} processors as the makespan is lower when using 2^{19} processors. With $g = 3$, instead, there is a reduction in average makespan when going from 2^{19} processors to 2^{20} processors for all our algorithms (the relative percentage reductions are between 14.58 and 18.81 %).

Based on the above, we conclude that for Exponential failures group replication can be useful when the checkpointing overhead is relatively large and/or when the scale of the execution is large. While large checkpointing overheads decrease parallel efficiency, the use of group replication makes it possible to limit this decrease or even to increase parallel efficiency at some scales. All our algorithms lead to comparable performance, with BESTPERIOD leading to good results, even though marginally outperformed by DPNEXTFAILURE in some instances. While these results are interesting, although Exponential failures have been studied in all previously published works, their relevance to practice is not clear given that real-world failures follow non-memoryless distributions. In the next section we present results for Weibull failure distributions.

4.4.4.2 Weibull Failures

Figure 4.5 show results for Weibull failures with $k = 0.7$ and $k = 0.5$, respectively. For low $R = C = 60$ s and for $k = 0.7$ (Fig. 4.5a), results are similar to those seen for Exponential failures: the use of multiple groups does not help, and all algorithms lead to sensibly the same performance. The gaps between the algorithms become larger for $k = 0.5$, i.e., when the failure distribution is farther from the Exponential distribution, with the advantage to BESTPERIOD (Fig. 4.5b). For instance, for $k = 0.5$, 2^{20} processors, and using $g = 2$ groups, BESTPERIOD leads to an average makespan lower than that of OPTEXP, OPTEXPGROUP, and DPNEXTFAILURE by 10.46, 51.04, and 2.08 %, respectively. A general observation in all the results for replication ($g > 1$) with Weibull failures, regardless of the value of $C = R$, is that OPTEXPGROUP leads to much poorer results than all the other algorithms. This is because the analytical development of Theorem 4.1 relies heavily on the Exponential failure assumption. As a result, OPTEXPGROUP is even outperformed by OPTEXP, even though this algorithm also assumes Exponential failures. In all that follows we no longer discuss the results for OPTEXPGROUP.

For $C = R = 600$ s and $k = 0.7$, and unlike the results for Exponential failures, at large scale the average makespan of the $g = 1$ executions increases sharply while the average makespans for $g > 1$ executions remain more stable (Fig. 4.5c). In other words, even when checkpointing overheads are moderate, group replication is useful for increasing parallel efficiency once the scale is large enough. This result is amplified when failures are further from being Exponential, i.e., for $k = 0.5$ (Fig. 4.5d). For $k = 0.5$, going from $g = 1$ to $g = 2$ groups is beneficial for OPTEXP starting at 2^{17} processors and for BESTPERIOD and DPNEXTFAILURE starting at 2^{18} processors. Going from $g = 2$ to $g = 3$ groups is beneficial for OPTEXP and BESTPERIOD starting at 2^{19} processors, and for DPNEXTFAILURE starting at 2^{20} processors. In terms of comparing the algorithms with each other, in Figure 4.5d all algorithms experience a makespan increase after the initial decrease. Only BESTPERIOD and DPNEXTFAILURE, when using $g = 3$ groups, have a decreasing makespan up to 2^{20} processors. When going to 2^{21} processors, these algorithms lead to relative increases in makespan of 18.50 and 14.99 %, and larger increases when going from 2^{21} to 2^{22} processors. Across the board, BESTPERIOD with $g = 3$ groups leads to the lowest average makespan, with DPNEXTFAILURE with $g = 3$ groups a close second. The average makespan of DPNEXTFAILURE is at most 15.66 % larger than that of BESTPERIOD, and in fact is shorter at low scales (for 2^{15} and 2^{16} processors).

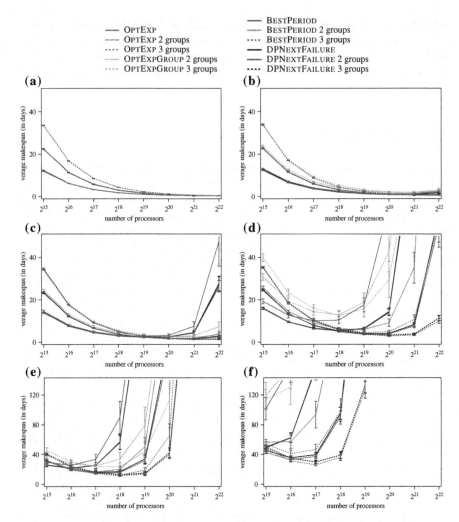

Fig. 4.5 Average makespan versus number of processors, Weibull failures, $k = 0.7$ or $k = 0.5$, $MTBF = 125$ years. **a** $C = R = 60$ s, $k = 0.7$. **b** $C = R = 60$ s, $k = 0.5$. **c** $C = R = 600$ s, $k = 0.7$. **d** $C = R = 600$ s, $k = 0.5$. **e** $C = R = 6000$ s, $k = 0.7$. **f** $C = R = 6000$ s, $k = 0.5$

Results for $C = R = 6000$ s show similar but accentuated trends. For $k = 0.7$ (Fig. 4.5e) the main results are similar to those obtained for $k = 0.5$ with $C = R = 600$ s. The best two algorithms are BESTPERIOD and DPNEXTFAILURE using $g = 3$ groups, but both algorithms show an increase in makespan starting at 2^{19} processors. For $k = 0.5$ (Fig. 4.5f) this increase occurs at 2^{18} processors and is sharper for DPNEXTFAILURE than BESTPERIOD. Even though group replication helps, with such large checkpointing overheads parallel efficiency cannot be maintained beyond 2^{17} processors.

We conclude that all our algorithms are more or less equivalent with Exponential failures (see Sect. 4.4.4.1), and show pronounced differences with Weibull failures. Overall, BESTPERIOD is the best algorithm. The only algorithm that leads to comparable makespans is DPNEXTFAILURE, but it never leads to a lower average makespan than BESTPERIOD at large scale. Even though DPNEXTFAILURE relies on a sophisticated DP approach, the brute-force but pragmatic approach used by BESTPERIOD turns out to be more effective. Even when using BESTPERIOD, our results show that application scalability is hindered by higher checkpoint overheads, which is expected, but also by lower k values, i.e., by less exponentially distributed failures.

4.4.4.3 Checkpointing Contention

The results presented so far are obtained assuming that the checkpointing overhead ($R = C$) does not depend on the number of groups. There are cases in which this assumption could give an unfair advantage to group replication. Consider an application with a given memory footprint V, in bytes, running on a platform with a total of q processors. With no replication ($g = 1$) the total volume of data involved in a checkpoint is V. Assuming that V is no larger than the aggregate RAM capacity of q/g processors, then group replication can be used with $g > 1$ groups. In this case, since each group executes the application, the total volume of data involved in a checkpoint at each group is also V. Since groups may checkpoint/recover at the same time, the amount of data involved can be up to $g \times V$, or a factor g larger than in the no-replication case.

To evaluate the impact of group replication on checkpointing overhead, we introduce a checkpointing contention model in our simulation. Whenever multiple checkpointing/recovery operations are concurrent, they receive a fair share of the checkpointing/recovery bandwidth. For instance, if n checkpointing operations begin at the same time, and no other checkpointing or recovery occurs over the next $n \times C$ time units, then all n checkpointing operations finish after $n \times C$ time units. More generally, considering that a checkpointing/recovery operations requires C units of activity, over a time interval Δt during which there are n ongoing such operations each operation performs $\frac{1}{n}/\Delta t$ units of activity (if one of these operations requires fewer units of work to complete, consider a shorter Δt interval).

Our objective in this section is to determine whether group replication can still be beneficial when considering checkpointing contention. We repeated all the experiments presented in Sects. 4.4.4.1 and 4.4.4.2. For $C = R = 60$ s, checkpointing contention has negligible impact on the results, and the impact for $C = R = 600$ s is lower than that for $C = R = 6000$ s. This is expected since the larger the checkpointing/recovery overhead, the more likely that more than one group is engaged in checkpointing or recovery at the same time. Thus, among all our results, those for $C = R = 6000$ s should be the most disadvantageous for group replication. These are the results presented in Fig. 4.6, which shows average makespan versus number of processors for BESTPERIOD without and with contention (denoted by

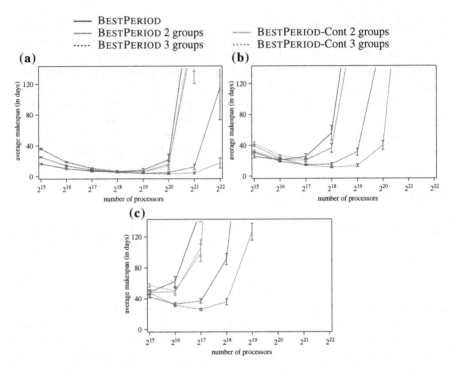

Fig. 4.6 Average makespan versus number of processors, $C = R = 6000$ s, $MTBF = 125$ years.
a Exponential failures. **b** Weibull failures, $k = 0.7$. **c** Weibull failures, $k = 0.5$

BESTPERIOD-Cont), for $g = 1, 2$, and 3, for $C = R = 6000$ s, for Exponential failures and for Weibull failures with $k = 0.7$ and $k = 0.5$.

As expected the average makespan of BESTPERIOD is increased due to check-pointing contention when multiple groups are used. However, even with contention, group replication outperforms the no-replication case at large scale. For Exponential failures, using $g = 2$ groups outperforms using $g = 1$ group as soon as the number of processors reaches 2^{18}, both with and without contention. Using $g = 3$ groups outperforms using $g = 2$ groups when there are either 2^{19} or 2^{20} processors with contention. The lowest average makespans with contention are achieved using either 2^{18} processors split in $g = 2$ groups, or 2^{19} processors split in $g = 3$ groups. For Weibull failures with $k = 0.7$, using $g = 2$ groups outperforms using $g = 1$ group starting at 2^{16} processors, with or without checkpointing contention. With contention, using $g = 3$ groups never outperforms using $g = 2$ groups, and ties its performance starting at 2^{18} processors. For Weibull failures with $k = 0.5$, using $g = 2$ groups outperforms using $g = 1$ group starting at 2^{15} processors with or without contention. With contention, using $g = 3$ groups is beneficial over using $g = 2$ groups when there are 2^{17} processors but the lowest makespan overall is achieved with $g = 2$ groups and 2^{15} processors.

We conclude that although checkpointing contention increases the makespan of group replication executions, the makespans of these executions are still shorter than that of no-replication execution at the same or slightly higher scales than when no contention takes place. One difference due to contention is that in our experiments using $g = 3$ groups is never worthwhile.

4.5 Process Replication

While in the previous section we replicate application instances, in this section we replicate processes within an instance with each process running on a distinct processor. Process replication was recently studied in [13], in which the authors propose to replicate each application process transparently on two processors. Only when both these processors fail must the job recover from the previous checkpoint. One replica performs redundant (thus wasteful) computations, but the probability that both replicas fail is much smaller than that of a single replica, thereby allowing for a drastic reduction of checkpoint frequency.

We consider the general case where each application process is replicated $g \geq 2$ times. We call *replica-group* the set of all the replicas of a given process, and we denote by n_{rg} the number of replica-groups. Altogether, if there are p available processors, there are $n_{rg} \times g \leq p$ processes running on the platform. We assume that when one of the g replicas of a replica-group fails it is not restarted, and the execution of the application proceeds as long as there is still at least one running replica in each of the replica-groups. In other words, for the whole application to fail, there must exist a replica-group whose g replicas have all been "hit" by a failure. One could envision a scenario where a failed replica is restarted based on the current state of the remaining replicas in its replica-group. This would increase application resiliency but would also be time-consuming. A certain amount of time would be needed to copy the state of one of the remaining replicas. Because all replicas of a same process must have a coherent state, the execution of the still running replicas would have to be paused during this copying. In a tightly coupled application, the execution of the whole application would be paused while copying. Consequently, restarting a failed replica would only be beneficial if the restarting cost were very small, when taking in consideration the frequency of failures and the checkpoint and restart costs. The benefit of such an approach is doubtful and, like [13], we do not consider it.

4.5.1 Theoretical Results

Two important quantities for evaluating the quality of an application execution, when replication is used, are: (i) the Mean Number of Failures To Interruption (*MNFTI*), i.e., the mean number of processor failures until application failure occurs; and

(ii) the Mean Time To Interruption (*MTTI*), i.e., the mean time elapsed until application failure occurs. In this section, we compute exact expressions of these two quantities. We first deal with the computation of *MNFTI* values in Sect. 4.5.1.1. Then we proceed to computing *MTTI* values, for Exponential failures in Sect. 4.5.1.2, and for arbitrary failures in Sect. 4.5.1.3. Note that the computation of *MNFTI* applies to any failure distribution, while that of *MTTI* is strongly distribution-dependent.

4.5.1.1 Computing *MNFTI*

We consider two options for "counting" failures. One option is to count each failure that hits any of the $g \cdot n_{rg}$ initial processors, including the processors *already hit* by a failure. Consequently, a failure that hits an already hit replica-group does not necessarily induce an application interruption. If the failure hits an already hit processor, whose replica had already been terminated due to an earlier failure, the application is not affected. If, on the contrary, the failure hits the other processor, in the case $g = 2$, then the whole application fails. This is the option chosen in [13]. Another option is to count only failures that hit *running processors*, and thus effectively kill replicas. This approach seems more natural as the running processors are the only ones that are important for the application execution.

We use *MNFTI*^{ah} to denote the *MNFTI* with the first option ("ah" stands for "already hit"), and *MNFTI*^{rp} to denote the *MNFTI* with the second option ("rp" stands for "running processors"). The following theorem gives a recursive expression for *MNFTI*^{ah} in the case $g = 2$ and for memoryless failure distributions.

Theorem 4.2 *If the failure inter-arrival times on the different processors are i.i.d. and independent from the failure history, then using process replication with $g = 2$, $MNFTI^{\text{ah}} = \mathbb{E}(NFTI^{\text{ah}}|0)$ where $\mathbb{E}(NFTI^{\text{ah}}|n_f) =$*

$$
\begin{cases}
2 & \text{if } n_f = n_{rg}, \\
\frac{2n_{rg}}{2n_{rg}-n_f} + \frac{2n_{rg}-2n_f}{2n_{rg}-n_f} \mathbb{E}\left(NFTI^{\text{ah}}|n_f + 1\right) & \text{otherwise.}
\end{cases}
$$

Note that Theorem 4.2 reproduces Proposition 1.3 in Sect. 1.4.2. We reproduce the result and proof for the convenience of the reader who studies this chapter independently.

Proof Let $\mathbb{E}(NFTI^{\text{ah}}|n_f)$ be the expectation of the number of failures needed for the whole application to fail, knowing that the application is still running and that failures have already hit n_f different replica-groups. Because each process initially has 2 replicas, this means that n_f different processes are no longer replicated, and that $n_{rg} - n_f$ are still replicated. Overall, there are $n_f + 2(n_{rg} - n_f) = 2n_{rg} - n_f$ processors still running.

The case $n_f = n_{rg}$ is the simplest. A new failure will hit an already hit replica-group, that is, a replica-group where one of the two initial replicas is still running. Two cases are then possible:

- The failure hits the running processor. This leads to an application failure, and in this case $\mathbb{E}(NFTI^{\text{ah}}|n_{rg}) = 1$.
- The failure hits the processor that has already been hit. Then the failure has no impact on the application. The $MNFTI^{\text{ah}}$ of this case is then: $\mathbb{E}(NFTI^{\text{ah}}|n_{rg}) = 1 + \mathbb{E}\left(NFTI^{\text{ah}}\,|n_{rg}\right)$.

The probability of failure is uniformly distributed between the two replicas, and thus between these two cases. Weighting the values by their probabilities of occurrence yields:

$$\mathbb{E}\left(NFTI^{\text{ah}}\,|n_{rg}\right) = \frac{1}{2} \times 1 + \frac{1}{2} \times \left(1 + \mathbb{E}\left(NFTI^{\text{ah}}\,|n_{rg}\right)\right) = 2.$$

For the general case $0 \le n_f \le n_{rg} - 1$, either the next failure hits a new replica-group, that is one with 2 replicas still running, or it hits a replica-group that has already been hit. The latter case leads to the same sub-cases as the $n_f = n_{rg}$ case studied above. As we have assumed that the failure inter-arrival times on the different processors are i.i.d. and *independent from the processor failure history* the failure probability is uniformly distributed among the $2n_{rg}$ processors, including the ones already hit. Hence the probability that the next failure hits a new replica-group is $\frac{2n_{rg} - 2n_f}{2n_{rg}}$. In this case, the expected number of failures needed for the whole application to fail is one (the considered failure) plus $\mathbb{E}\left(NFTI^{\text{ah}}|n_f + 1\right)$. Altogether we have:

$$\mathbb{E}\left(NFTI^{\text{ah}}|n_f\right) = \frac{2n_{rg} - 2n_f}{2n_{rg}} \times \left(1 + \mathbb{E}\left(NFTI^{\text{ah}}|n_f + 1\right)\right)$$
$$+ \frac{2n_f}{2n_{rg}} \times \left(\frac{1}{2} \times 1 + \frac{1}{2}\left(1 + \mathbb{E}\left(NFTI^{\text{ah}}|n_f\right)\right)\right).$$

Therefore, $\mathbb{E}\left(NFTI^{\text{ah}}|n_f\right) = \frac{2n_{rg}}{2n_{rg} - n_f} + \frac{2n_{rg} - 2n_f}{2n_{rg} - n_f}\mathbb{E}\left(NFTI^{\text{ah}}|n_f + 1\right)$.

We obtain a very similar recursive formula for $MNFTI^{\text{rp}}$.

Theorem 4.3 *If the failure inter-arrival times on the different processors are independent and identically distributed, then under the process replication scheme, with $g = 2$, we have $MNFTI^{\text{rp}} = \mathbb{E}(NFTI^{\text{rp}}|0)$ where*

$$\mathbb{E}(NFTI^{\text{rp}}|n_f) = \begin{cases} 1 & \text{if } n_f = n_{rg}, \\ 1 + \frac{2n_{rg} - 2n_f}{2n_{rg} - n_f}\mathbb{E}(NFTI^{\text{rp}}|n_f + 1) & \text{otherwise.} \end{cases}$$

It turns out that there is a simple (and quite unexpected) relationship between both failure models:

Proposition 4.4 *If the failure inter-arrival times on the different processors are i.i.d. and independent from the processor failure history then, for $g = 2$,*

$$MNFTI^{\text{ah}} = 1 + MNFTI^{\text{rp}}.$$

Proof We prove by induction that $\mathbb{E}(NFTI^{\text{ah}}|n_f) = 1 + \mathbb{E}(NFTI^{\text{rp}}|n_f)$, for any $n_f \in [0, n_{rg}]$. The base case is for $n_f = n_{rg}$ and the induction uses nonincreasing values of n_f.

For the base case, we have $\mathbb{E}(NFTI^{\text{rp}}|n_{rg}) = 1$ and $\mathbb{E}(NFTI^{\text{ah}}|n_{rg}) = 2$. Hence the property is true for $n_f = n_{rg}$. Consider a value $n_f < n_{rg}$, and assume to have proven that $\mathbb{E}(NFTI^{\text{ah}}|i) = 1 + \mathbb{E}(NFTI^{\text{rp}}|i)$, for any value of $i \in [1 + n_f, n_{rg}]$. We now prove the equation for n_f. According to Theorem 4.2, we have:

$$\mathbb{E}(NFTI^{\text{ah}}|n_f) = \frac{2n_{rg}}{2n_{rg} - n_f} + \frac{2n_{rg} - 2n_f}{2n_{rg} - n_f}\mathbb{E}\left(NFTI^{\text{ah}}|n_f + 1\right).$$

Therefore, using the induction hypothesis, we have:

$$\mathbb{E}(NFTI^{\text{ah}}|n_f) = \frac{2n_{rg}}{2n_{rg}-n_f} + \frac{2n_{rg}-2n_f}{2n_{rg}-n_f}\left(1 + \mathbb{E}\left(NFTI^{\text{rp}}|n_f + 1\right)\right)$$
$$= 2 + \frac{2n_{rg}-2n_f}{2n_{rg}-n_f}\mathbb{E}\left(NFTI^{\text{rp}}|n_f + 1\right) = 1 + \mathbb{E}\left(NFTI^{\text{rp}}|n_f\right)$$

the last equality being established using Theorem 4.3. Therefore, we have proved by induction that $\mathbb{E}(NFTI^{\text{ah}}|0) = 1 + \mathbb{E}(NFTI^{\text{rp}}|0)$. To conclude, we remark that $\mathbb{E}(NFTI^{\text{ah}}|0) = MNFTI^{\text{ah}}$ and $\mathbb{E}(NFTI^{\text{rp}}|0) = MNFTI^{\text{rp}}$.

We now show that Theorems 4.2 and 4.3 can be generalized to $g > 2$. Because the proofs are very similar, we only give the one for the $MNFTI^{\text{rp}}$ accounting approach (failures on running processors only), as it does not make any assumption on failures besides the i.i.d. assumption.

Proposition 4.5 *If the failure inter-arrival times on the different processors are i.i.d. then using process replication for $g \geq 2$,* $MNFTI^{\text{rp}} = \mathbb{E}\left(NFTI^{\text{rp}}|\underbrace{0, \dots, 0}_{g-1\ zeros}\right)$ *where:*

$$\mathbb{E}\left(NFTI^{\text{rp}}|n_f^{(1)}, \dots, n_f^{(g-1)}\right) =$$
$$1 + \frac{g \cdot \left(n_{rg} - \sum_{i=1}^{g-1} n_f^{(i)}\right)}{g \cdot n_{rg} - \sum_{i=1}^{g-1} i \cdot n_f^{(i)}} \cdot \mathbb{E}\left(NFTI^{\text{rp}}|n_f^{(1)}, n_f^{(2)}, \dots, n_f^{(g-1)}\right)$$
$$+ \sum_{i=1}^{g-2} \frac{(g-i) \cdot n_f^{(i)}}{g \cdot n_{rg} - \sum_{i=1}^{g-1} i \cdot n_f^{(i)}} \cdot \mathbb{E}\left(NFTI^{\text{rp}}|n_f^{(1)}, \dots, n_f^{(i-1)}, n_f^{(i)} - 1, n_f^{(i+1)} + 1, n_f^{(i+2)}, \dots, n_f^{(g-1)}\right)$$

Proof Let $\mathbb{E}\left(NFTI^{\text{rp}}|n_f^{(1)}, \dots, n_f^{(g-1)}\right)$ be the expectation of the number of failures needed for the whole application to fail, knowing that the application is still running and that, for $i \in [1..g - 1]$, there are $n_f^{(i)}$ replica-groups that have already been hit by exactly i failures. Note that a replica-group hit by i failures still contains exactly $g - i$ running replicas. Therefore, in a system where $n_f^{(i)}$ replica-groups have been hit by exactly i failures, there are still overall exactly $g \cdot n_{rg} - \sum_{i=1}^{g-1} i \cdot n_f^{(i)}$ running replicas, $g \cdot \left(n_{rg} - \sum_{i=1}^{g-1} n_f^{(i)} \right)$ of which are in replica-groups that have not yet been hit by any failure. Now, consider the next failure to hit the system. There are three cases to consider.

1. The failure hits a replica-group that has not been hit by any failure so far. This happens with probability:

$$\frac{g \cdot \left(n_{rg} - \sum_{i=1}^{g-1} n_f^{(i)} \right)}{g \cdot n_{rg} - \sum_{i=1}^{g-1} i \cdot n_f^{(i)}}$$

and, in that case, the expected number of failures needed for the whole application to fail is one (the studied failure) plus $\mathbb{E}\left(NFTI^\text{rp} | 1 + n_f^{(1)}, n_f^{(2)}, \ldots, n_f^{(g-1)} \right)$. Remark that we should have conditioned the above expectation with the statement "if $n_{rg} > \sum_{i=1}^{g-1} n_f^{(i)}$." In order to keep equations as simple as possible we rather do not explicitly state the condition and use the following abusive notation:

$$\frac{g \cdot \left(n_{rg} - \sum_{i=1}^{g-1} n_f^{(i)} \right)}{g \cdot n_{rg} - \sum_{i=1}^{g-1} i \cdot n_f^{(i)}} \cdot \left(1 + \mathbb{E}\left(NFTI^\text{rp} | 1 + n_f^{(1)}, n_f^{(2)}, \ldots, n_f^{(g-1)} \right) \right)$$

considering than when $n_{rg} = \sum_{i=1}^{g-1} n_f^{(i)}$ the first term is null and thus that it does not matter that the second term is not defined.
2. The failure hits a replica-group that has already been hit by $g - 1$ failures. Such a failure leads to a failure of the whole application. As there are $n_f^{(g-1)}$ such groups, each containing exactly one running replica, this event happens with probability:

$$\frac{n_f^{(g-1)}}{g \cdot n_{rg} - \sum_{i=1}^{g-1} i \cdot n_f^{(i)}}.$$

In this case, the expected number of failures needed for the whole application to fail is exactly equal to one (the considered failure).
3. The failure hits a replica-group that had already been hit by at least one failure, and by at most $g - 2$ failures. Let i be any value in $[1..g - 2]$. The probability that the failure hits a group that had previously been the victim of exactly i failures is equal to:

$$\frac{(g - i) \cdot n_f^{(i)}}{g \cdot n_{rg} - \sum_{i=1}^{g-1} i \cdot n_f^{(i)}}$$

as there are $n_f^{(i)}$ such replica-groups and that each contains exactly $g - i$ still running replicas. In this case, the expected number of failures needed for the whole application to fail is one (the studied failure) plus $\mathbb{E}\left(NFTI^{\text{rp}}|n_f^{(1)}, \ldots, n_f^{(i-1)},\right.$ $n_f^{(i)} - 1, n_f^{(i+1)} + 1, \ n_f^{(i+2)}, \ldots, n_f^{(g-1)}\left.\right)$ as there is one less replica-group hit by exactly i failures and one more hit by exactly $i + 1$ failures.

We aggregate all the cases to obtain:

$$
\mathbb{E}\left(NFTI^{\text{rp}}|n_f^{(1)}, \ldots, n_f^{(g-1)}\right) =
$$

$$
\frac{g \cdot \left(n_{rg} - \sum_{i=1}^{g-1} n_f^{(i)}\right)}{g \cdot n_{rg} - \sum_{i=1}^{g-1} i \cdot n_f^{(i)}} \cdot \left(1 + \mathbb{E}\left(NFTI^{\text{rp}}|1 + n_f^{(1)}, n_f^{(2)}, \ldots, n_f^{(g-1)}\right)\right)
$$

$$
+ \sum_{i=1}^{g-2} \frac{(g-i) \cdot n_f^{(i)}}{g \cdot n_{rg} - \sum_{i=1}^{g-1} i \cdot n_f^{(i)}} \left(1 + \mathbb{E}\left(NFTI^{\text{rp}}|n_f^{(1)}, \ldots, n_f^{(i-1)}, n_f^{(i)} - 1, n_f^{(i+1)} + 1, n_f^{(i+2)}, \ldots, n_f^{(g-1)}\right)\right)
$$

$$
+ \frac{n_f^{(g-1)}}{g \cdot n_{rg} - \sum_{i=1}^{g-1} i \cdot n_f^{(i)}} \cdot 1
$$

which is equivalent to the target equation.

Following the construction used to establish Proposition 4.5, here is the recursion to compute $MNFTI^{\text{ah}}$ for $g = 3$:

Proposition 4.6 *If the failure inter-arrival times on the different processors are i.i.d. and independent from the failure history, then using process replication with $g = 3$, $MNFTI^{\text{ah}} = \mathbb{E}(NFTI^{\text{ah}}|0, 0)$ where*

$$
\mathbb{E}\left(NFTI^{\text{ah}}|n_2, n_1\right) =
$$

$$
\frac{1}{3n_{rg} - n_2 - 2n_1} \left(3n_{rg} + 3(n_{rg} - n_1 - n_2)\mathbb{E}\left(NFTI^{\text{ah}}|n_2 + 1, n_1\right)\right.
$$

$$
+ 2n_2 \mathbb{E}\left(NFTI^{\text{ah}}|n_2 - 1, n_1 + 1\right)\left.\right)
$$

One can solve this recursion using a dynamic programming algorithm of quadratic cost $O(p^2)$ (and linear memory space $O(p)$).

Proposition 4.7 *If the failure inter-arrival times on the different processors are i.i.d. and independent from the failure history, then using process replication with $g = 3$, $MNFTI^{\text{rp}} = \mathbb{E}(NFTI^{\text{rp}}|0, 0)$ where*

$$\mathbb{E}\left(NFTI^{rp}|n_2, n_1\right) =$$

$$1 + \frac{1}{3n_{rg} - n_2 - 2n_1}\left(3(n_{rg} - n_1 - n_2)\mathbb{E}\left(NFTI^{rp}|n_2 + 1, n_1\right)\right.$$

$$\left. + 2n_2\mathbb{E}\left(NFTI^{rp}|n_2 - 1, n_1 + 1\right)\right)$$

Given the simple additive relationship that exists between $MNFTI^{ah}$ and $MNFTI^{rp}$ for $g = 2$ (Proposition 4.4), one may expect a similar relationship for large g. Table 4.1 shows $MNFTI^{ah}$ and $MNFTI^{rp}$ values and the difference between them for $g = 3$. The difference is not constant and increases as n_{rg} increases, and no simple relationship seems to exist between $MNFTI^{ah}$ and $MNFTI^{rp}$.

We can now evaluate our approach for computing the $MNFTI$ value and compare it to that in [13]. The authors therein observe that the generalized birthday problem is related to the problem of determining the number of processor failures needed to induce an application failure. The generalized birthday problem asks the following question: what is the expected number of balls $BP(m)$ to randomly put into m (originally empty) bins so that there is a bin with two balls? This problem has a well-known closed-form solution [14]. In the context of process replication, it is tempting to consider each replica group as a bin, and each ball as a processor failure, thus computing $MNFTI = BP(n_{rg})$. Unfortunately, this analogy is incorrect because processors in a replica group are distinguished. Let us consider the case $g = 2$, i.e., two replicas per replica group, and the two failure models described in Sect. 4.5.1.1. In the "already

Table 4.1 $MNFTI^{ah}$ and $MNFTI^{rp}$ computed using Propositions 4.6 and 4.7 and the difference between them, for $n_{rg} = 2^0, \ldots, 2^{20}$, with $g = 3$

n_{rg}	2^0	2^1	2^2	2^3	2^4	2^5	2^6
$MNFTI^{ah}$	5.5	7.3	10.1	14.6	21.6	32.4	49.4
$MNFTI^{rp}$	3.0	4.5	6.9	10.9	17.1	27.1	42.9
($MNFTI^{ah}$ $-$ $MNFTI^{rp}$)	2.5	2.8	3.2	3.7	4.4	5.3	6.4
n_{rg}	2^7	2^8	2^9	2^{10}	2^{11}	2^{12}	2^{13}
$MNFTI^{ah}$	75.9	117.6	183.3	286.8	450.2	708.5	1117.0
$MNFTI^{rp}$	68.1	108.0	171.5	272.2	432.1	685.8	1088.7
($MNFTI^{ah}$ $-$ $MNFTI^{rp}$)	7.8	9.6	11.8	14.6	18.2	22.7	28.3
n_{rg}	2^{14}	2^{15}	2^{16}	2^{17}	2^{18}	2^{19}	2^{20}
$MNFTI^{ah}$	1763.5	2787.6	4410.2	6982.3	11060.6	17528.6	27788.6
$MNFTI^{rp}$	1728.1	2743.2	4354.6	6912.5	10972.9	17418.4	27650.1
($MNFTI^{ah}$ $-$ $MNFTI^{rp}$)	35.4	44.3	55.6	69.8	87.7	110.2	138.6

hit" model, which is used in [13], if a failure hits a replica group after that replica group has already been hit once (i.e., a second ball is placed in a bin) an application failure does not necessarily occur. This is unlike the birthday problem, in which the stopping criterion is for a bin to contain two balls, thus breaking the analogy. In the "running processor" model, the analogy also breaks down. Consider that one failure has already occurred. The replica group that has suffered that first failure is now twice less likely to be hit by another failure as all the other replica groups as it contains only one replica. Since probabilities are no longer identical across replica groups, i.e., bins, the problem is not equivalent to the generalized birthday problem. However, there is a direct and valid analogy between the process replication problem and another version of the birthday problem with distinguished types, which asks: what is the expected number of randomly drawn red or white balls $BT(m)$ to randomly put into m (originally empty) bins so that there is a bin that contains at least one red ball and one white ball? Unfortunately, there is no known closed-form formula for $BT(m)$, even though the results in Sect. 4.5.1.1 provide a recursive solution.

In spite of the above, [13] uses the solution of the generalized birthday problem to compute $MNFTI$. According to [25], a previous article by the authors of [13], it would seem that the value $BP(n_{rg})$ is used. While [13] does not make it clear which value is used, a recent research report by the same authors states that they use $BP(g \cdot n_{rg})$. For completeness, we include both values in the comparison hereafter.

Table 4.2 shows the $MNFTI^{ah}$ values computed as $BP(n_{rg})$ or as $BP(g \cdot n_{rg})$, as well as the exact value computed using Theorem 4.2, for various values of n_{rg} and for $g = 2$. (Recall that in this case, $MNFTI^{ah}$ and $MNFTI^{rp}$ differ only by 1). The percentage relative differences between the two BP values and the exact value are included in the table as well. We see that the $BP(n_{rg})$ value leads to relative differences with the exact value between 29 and 33 %. This large difference seems easily explained due to the broken analogy with the generalized birthday problem. The unexpected result is that the relative difference between the $BP(g \cdot n_{rg})$ value and the exact value is below 16 % and, more importantly, decreases and approaches zero as n_{rg} increases. The implication is that using $BP(g \cdot n_{rg})$ is an effective heuristic for computing $MNFTI^{ah}$ even though the birthday problem is not analogous to the process replication problem! These results thus provide an empirical, if not theoretical, justification for the approach in [13], whose validity was not assessed experimentally therein.

4.5.1.2 Computing *MTTI* for Exponential Failures

With the "already hit" assumption, and assuming Exponential failures, the *MTTI* can be computed easily as

$$MTTI = systemMTBF(g \times n_{rg}) \times MNFTI^{ah} \qquad (4.7)$$

where $systemMTBF(p)$ denotes the mean time between failures of a platform with p processors and $MNFTI^{ah}$ is given by Theorem 4.2. Recall that $systemMTBF(p)$

Table 4.2 $MNFTT^{ah}$ computed as $BP(n_{rg})$, $BP(g \cdot n_{rg})$, and using Theorem 4.2, for $n_{rg} = 2^0, \ldots, 2^{20}$, with $g = 2$

n_{rg}	2^0	2^1	2^2	2^3	2^4	2^5	2^6
Theorem 4.2	3.0	3.7	4.7	6.1	8.1	11.1	15.2
$BP(n_{rg})$	2.0 (−33.3 %)	2.5 (−31.8 %)	3.2 (−30.9 %)	4.2 (−30.3 %)	5.7 (−30.0 %)	7.8 (−29.7 %)	10.7 (−29.6 %)
$BP(g \cdot n_{rg})$	2.5 (−16.7 %)	3.2 (−12.2 %)	4.2 (−8.8 %)	5.7 (−6.4 %)	7.8 (−4.6 %)	10.7 (−3.3 %)	14.9 (−2.3 %)

n_{rg}	2^7	2^8	2^9	2^{10}	2^{11}	2^{12}	2^{13}
Theorem 4.2	21.1	29.4	41.1	57.7	81.2	114.4	161.4
$BP(n_{rg})$	14.9 (−29.5 %)	20.7 (−29.4 %)	29.0 (−29.4 %)	40.8 (−29.4 %)	57.4 (−29.3 %)	80.9 (−29.3 %)	114.1 (−29.3 %)
$BP(g \cdot n_{rg})$	20.7 (−1.6 %)	29.0 (−1.2 %)	40.8 (−0.8 %)	57.4 (−0.6 %)	80.9 (−0.4 %)	114.1 (−0.3 %)	161.1 (−0.2 %)

n_{rg}	2^{14}	2^{15}	2^{16}	2^{17}	2^{18}	2^{19}	2^{20}
Theorem 4.2	227.9	321.8	454.7	642.7	908.5	1284.4	1816.0
$BP(n_{rg})$	161.1 (−29.3 %)	227.5 (−29.3 %)	321.5 (−29.3 %)	454.4 (−29.3 %)	642.4 (−29.3 %)	908.2 (−29.3 %)	1284.1 (−29.3 %)
$BP(g \cdot n_{rg})$	227.5 (−0.1 %)	321.5 (−0.1 %)	454.4 (−0.1 %)	642.4 (−0.1 %)	908.2 (−0.04 %)	1284.1 (−0.03 %)	1815.7 (−0.02 %)

is simply equal to the MTBF of an individual processor divided by p (cf. Proposition 1.2). A recursive expression for *MTTI* can also be obtained directly. While the *MTTI* value should not depend on the way to count failures, it would be interesting to compute it with the "running processor" assumption as a sanity check. It turns out that there is no equivalent to Eq. (4.7) for linking *MTTI* and $MNFTI^{rp}$. The reason is straightforward. While $systemMTBF(2n_{rg})$ is the expectation of the date at which the first failure will happen, it is not the expectation of the inter-arrival time of the first and second failures when only considering failures on processors still running. Indeed, after the first failure, there only remain $2n_{rg} - 1$ running processors. Therefore, the inter-arrival time of the first and second failures has an expectation of $systemMTBF(2n_{rg} - 1)$. We can, however, use a reasoning similar to that in the proof of Theorem 4.3 and obtain a recursive expression for *MTTI*:

Theorem 4.4 *If the failure inter-arrival times on the different processors follow an Exponential distribution of parameter λ then, when using process replication with $g = 2$, $MTTI = \mathbb{E}(TTI|0)$ where $\mathbb{E}(TTI|n_f) =$*

$$
\begin{cases}
\frac{1}{n_{rg}}\frac{1}{\lambda} & \text{if } n_f = n_{rg} \\
\frac{1}{(2n_{rg}-n_f)}\frac{1}{\lambda} + \frac{2n_{rg}-2n_f}{2n_{rg}-n_f}\mathbb{E}(TTI|n_f+1) & \text{otherwise}
\end{cases}
$$

Proof We denote by $\mathbb{E}(TTI|n_f)$ the expectation of the time an application will run before failing, knowing that the application is still running and that failures have already hit n_f different replica-groups. Since each process initially has 2 replicas, this means that n_f different processes are no longer replicated and that $n_{rg} - n_f$ are still replicated. Overall, there are thus still $n_f + 2(n_{rg} - n_f) = 2n_{rg} - n_f$ running processors.

The case $n_f = n_{rg}$ is the simplest: a new failure will hit an already hit replica-group and hence leads to an application failure. As there are exactly n_{rg} remaining running processors, the inter-arrival times of the n_{rg}th and $(n_{rg} + 1)$th failures is equal to $\frac{1}{\lambda n_{rg}}$ (minimum of n_{rg} Exponential laws). Hence:

$$
\mathbb{E}\left(TTI \,\middle|\, n_{rg}\right) = \frac{1}{\lambda n_{rg}}.
$$

For the general case, $0 \le n_f \le n_{rg} - 1$, either the next failure hits a replica-group with still 2 running processors, or it strikes a replica-group that had already been victim of a failure. The latter case leads to an application failure; then, after n_f failures, the expected application running time before failure is equal to the inter-arrival times of the n_fth and $(n_f + 1)$th failures, which is equal to $\frac{1}{(2n_{rg}-n_f)\lambda}$. The failure probability is uniformly distributed among the $2n_{rg} - n_f$ running processors, hence the probability that the next failure strikes a new replica-group is $\frac{2n_{rg}-2n_f}{2n_{rg}-n_f}$. In this case, the expected application running time before failure is equal to the inter-arrival times of the n_fth and $(n_f + 1)$th failures plus $\mathbb{E}\left(TTI|n_f + 1\right)$. We derive that:

$$\mathbb{E}\left(TTI|n_f\right) = \frac{2n_{rg} - 2n_f}{2n_{rg} - n_f} \times \left(\frac{1}{(2n_{rg} - n_f)\lambda} + \mathbb{E}\left(TTI|n_f + 1\right) \right)$$
$$+ \frac{n_f}{2n_{rg} - n_f} \times \frac{1}{(2n_{rg} - n_f)\lambda}.$$

Therefore,

$$\mathbb{E}\left(TTI|n_f\right) = \frac{1}{(2n_{rg} - n_f)\lambda} + \frac{2n_{rg} - 2n_f}{2n_{rg} - n_f}\mathbb{E}\left(TTI|n_f + 1\right).$$

The above results can be generalized to $g \geq 2$. To compute $MTTI$ under the "already hit" assumption one can use Eq. (4.7) replacing $MNFTI^{ah}$ by the value given by Proposition 4.6 or its generalization for a higher value of g. To compute $MNFTI^{rp}$ under the "running processors," Theorem 4.4 can be generalized using the same proof technique as when proving Proposition 4.5.

The linear relationship between $MNFTI^{ah}$ and $MTTI$, seen in Eq. (4.7), allows us to use the results in Table 4.2 to compute $MTTI$ values. To quantify the potential benefit of replication, Table 4.3 shows these values as the total number of processors increases. For a given total number of processors, we show results for $g = 1, 2,$ and 3. As a safety check, we have compared these predicted values with those computed through simulations, using an individual processor MTBF equal to 125 years. For each value of n_{rg} in Table 4.3, we have generated 1,000,000 random failure dates, computed the Time To application Interruption for each instance, and computed the mean of these values. This *simulated MTTI*, is in full agreement with the predicted $MTTI$ in Table 4.3.

The main and expected observation in Table 4.3 is that increasing g, i.e., the level of replication, leads to increased $MTTI$. The improvement in $MTTI$ due to replication increases as n_{rg} increases, and increases when the level of replication, g, increases. Using $g = 2$ leads to large improvement over using $g = 1$, with an $MTTI$ up to 3 orders of magnitude larger for $n_{rg} = 2^{20}$. Increasing the replication level to $g = 3$ leads to more moderate improvement over $g = 2$, with an $MTTI$ only about 10 times larger for $n_{rg} = 2^{20}$. Overall, these results show that, at least in terms of $MTTI$, replication is beneficial. Although these results are for a particular MTBF value, they lead us to believe that moderate replication levels, namely $g = 2$, are sufficient to achieve drastic improvements in fault tolerance.

4.5.1.3 Computing $MTTI$ for Arbitrary Failures

The approach that computes $MTTI$ from $MNFTI^{ah}$ is limited to memoryless (i.e., Exponential) failure distributions. To encompass arbitrary distributions, we use another approach based on the failure distribution density at the platform level. Theorem 4.5 quantifies the probability of successfully completing an amount of work of size $\mathcal{W}$ when using process replication for any failure distribution, which makes it possible to compute $MTTI$ via numerical integration:

Table 4.3 *MTTI* values achieved for Exponential failures and a given number of processors using different replication factors (total of $p = 2^0, \ldots, 2^{20}$ processors, with $g = 1, 2,$ and 3)

p	2^0	2^1	2^2	2^3	2^4	2^5	2^6
$g = 1$	10,95,000	5,47,500	2,73,750	1,36,875	68,438	34,219	17,109
$g = 2$		16,42,500	10,03,750	6,37,446	4,16,932	2,78,726	1,89,328
$g = 3$			15,05,625	9,99,188	7,78,673	5,65,429	4,32,102
p	2^7	2^8	2^9	2^{10}	2^{11}	2^{12}	2^{13}
$g = 1$	8555	4277	2139	1069	535	267	134
$g = 2$	1,30,094	90,135	62,819	43,967	30,864	21,712	15,297
$g = 3$	3,26,569	2,51,589	1,94,129	1,51,058	1,17,905	92,417	72,612
p	2^{14}	2^{15}	2^{16}	2^{17}	2^{18}	2^{19}	2^{20}
$g = 1$	66.8	33.4	16.7	8.35	4.18	2.09	1.04
$g = 2$	10,789	7615	5378	3799	2685	1897	1341
$g = 3$	57,185	45,106	35,628	28,169	22,290	17,649	13,982

The individual processor MTBF is 125 years, and MTTIs are expressed in hours

Theorem 4.5 *Consider an application with n_{rg} processes, each replicated g times using process replication, so that processor P_i, $1 \leq i \leq g \cdot n_{rg}$, executes a replica of process $\left\lceil \frac{i}{g} \right\rceil$. Assume that the failure inter-arrival times on the different processors are i.i.d, and let τ_i denote the time elapsed since the last failure of processor P_i. Let F denote the cumulative distribution function of the failure probability, and $F(t|\tau)$ be the probability that a processor fails in the next t units of time, knowing that its last failure happened τ units of time ago. Then the probability that the application will still be running after t units of time is:*

$$R(t) = \prod_{j=1}^{n_{rg}} \left(1 - \prod_{i=1}^{g} F\left(t|\tau_{i + g(j-1)}\right) \right). \tag{4.8}$$

Let f denote the probability density function of the entire platform (f is the derivative of the function $1 - R$): the MTTI is given by:

$$MTTI = \int_0^{+\infty} tf(t)dt = \int_0^{+\infty} R(t)dt = \int_0^{+\infty} \prod_{j=1}^{n_{rg}} \left(1 - \prod_{i=1}^{g} F\left(t|\tau_{i + g(j-1)}\right) \right) dt. \tag{4.9}$$

This theorem can then be used to obtain a closed-form expression for *MTTI* when the failure distribution is Exponential (Theorem 4.6) or Weibull (Theorem 4.7):

Theorem 4.6 *Consider an application with n_{rg} processes, each replicated g times using process replication. If the probability distribution of the time to failure of each processor is Exponential with parameter λ, then the MTTI is given by:*

$$MTTI = \frac{1}{\lambda} \sum_{i=1}^{n_{rg}} \sum_{j=1}^{i \cdot g} \left(\frac{\binom{n_{rg}}{i} \binom{i \cdot g}{j} (-1)^{i+j}}{j} \right).$$

The following corollary gives a simpler expression for the case $g = 2$:

Corollary 4.1 *Consider an application with n_{rg} processes, each replicated 2 times using process replication. If the probability distribution of the time to failure of each processor is Exponential with parameter λ, then the MTTI is given by:*

$$MTTI = \frac{1}{\lambda} \sum_{i=1}^{n_{rg}} \sum_{j=1}^{i \cdot 2} \left(\frac{\binom{n_{rg}}{i} \binom{i \cdot 2}{j} (-1)^{i+j}}{j} \right) = \frac{2^{n_{rg}}}{\lambda} \sum_{i=0}^{n_{rg}} \left(\frac{-1}{2} \right)^{i} \frac{\binom{n_{rg}}{i}}{(n_{rg} + i)}.$$

Theorem 4.7 *Consider an application with n_{rg} processes, each replicated g times using process replication. If the probability distribution of the time to failure of each processor is Weibull with scale parameter λ and shape parameter k, then the MTTI is given by:*

$$MTTI = \frac{\lambda}{k} \Gamma \left(\frac{1}{k} \right) \sum_{i=1}^{n_{rg}} \sum_{j=1}^{i \cdot g} \frac{\binom{n_{rg}}{i} \binom{i \cdot g}{j} (-1)^{i+j}}{j^{\frac{1}{k}}}.$$

While Theorem 4.6 is yet another approach to computing the *MTTI* for Exponential distributions, Theorem 4.7 is the first analytical result (to the best of our knowledge) for Weibull distributions. Unfortunately, the formula in Theorem 4.7 is not numerically stable for large values of n_{rg}. As a result, we resort to simulation to compute *MTTI* values. Table 4.4, which is the counterpart of Table 4.3 for Weibull failures, show *MTTI* results obtained as averages computed on the first 100,000 application failures of each simulated scenario. The results are similar to those in Table 4.3. The *MTTI* with $g = 2$ is much larger than that using $g = 1$, up to more than 3 orders of magnitude at large scale ($n_{rg} = 2^{20}$). The improvement in *MTTI* with $g = 3$ compared to $g = 2$ is more modest, reaching about a factor 10. The conclusions are thus similar: replication leads to large improvements, and a moderate replication level ($g = 2$) may be sufficient.

4.5.2 Empirical Evaluation

In the previous section, we have obtained exact expressions for the *MNFTI* and *MTTI* quantities, which are of direct relevance to the performance of the application and are amenable to analytical derivations. The main performance metric of interest to end-users, however, is the application *makespan*, i.e., the time elapsed between the launching of the application and its successful completion. But since it is not tractable to derive a closed-form expression of the expected makespan, in this section we

Table 4.4 Simulated *MTTI* values achieved for Weibull failures with shape parameter 0.7 and a given number of processors p using different replication factors (total of $p = 2^0, \ldots, 2^{20}$ processors, with $g = 1, 2,$ and 3)

p	2^0	2^1	2^2	2^3	2^4	2^5	2^6
$g = 1$	10,91,886	5,49,031	2,74,641	1,37,094	68,812	34,383	17,202
$g = 2$		20,81,689	12,43,285	7,69,561	4,91,916	3,21,977	2,14,795
$g = 3$			28,10,359	18,11,739	10,83,009	7,63,629	5,39,190
p	2^7	2^8	2^9	2^{10}	2^{11}	2^{12}	2^{13}
$g = 1$	8603	4275	2132	1060	525	260	127
$g = 2$	1,44,359	98,660	67,768	46,764	32,520	22,496	15,767
$g = 3$	3,98,410	2,96,301	2,23,701	1,70,369	1,31,212	1,01,330	78,675
p	2^{14}	2^{15}	2^{16}	2^{17}	2^{18}	2^{19}	2^{20}
$g = 1$	60.1	27.9	12.2	5.09	2.01	0.779	0.295
$g = 2$	11,055	7766	5448	3843	2708	1906	1345
$g = 3$	61,202	47,883	37,558	29,436	23,145	18,249	14,391

The individual processor MTBF is 125 years, and *MTTIs* are expressed in hours

compute the makespan empirically via simulation experiments. One of our goals here is to verify that the performance advantage of process replication seen in Sects. 4.5.1.2 and 4.5.1.3 in terms of *MTTI* are also seen when considering the makespan.

4.5.2.1 Simulation Framework and Models

In this section we provide details on our simulation methodology for evaluating the benefits of process replication.

Failure distributions and failure scenarios—We use the methodology described in Sect. 4.4.3.2 to generate failure distributions and the methodology described in Sect. 4.4.3.4 to generate failure scenarios,

Checkpointing policy—Replication dramatically reduces the number of application failures, so that standard periodic checkpointing strategies can be used. The checkpointing period can be computed based on the *MTTI* value using Young's approximation [36] or Daly's first-order approximation [7], the latter being used in [13]. We use Daly's approximation in this work because it is classical, often used in practice, and used in previous work [13]. It would be also interesting to present results obtained with the optimal checkpointing period, so as to evaluate the impact of the choice of the checkpointing period on our results. However, deriving the optimal period is not tractable. However, since our experiments are in simulation, we can search numerically for the best period among a sensible set of candidate periods. To build the candidate periods, we use the period computed in [3] (called OPTEXP) as a starting point. We then multiply and divide this period by $1 + 0.05 \times i$ with $i \in \{1, \ldots, 180\}$, and by 1.1^j with $j \in \{1, \ldots, 60\}$ and pick among these the value that leads to the lowest makespan. For a given replication level ($g = x$), we present results with the period computed using Daly's approximation (DALY-$g = x$) and with the best candidate period found numerically (BESTPERIOD-$g = x$).

Replication overhead—In [13], the authors consider that the communication overhead due to replication is proportional to the application's communication demands. Arguing that, to be scalable, an application must have sub-linear communication costs with respect to increasing processor counts, they consider an approximate logarithmic model for the percentage replication overhead: $\frac{log(p)}{10} + 3.67$, where p is the number of processors. The parameters to this model are instantiated from the application in [13] that has the highest replication overhead. When $g = 2$, we use the same logarithmic model to augment our first two parallel job models in Sect. 4.3:

- **Perfectly parallel jobs**: $W(p) = \frac{W}{p} \times (1 + \frac{1}{100} \times (\frac{log(p)}{10} + 3.67))$.
- **Generic parallel jobs**: $W(p) = (\frac{W}{p} + \gamma W) \times (1 + \frac{1}{100} \times (\frac{log(p)}{10} + 3.67))$.

For the numerical kernel job model, we can use a more accurate overhead model that does not rely on the above logarithmic approximation. Our original model in Sect. 4.3 comprises a computation component and a communication component. Using replication ($g = 2$), for each point-to-point communication between two original application processes, now a communication occurs between each process pair, considering both original processors and replicas, for a total of 4 communications. We can thus simply multiply the communication component of the model by a factor 4 and obtain the augmented model:

- **Numerical kernels**: $W(p) = \frac{W}{p} + \frac{\gamma \times W^{\frac{2}{3}}}{\sqrt{p}} \times 4$.

When $g = 3$, we (somewhat arbitrarily) multiply by 9/4 the overhead for perfectly parallel and generic parallel jobs, because the number and volume of communications are multiplied by 4 when $g = 2$ and by 9 when $g = 3$. When $g = 3$, we multiply the communication component by a factor 9 for numerical kernels.

Parameter values—Following Sect. 4.4.3.3, we use the following default parameter values to instantiate the simulations: $C = R = 600$ s, $D = 60$ s and $\mathcal{W} = 10,000$ years.

4.5.2.2 Choice of the Checkpointing Period

Our first set of experiments aims at determining whether using Daly's approximation for computing the checkpointing period, as done in [13], is a reasonable idea when replication is used. In the $g = 2$ case (two replicas per application process), we compute this period using the exact *MTTI* expression from Corollary 4.1. Given a failure distribution and a parallel job model, we compute the average makespan over 100 sample simulated application executions for a range of numbers of processors. Each sample is obtained using a different seed for generating random failure events based on the failure distribution. We present results using the best period found via a numerical search in a similar manner. In addition to the $g = 2$ and $g = 3$ results, we also present results for $g = 1$ (no replication) as a baseline, in which case the *MTTI* is simply the processor MTBF. In the three options the total number of processors is the same, i.e., $g \times n/g$.

We show experimental results for three failure distributions: (i) Exponential with a 125-year MTBF; (ii) Weibull with a 125-year MTBF and shape parameter $k = 0.70$; and (iii) Weibull with a 125-year MTBF and shape parameter $k = 0.50$. For each failure distribution, we use five parallel job models as described in Sect. 4.3, augmented with the replication overhead model described in Sect. 4.5.2.1: (i) perfectly parallel; (ii) generic parallel jobs with $\gamma = 10^{-6}$; (iii) numerical kernels with $\gamma = 0.1$; (iv) numerical kernels with $\gamma = 1$; and (v) numerical kernels with $\gamma = 10$. We thus have $5 \times 5 = 25$ sets of results.

Figures 4.7, 4.8 and 4.9 show average makespan versus number of processors. It turns out that, for a given failure distribution, all results follow the same trend regardless of the job model, as illustrated in Fig. 4.9 for Weibull failures with $k = 0.7$. But for Fig. 4.9 we show results only for generic parallel jobs.

Figures 4.7 and 4.8 show average makespan versus number of processors for generic parallel jobs subject to each of the three considered failure distributions. We first note that, without replication, and except for Exponential failures, the minimum makespan is not achieved on the largest platform. The fact that in most cases the makespan with 2^{19} processors is lower than the makespan with 2^{20} processors suggests that duplicating processes should be beneficial. This is indeed always the case

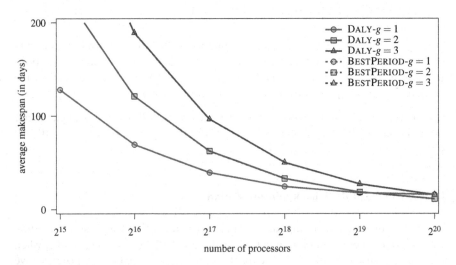

Fig. 4.7 Average makespan versus number of processors for two choices of the checkpointing period, without process replication (DALY-$g = 1$ and BESTPERIOD-$g = 1$) and with process replication (DALY-$g = 2$ or 3 and BESTPERIOD-$g = 2$ or 3), for generic parallel jobs subject to Exponential failures (*MTBF = 125 years*)

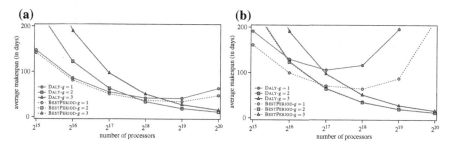

Fig. 4.8 Same as Fig. 4.7 (generic parallel jobs) but for Weibull failures ($MTBF = 125$ years). **a** $k = 0.70$. **b** $k = 0.50$

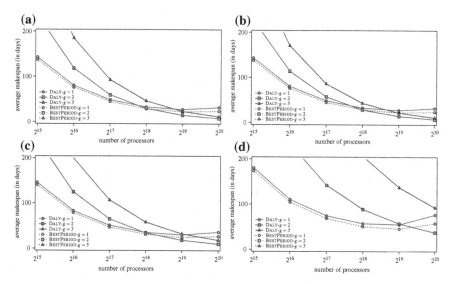

Fig. 4.9 Same as Fig. 4.8a (Weibull failures with $k = 0.7$, $MTBF = 125$ years) but for other job types. **a** Perfectly parallel jobs. **b** Numerical kernels ($\gamma = 0.1$). **c** Numerical kernels ($\gamma = 1$). **d** Numerical kernels ($\gamma = 10$)

for the largest platforms: when using 2^{20} processors, the makespan without replication is always larger than the makespan with replication, the replication factor being either $g = 2$ or $g = 3$. However, in none of the configurations, using a replication of $g = 3$ is more beneficial than with $g = 2$. More importantly, in each configuration, the minimum makespan is always achieved while duplicating the processes ($g = 2$) and using the maximum number of processors.

The two curves for $g = 1$ are exactly superposed in Fig. 4.7. For $g = 2$ and for $g = 3$ the two curves are exactly superposed in all three figures. Results for the case $g = 1$ (no replication) show that Daly's approximation achieves the same performance as the best periodic checkpointing policy for Exponential failures. For Weibull failures, however, Daly's approximation leads to significantly suboptimal results that worsen

Table 4.5 Number of application failures and fraction of processor failures that cause application failures with process replication ($g = 2$) assuming Weibull failure distributions ($k = 0.7$ or 0.5) for various numbers of processors and $C = 600$ s

# of proc.	# of app. failures		% of proc. failures	
	$k = 0.7$	$k = 0.5$	$k = 0.7$	$k = 0.5$
2^{14}	1.95	4.94	0.35	0.39
2^{15}	1.44	3.77	0.25	0.28
2^{16}	0.88	2.61	0.15	0.19
2^{17}	0.45	1.67	0.075	0.12
2^{18}	0.20	1.11	0.034	0.076
2^{19}	0.13	0.72	0.022	0.049
2^{20}	0.083	0.33	0.014	0.023

Results are averaged over 100 experiments

as k decreases (as expected and already reported in [3]). What is perhaps less expected is that in the cases $g = 2$ and $g = 3$, using Daly's approximation leads to virtually the same performance as using the best period even for Weibull failures. With replication, application makespan is simply not sensitive to the checkpointing period, at least in a wide neighborhood around the best period. This is because application failures and recoveries are infrequent, i.e., the MTBF of a pair of replicas is large. To quantify the frequency of application failures Table 4.5 shows the percentage of processor failures that actually lead to failure recoveries when using process replication. Results are shown in the case of Weibull failures for $k = 0.5$ and $k = 0.7$, $C = 600$ s, and for various numbers of processors. We see that very few application failures, and thus recoveries, occur throughout application execution (recall that makespans are measured in days in our experiments). This is because a very small fraction of processor failures manifest themselves as application failures (below 0.4 % in our experiments). This also explains why using $g = 3$ replicas does not lead to any further performance improvements (recall that the expectation was that further improvement would be low anyway given the results in Tables 4.3 and 4.4). While this low number of application failures demonstrates the benefit of process replication, the interesting result is that it also makes the choice of the checkpointing period not critical.

When setting the processor MTBF to a lower value so that the MTBF of a pair of replicas is not as large, then the choice of the checkpointing period matters. Consider for instance a process replication scenario with Weibull failures of shape parameters $k = 0.7$, a generic parallel job, and a platform with 2^{20} processors. When setting the MTBF to an unrealistic 0.1 year, using Daly's approximation yields an average makespan of 22.7 days, as opposed to 19.1 days (an increase of more than 18 %) when using the best period. Similar cases can be found for Exponential failures.

We summarize our findings so far as follows. Without replication, a poor choice of checkpointing period produces significantly suboptimal performance. When using replication, a poor choice can also theoretically lead to poor results, but this is very unlikely in practice because replication drastically reduces the number of failures.

In fact, in practical settings, the choice of the checkpointing period is simply not critical when replication is used. Consequently, setting the checkpointing period based on an approximation, Daly's being the most commonplace and oft referenced, is appropriate.

4.5.2.3 When Is Process Replication Beneficial?

In this section we determine under which conditions process replication is beneficial, i.e., leads to a lower makespan, when compared to a standard application execution that uses only checkpoint–recovery. We restrict our study to duplication ($g = 2$) as we have seen that the case $g = 3$ was never beneficial with respect to the case $g = 2$.

In a 2-D plane defined by the processor MTBF and the number of processors, and given a checkpointing overhead, simulation results can be used to construct a curve that divides the plane into two regions. Points above the curve correspond to cases in which process replication is beneficial. Points below the curve correspond to cases in which process replication is detrimental, i.e., the resource waste due to replication is not worthwhile because the processor MTBF is too large or the number of processors is too low. Several such curves are shown in [13] (Fig. 9 therein) for different checkpointing overheads, and, as expected, the higher the overhead the more beneficial it is to use process replication.

One question when comparing the replication and the no-replication cases is that of the checkpointing period. We have seen in the previous section that when using process replication the choice of the period has little impact and that Daly's approximation can be used safely. In the no-replication case, however, Daly's approximation should only be used in the case of exponentially distributed failures as it leads to poor results when the failure distribution is Weibull (see the $g = 1$ curves in Fig. 4.8). Furthermore, there is evidence that, in general, failure distributions are well approximated by Weibull distributions [17, 18, 22, 29], while not at all by exponential distributions. Most recently, in [18], the authors show that failures observed on a production cluster, over a cumulative 42-month time period, are modeled well by a Weibull distribution with shape parameter $k < 0.5$. In other words, the failure distribution is far from being Exponential and thus Daly's approximation would be far from the best period (compare Fig. 4.8a for $k = 0.7$ to Fig. 4.8b for $k = 0.5$).

Given the above, comparing the replication case to the no-replication case with Weibull failure distributions and using Daly's approximation as the checkpointing period gives an unfair advantage to process replication. To isolate the effect of replication from checkpointing period effects, we opt for the following method: we always use the best checkpointing period for each simulated application execution, as computed by a numerical search over a range of simulated executions each with a different checkpointing period. These results, for $g = 2$, are shown as solid curves in Fig. 4.10, for Weibull failures with $k = 0.7$ and $k = 0.5$, each curve corresponding to a different checkpointing overhead (C) value.

Each curve corresponds to the break-even point and the area above the curve corresponds to settings for which replication is beneficial. As expected, replication

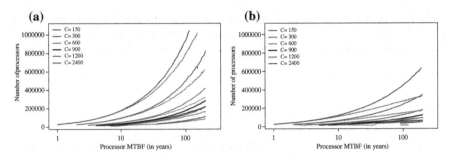

Fig. 4.10 Break-even point curves for replication ($g = 2$) versus no-replication for various check-pointing overheads, as computed using the best checkpointing periods (*solid lines*) and Daly's approximation (*dashed lines*), assuming Weibull failure distributions. **a** $k = 0.70$. **b** $k = 0.50$

becomes detrimental when the number of processors is too small, when the check-pointing overhead is too low, and/or when the processor MTBF is too large. For comparison purposes, the figure also shows a set of dashed curves that correspond to results obtained when using Daly's approximation as the checkpointing period instead of using our numerical search for the best such period. We see that, as expected, using Daly's approximation gives an unfair advantage to process repli-cation. This advantage increases as k decreases, since the Weibull distribution is then further away from the Exponential distribution. (For exponential distributions, all curves match.) For instance, for $k = 0.5$ (Fig. 4.10b), the break-even curve for $C = 600$ s as obtained using Daly's approximation is in fact, for most values of the MTBF, below the break-even curve for $C = 900$ s as obtained using the best check-pointing period. Note that the results presented in [13] are obtained using Daly's approximation as the checkpointing period.

4.6 Conclusion

In this chapter we have presented a rigorous study of replication techniques for large-scale platforms. These platforms are subject to failures, the frequencies of which increase dramatically with platform scale. We have investigated replication as a technique to better use all the resources provided by the platform. Replication comes in two flavors, *group replication* and *process replication*. Group replication consists in partitioning the platform into several groups, which each executes an instance of the application concurrently in phases. All groups synchronize as soon as one of them completes a phase. Instead, process replication replicates each application process onto several processors (a replica group), thereby reducing the need to recover from a failure only when all processors in a replica group have failed. Process replication is the approach followed in [13] with two processors per replica group.

While both replication techniques improve reliability, they have very different characteristics. Group replication can be used for any kind of parallel application, while process replication requires a replication-aware runtime (e.g., an MPI implementation with process replication support). Also, with process replication the total communication volume is increased by a factor proportional to the square of the replication degree, while the increase is only linear with group replication.

We have provided a detailed analysis of group replication for large-scale platforms. Given an execution protocol, ASAP, it is possible to derive a bound on the expected application makespan using this protocol when failures are exponentially distributed, which suggests a checkpointing period that can be used in practice. We have also proposed two approaches to minimize application makespan that are applicable regardless of the failure distribution: (i) a brute-force search for a checkpointing period, called BESTPERIOD; and (ii) a Dynamic Programming algorithm, called DPNEXTFAILURE. Using simulation we have obtained the following main findings: (i) when considering realistic non-Exponential failures group replication can significantly lower application makespan on large-scale platforms; (ii) our pragmatic BESTPERIOD approach outperforms the more sophisticated DPNEXTFAILURE Dynamic Programming approach; (iii) even when accounting for the contention due to concurrent checkpointing/recovery by multiple groups, group replication remains beneficial at large scale. Group replication leads to particularly good results when failures are far from being exponentially distributed, which several studies have shown to be the case in production platforms [17, 18, 22, 29].

We have also provided a detailed analysis of process replication for large-scale platforms. We have obtained recursive expressions for *MNFTI*, and analytical expressions for *MTTI* with arbitrary distributions, which lead to closed-form expressions for Exponential and Weibull distributions. It turns out that there is an unexpected relationship between two natural failure models (*already hit* and *running processors*) in the case of process duplication ($g = 2$). Simulation results show that, although the choice of a good checkpointing period can be important in the no-replication case (e.g., for non-Exponential failures) this choice is not critical when process replication is used. This is because with process replication few processor failures lead to application failures (i.e., rollback and recovery). This effect is essentially the reason why process replication was proposed in the first place. But a surprising and interesting side effect is that choosing a good checkpointing period is no longer challenging. Finally, we have determined the break-even point between replication and no-replication for Weibull failures in a way that is agnostic to the choice of the checkpointing period. These results identify relevant scenarios in which replication is worthwhile when compared to the no-replication case.

An interesting direction for future work on group replication would be to compare the checkpoints saved by multiple groups as a way to detect silent errors or corrupted data. This would require modifying the group replication approach so that at least 2 groups among $g > 2$ groups compute a chunk of work successfully, thereby trading off performance for reliability. Another direction is to generalize the work on group replication beyond the case of coordinated checkpointing, for instance to deal with hierarchical checkpointing schemes based on message logging, or with containment domains [6]. Both these techniques alleviate the cost of checkpointing and recovery, and would dramatically decrease checkpointing contention costs.

Further work could investigate the impact of *partial* replication instead of full replication. In this approach, replication would be used only for critical components (e.g., message loggers in uncoordinated checkpoint protocols), while traditional checkpointing would be used for noncritical components. The goal would be to reduce the overhead of replication while still achieving some of its benefits in terms of resilience.

Another direction for future work is to study the impact of resilience techniques on energy consumption. Together with fault tolerance, energy consumption is expected to be a major challenge for exascale machines [8, 27]. A promising next step in this search is the study of the interplay between checkpointing, replication, and energy consumption. By definition, both checkpointing and replication induce additional power consumption, but both techniques lead to faster executions in expectation. There are thus various energy trade-offs to achieve. The key question is to determine the best execution strategy given both an energy budget and a maximum admissible application makespan.

References

1. Amdahl G (1967) The validity of the single processor approach to achieving large scale computing capabilities. In: AFIPS conference proceedings, vol 30. AFIPS Press, pp 483–485
2. Blackford LS, Choi J, Cleary A, D'Azevedo E, Demmel J, Dhillon I, Dongarra J, Hammarling S, Henry G, Petitet A, Stanley K, Walker D, Whaley RC (1997) ScaLAPACK users' guide. SIAM
3. Bougeret M, Casanova H, Rabie M, Robert Y, Vivien F (2011) Checkpointing strategies for parallel jobs. In: Proceedings of 2011 international conference high performance computing, networking, storage and analysis SC'11. ACM Press
4. Bouguerra M-S, Gautier T, Trystram D, Vincent J-M (2010) A flexible checkpoint/restart model in distributed systems. In: PPAM, vol 6067. LNCS, pp 206–215
5. Castelli V, Harper RE, Heidelberger P, Hunter SW, Trivedi KS, Vaidyanathan K, Zeggert WP (2001) Proactive management of software aging. IBM J Res Dev 45(2):311–332
6. Chung J, Lee I, Sullivan M, Ryoo JH, Kim DW, Yoon DH, Kaplan L, Erez M (2012) Containment domains: a scalable, efficient, and flexible resilience scheme for exascale systems. In: Proceedings of international conference on high performance computing, networking, storage and analysis SC'12. ACM Press
7. Daly JT (2004) A higher order estimate of the optimum checkpoint interval for restart dumps. Future Gener Comput Syst 22(3):303–312
8. Dongarra J, Beckman P, Aerts P, Cappello F, Lippert T, Matsuoka S, Messina P, Moore T, Stevens R, Trefethen A, Valero M (2009) The international exascale software project: a call to cooperative action by the global high-performance community. Int J High Perform Comput Appl 23(4):309–322
9. Elliott J, Kharbas K, Fiala D, Mueller F, Ferreira K, Engelmann C (2012) Combining partial redundancy and checkpointing for HPC. In: ICDCS'12. IEEE
10. Elnozahy E, Plank J (2004) Checkpointing for peta-scale systems: a look into the future of practical rollback-recovery. IEEE Trans Dependable Secur Comput 1(2):97–108
11. Engelmann C, Swen B (2011) Redundant execution of HPC applications with MR-MPI. In: PDCN. IASTED

12. Engelmann C, Ong HH, Scorr SL (2009) The case for modular redundancy in large-scale high performance computing systems. In: Proceedings of the 8th IASTED international conference on parallel and distributed computing and networks (PDCN), pp 189–194

13. Ferreira K, Stearley J, Laros JHI, Oldfield R, Pedretti K, Brightwell R, Riesen R, Bridges PG, Arnold D (2011) Evaluating the viability of process replication reliability for exascale systems. In: Proceedings of 2011 international conference on high performance computing, networking, storage and analysis SC'11. ACM Press

14. Flajolet P, Grabner PJ, Kirschenhofer P, Prodinger H (1995) On Ramanujan's Q-function. J Comput Appl Math 58:103–116

15. George C, Vadhiyar SS (2012) AdFT: an adaptive framework for fault tolerance on large scale systems using application malleability. Procedia Comput Sci 9:166–175

16. Gärtner F (1999) Fundamentals of fault-tolerant distributed computing in asynchronous environments. ACM comput Surv 31(1):1–26

17. Heath T, Martin RP, Nguyen TD (2002) Improving cluster availability using workstation validation. SIGMETRICS Perf Eval Rev 30(1):217–227

18. Heien R, Kondo D, Gainaru A, LaPine D, Kramer B, Cappello F (2011) Modeling and tolerating heterogeneous failures on large parallel system. In: Proceedings of the IEEE/ACM supercomputing conference (SC)

19. Jones W, Daly J, DeBardeleben N (2010) Impact of sub-optimal checkpoint intervals on application efficiency in computational clusters. In: HPDC'10. ACM, pp 276–279

20. Kolettis N, Fulton ND (1995) Software rejuvenation: analysis, module and applications. In: FTCS'95. IEEE CS, Washington, p 381

21. Leblanc T, Anand R, Gabriel E, Subhlok J (2009) VolpexMPI: an MPI library for execution of parallel applications on volatile nodes. In: 16th European PVM/MPI users' group meeting. Springer, pp 124–133

22. Liu Y, Nassar R, Leangsuksun C, Naksinehaboon N, Paun M, Scott S (2008) An optimal checkpoint/restart model for a large scale high performance computing system. In: IPDPS 2008. IEEE, pp 1–9

23. Oldfield RA, Arunagiri S, Teller PJ, Seelam S, Varela MR, Riesen R, Roth PC (2007) Modeling the impact of checkpoints on next-generation systems. In: Proceedings of the 24th IEEE conference on mass storage systems and technologies, pp 30–46

24. Pinedo M (2008) Scheduling: theory, algorithms, and systems, 3rd edn. Springer, New York

25. Riesen R, Ferreira K, Stearley J (2010) See applications run and throughput jump: the case for redundant computing in HPC. In: Proceedings of the dependable systems and networks workshops, pp 29–34

26. Ross SM (2009) Introduction to probability models, 11th edn. Academic Press, New York

27. Sarkar V, Harrod W, Snavely A (2009) Software challenges in extreme scale systems. J Phys Conf Ser 180(1):012045

28. Schroeder B, Gibson G (2007) Understanding failures in petascale computers. J Phys Conf Ser 78(1):012022

29. Schroeder B, Gibson GA (2006) A large-scale study of failures in high-performance computing systems. In: Proceedings of DSN, pp 249–258

30. Schroeder B, Gibson GA (2007) Understanding failures in petascale computers. J Phys Conf Ser 78(1):188–198

31. Stearley J, Ferreira KB, Robinson DJ, Laros J, Pedretti KT, Arnold D, Bridges PG, Riesen R (2012) Does partial replication pay off? In FTXS (a DSN workshop). IEEE

32. Venkatesh K (2010) Analysis of dependencies of checkpoint cost and checkpoint interval of fault tolerant MPI applications. Analysis 2(08):2690–2697

33. Wang L, Karthik P, Kalbarczyk Z, Iyer R, Votta L, Vick C, Wood A (2005) Modeling coordinated checkpointing for large-scale supercomputers. In: Proceedings of the international conference on dependable systems and networks, pp 812–821

34. Yang X-J, Wang Z, Xue J, Zhou Y (2012) The reliability wall for exascale supercomputing. IEEE Trans Comput 61(6):767–779

35. Yi S, Kondo D, Kim B, Park G, Cho Y (2010) Using replication and checkpointing for reliable task management in computational grids. In: Proceedings of the international conference on high performance computing and simulation
36. Young JW (1974) A first order approximation to the optimum checkpoint interval. Commun ACM 17(9):530–531
37. Zheng G, Ni X, Kale L (2012) A scalable double in-memory checkpoint and restart scheme towards exascale. In: Dependable systems and networks workshops (DSN-W)
38. Zheng Z, Lan Z (2009) Reliability-aware scalability models for high performance computing. In: Proceedings of the IEEE conference on cluster computing

Chapter 5
Energy-Aware Checkpointing Strategies

**Guillaume Aupy, Anne Benoit, Mohammed El Mehdi Diouri,
Olivier Glück and Laurent Lefèvre**

Abstract Future extreme-scale supercomputers will gather several millions of cores. The main problem that we address in this chapter is the energy consumption of these systems. Fault-tolerant methods must be deployed in such extreme-scale systems and these methods have a dramatic impact on total energy consumption. Fault-tolerant protocols have different energy consumption rates, depending on parameters such as platform characteristics, application features, and number of processes used in the execution. Currently, in order to evaluate the power consumption of fault-tolerant protocols in a given execution context, the only approach is to run the application with the different versions of fault-tolerant protocols and to monitor energy consumption. In order to avoid this time and energy consuming process, we describe in this chapter a methodology to estimate the energy consumption of the fault-tolerant protocols used for HPC applications. This methodology relies on an energy calibration of the supercomputer and a user description of the execution setting. We evaluate the accuracy of the estimations with applications and scenarios executed on a real platform with energy consumption monitoring. Results show that the energy estimations provided before the execution are highly accurate, and allow users to select the less energy consuming fault-tolerant protocol without pre-running their applications.

G. Aupy · A. Benoit (✉)
Ecole Normale Supérieure de Lyon, Lyon Cedex 07, France
e-mail: guillaume.aupy@ens-lyon.fr

A. Benoit
e-mail: anne.benoit@ens-lyon.fr

M.E.M. Diouri
IGA Casablanca, Casablanca, Morocco
e-mail: mehdi.diouri@iga-casablanca.ma

O. Glück
Université Claude Bernard de Lyon, Villeurbanne, France
e-mail: olivier.gluck@univ-lyon1.fr

L. Lefèvre
INRIA & Ecole Normale Supérieure de Lyon, Lyon Cedex 07, France
e-mail: laurent.lefevre@inria.fr

© Springer International Publishing Switzerland 2015
T. Herault and Y. Robert (eds.), *Fault-Tolerance Techniques
for High-Performance Computing*, Computer Communications and Networks,
DOI 10.1007/978-3-319-20943-2_5

5.1 Introduction

For decades, the computer science research community exclusively focused on performance, which resulted in highly powerful, but in turn, low efficient systems with a very high total cost of ownership (TCO)[31]. A significant research effort is focusing on the characteristics, features, and challenges of High-Performance Computing (HPC) systems capable of reaching the Exaflop performance mark[21, 42]. The portrayed Exascale systems will necessitate billion way parallelism, resulting not only in a massive increase in the number of processing units (cores), but also in terms of computing nodes.

Considering the relative slopes describing the evolution of the reliability of individual components on one side, and the evolution of the number of components on the other side, the reliability of the entire platform is expected to decrease, due to probabilistic amplification (see Sects. 1.1.1 and 1.3.2). Even if each independent component is quite reliable, the Mean Time Between Failures (MTBF) is expected to drop drastically. Executions of large parallel applications on these systems will have to tolerate a higher degree of errors and failures than in current systems. The de-facto general-purpose error recovery technique in high-performance computing is checkpoint and rollback recovery. Such protocols employ checkpoints to periodically save the state of a parallel application, so that when an error strikes some process, the application can be restored into one of its former states. The most widely used protocol is coordinated checkpointing, where all processes periodically stop computing and synchronize to write critical application data onto stable storage. Coordinated checkpointing is well understood, at least in its blocking form (when no computing activity takes place during checkpoints), and good approximations of the optimal checkpoint interval exist; they are known as Young's and Daly's formula[10, 44] (see also Sect. 1.3.1). While the future Exascale applications are not yet designed and developed, it is anticipated, from the current knowledge and observations of existing large systems, that fault tolerance is unavoidable at the post-Petascale era, since Exascale systems will experience various kind of faults many times per day[8].

While reliability is a major concern for Exascale, another key challenge is to minimize energy consumption, both for economic and environmental reasons. The HPC community has recently acknowledged that the energy efficiency of HPC systems is a major concern in designing future Exascale systems for the end of the decade [20, 25]. One of the most power-consuming components of today's systems is the processor: even when idle, it dissipates a significant fraction of the total power. However, for future Exascale systems, the power dissipated to execute I/O transfers is likely to play an even more important role, because the relative cost of communication is expected to dramatically increase, both in terms of latency and consumed energy[43]. An Exascale supercomputer will gather several millions of CPU cores running up to a billion trends to achieve a performance of 10^{18} FLoat Operations Per Second, and it will consume several megawatts. The energy consumption issue at the Exascale becomes even more worrying when we know that we already reach power consumptions higher than 17 MW at the Petascale, while DARPA has set to

20 MW the threshold for Exascale supercomputers [42]. Hence, reducing the energy consumption of high-performance computing infrastructures is a major challenge for the next years in order to be able to move to the Exascale era. Nowadays, there exists a strong research effort toward energy-efficient supercomputers. Hardware provides part of the solution by exposing unceasingly more energy-efficient devices which also provide abilities that current operating systems can successfully leverage to save energy [33]. Mechanisms such as dynamic voltage scaling (DVFS) or P-state management have also been used to develop power-aware user-level software [1, 2, 33].

Hence, dealing with errors and minimizing the energy consumption are two main challenges that should be addressed. However, fault tolerance and energy consumption are interrelated: fault tolerance consumes energy and some energy reduction techniques can increase error and failure rates [23].

Very few papers consider the general problem of the interplay between energy consumption and fault tolerance. Aupy et al. [3] discuss energy-aware checkpointing strategies for divisible tasks, using DVFS to reduce the energy consumption. Given a workload, they show how to decide how many chunks to use, what are the sizes of these chunks, and at which speed each chunk is executed. Tackling HPC platforms, Diouri et al. [16] present the energy consumption of the three most important parts of fault tolerance: message logging, checkpointing, and task coordination. Their first result is that task coordination is the most energy consuming part of fault-tolerant protocols. They also show that while it involves more power to store data on RAM, HDD logging is more energy consuming than RAM logging because of the logging duration. In a second paper, Diouri et al. [18] extend these results into a framework that predicts the energy consumption of a fault-tolerant protocol, allowing the user to choose amongst three fault-tolerant protocols: coordinated, uncoordinated and hierarchical, depending on the application running on the platform. We detail the coordinated and uncoordinated protocols below. Finally, Meneses et al. [35] study the energy consumption of the coordinated periodic checkpointing protocol as a function of $\mathscr{P}_{Static}$ (the base power consumed when the platform is switched on) and $\mathscr{P}_{Cal}$ (the CPU overhead when the platform is active).

We identify two classes of fault-tolerant protocols: coordinated and uncoordinated protocols. Both coordinated and uncoordinated protocols rely on checkpointing regularly (each checkpoint interval) the global state of the application in order to restart it in case of failure from the last checkpoint instead of re-executing the whole application. The problem of checkpointing is to ensure a global coherent state of the system. A global state is considered as coherent if it does not contain messages that are received but that were not sent. Coordinated protocols (already discussed above) are currently the most used fault-tolerant protocols in high-performance computing applications. In order to ensure the global coherent state of the system, the coordinated protocol relies on a coordination that consists of synchronizing all the processes before checkpointing [36]. Coordination may result in a huge waste in terms of performance. Indeed in order to synchronize all the processes, it is necessary to wait for all the inflight messages to be transmitted and received. Moreover, in case of failure with the coordinated protocol, all the processes have to be restarted from the last checkpoint even if a single process has crashed. This results in a huge waste in terms of energy consumption since all the processes even the non-crashed ones have to redo all the computations and the communications from the last checkpoint. The uncoordinated protocol with message logging addresses this issue by restarting only the failed processes. Thus, the power consumption in recovery is supposed to be much smaller than for coordinated checkpointing. However, in order to ensure a global coherent state of the system, all message logging protocols need to log all messages sent by all processes during the whole execution and this impacts the performance [5]. Hence, in case of failure, the non-crashed processes send to the crashed ones the messages that they have logged.

A first important result of this chapter is to determine the optimal checkpointing interval in terms of energy consumption, for coordinated checkpointing. Section 5.2 presents a detailed analysis to compute this optimal checkpointing interval, considering two distinct objectives: minimizing execution time or energy consumption. Then, according to the determined checkpointing interval, the next results presented in this chapter allow supercomputer users to choose between the coordinated and the uncoordinated protocols before pre-executing the HPC application in a given execution context. To this end, we rely on a methodology that estimates the energy consumption of fault-tolerant protocols relying on an energy calibration of the execution platform and a description of execution parameters. Section 5.3 presents this methodology and shows how it enables supercomputer users to select the less energy consuming fault-tolerant protocol without pre-running the application. We conclude the chapter in Sect. 5.4.

5.2 Optimal Checkpointing Period: Time versus Energy

This section deals with parallel scientific applications using non-blocking and periodic coordinated checkpointing to enforce resilience. We provide a model and detailed formulas for total execution time and consumed energy. We characterize the optimal period for both objectives, and we assess the range of time/energy trade-offs to be made by instantiating the model with a set of realistic scenarios for Exascale systems. We give a particular emphasis to I/O transfers, because the relative cost of communication is expected to dramatically increase, both in terms of latency and consumed energy, for future Exascale platforms.

This section extends the analysis of the waste presented in Sect. 1.3.1. We investigate trade-offs between execution time and energy consumption for the execution of parallel applications on future Exascale systems. The optimal period $\mathcal{T}_{\text{Time}}^{\text{opt}}$ given by Young's and Daly's formula [10, 44] will minimize (expected) execution time. However, this period $\mathcal{T}_{\text{Time}}^{\text{opt}}$ will not minimize energy consumption, mainly because the fraction of power $\mathcal{P}_{\text{Cal}}$ spent when computing (by the CPUs) is not the same as the fraction of power $\mathcal{P}_{\text{I/O}}$ spent when checkpointing. In particular, we revisit the work of Meneses et al. [35] for checkpoint/restart, where formulas are given to compute the time-optimum and energy-optimum periods. However, our model is more precise: (i) we carefully assess the impact of the power consumption required for I/O activity, which is likely to play a key role at the Exascale; (ii) we consider non-blocking checkpointing that can be partially overlapped with computations; (iii) we give a more accurate analysis of the consumed energy. To state this more precisely:

- We provide a refined analytical model to compute both the execution time and the consumed energy with a given checkpoint period. The model handles the case where checkpointing activity can be non-blocking, i.e., partially overlapped with computations.
- We provide analytical formulas to approximate the optimal period for time $\mathscr{T}_{\text{Time}}^{\text{opt}}$ as well as the optimal period for energy $\mathscr{T}_{\text{Energy}}^{\text{opt}}$, thereby refining and extending Daly [10] and Meneses, Sarood, and Kalé [35] results to non-blocking checkpoints.
- We assess the range of time/energy trade-offs to be made by instantiating the model with a set of realistic scenarios for Exascale systems.

5.2.1 Model

In this section, we introduce all model parameters. We start with parameters related to resilience (checkpointing) before moving to parameters related to energy consumption.

5.2.1.1 Checkpointing

We model coordinated checkpointing [9] where checkpoints are taken at regular intervals, after some fixed amount of work units have been performed. This corresponds to an execution partitioned into periods of duration T. Every period, a checkpoint of length C is taken.

An important question is whether checkpoints are blocking or not. On some architectures, we may have to stop executing the application before writing to the stable storage where the checkpoint data is saved; in that case checkpoint is fully blocking. On other architectures, checkpoint data can be saved on the fly into a local memory before the checkpoint is sent to the stable storage, while computation can resume progress; in that case, checkpoints can be fully overlapped with computations. To deal with all situations, we introduce a slowdown factor ω: during a checkpoint of duration C, the work that is performed is ωC work units. In other words, $(1 - \omega)C$ work units are wasted due to checkpoint jitter disrupting the progress of computation. Here, $0 \leq \omega \leq 1$ is an arbitrary parameter. The case $\omega = 0$ corresponds to a fully blocking checkpoint, while $\omega = 1$ corresponds to a checkpoint totally overlapped with computations. All intermediate situations can be represented.

Next we have to account for failures. During t time units of execution, the expectation of the number of failures is $\frac{t}{\mu}$, where μ is the MTBF (Mean Time Between Failures) of the platform. Note that if the platform if made of N identical resources whose individual mean time between failures is μ_{ind}, then $\mu = \frac{\mu_{\text{ind}}}{N}$. This relation is agnostic of the granularity of the resources, which can be anything from a single CPU to a complex multi-core socket. When a failure strikes, there is a downtime of length D (time to reboot the resource or set up a spare), and then a recovery of length R (time to read the last stored checkpoint). The work executed by the application since the last checkpoint and before the failure needs to be re-executed. Clearly, the shorter the period T, the less work to re-execute, but also the more overhead due to frequent checkpoints in a failure-free execution. The best trade-off when $\omega = 0$ (blocking checkpoint) is achieved for $T = \sqrt{2C\mu} + C$ (Young's formula [44]) or $T = \sqrt{2C(\mu + D + R)} + C$ (Daly's formula [10]). Both formulas are first-order approximations and valid only if all checkpoint parameters C, D, and R are small in front of μ (and these formulas collapse if they become negligible). In Sect. 5.2.2, we show how to extend these formulas to the case of non-blocking checkpoints (see also [4] for more details).

5.2.1.2 Energy

To compute the energy consumption of the application, we need to consider the energy consumption of the different phases, and hence the power consumption at each time-step. To this purpose, we define:

- $\mathscr{P}_{\text{Static}}$: this is the base power consumed when the platform is switched on.
- $\mathscr{P}_{\text{Cal}}$: when the platform is active, we have to consider the CPU overhead in addition to the static power $\mathscr{P}_{\text{Static}}$.
- $\mathscr{P}_{\text{I/O}}$: similarly, this is the power overhead due to file I/O. This supplementary power consumption is induced by checkpointing, or when recovering from a failure.
- $\mathscr{P}_{\text{Down}}$: for coordinated checkpointing, when one processor fails, the rest of the machine stays idle. $\mathscr{P}_{\text{Down}}$ is the power consumption overhead when one machine is down, that may be incurred for instance by rebooting the machine. In general, we let $\mathscr{P}_{\text{Down}} = 0$.

Meneses et al. [35] have a simpler model with two parameters, namely L, the base power (corresponding to $\mathscr{P}_{\text{Static}}$ with our notations), and H, the maximum power (corresponding to $\mathscr{P}_{\text{Static}} + \mathscr{P}_{\text{Cal}}$ with our notations). They use $\mathscr{P}_{\text{I/O}} = \mathscr{P}_{\text{Down}} = 0$.

In Sect. 5.2.2, we show how to compute the optimal period that minimizes the energy consumption. In Sect. 5.2.3, we instantiate the model with expected values for power consumption of Exascale platforms.

5.2.2 Optimal Checkpointing Period

We consider a parallel application whose execution time is $\mathscr{T}_{\text{base}}$ without any overhead due to the resilience method or the occurrence of failures. We compute the expectation $\mathscr{T}_{\text{final}}$ of the total execution time (accounting both for checkpointing and for failures) in Sect. 5.2.2.1, and the expectation $\mathscr{E}_{\text{final}}$ of the total energy consumed during this execution of length $\mathscr{T}_{\text{final}}$ in Sect. 5.2.2.2. We will compute the optimal period T that minimizes the objective, either $\mathscr{T}_{\text{final}}$ or $\mathscr{E}_{\text{final}}$.

5.2.2.1 Execution Time

The total execution time $\mathscr{T}_{\text{final}}$ of the application depends on two sources of overhead. We first compute $\mathscr{T}_{\text{ff}}$, the time taken by a fault-free execution, thereby accounting only for the overhead due to periodic checkpointing. Then we compute $\mathscr{T}_{\text{fails}}$, the time lost due to failures. Finally, $\mathscr{T}_{\text{final}} = \mathscr{T}_{\text{ff}} + \mathscr{T}_{\text{fails}}$. We detail here both computations:

- The reasoning to derive $\mathscr{T}_{\text{ff}}$ is simple. We need to execute a total amount of work equal to $\mathscr{T}_{\text{base}}$. During each period of length T, there is an amount of time $T - C$ where only computations take place, and an amount of time C of checkpointing, where only a work ωC is done. Therefore, the total number of work units executed during a period of length T is $T - C + \omega C = T - (1 - \omega)C$, and

$$\mathscr{T}_{\text{ff}} = \mathscr{T}_{\text{base}} \frac{T}{T - (1 - \omega)C}.$$

- The reasoning to compute $\mathscr{T}_{\text{fails}}$ is the following. Since the mean time between two failures is μ, the average number of failures during execution is $\frac{\mathscr{T}_{\text{final}}}{\mu}$. For each failure, the time lost is expressed as:

  - $D + R$ for downtime and recovery;
  - a time ωC for the work that was done during the previous checkpoint and that has to be redone because it was not checkpointed (because of the failure);
  - with probability $\frac{T-C}{T}$, the failure happens while we are not checkpointing, and the time lost is on average $A = \frac{T-C}{2}$;
  - otherwise, with probability $\frac{C}{T}$, the failure happens while we are checkpointing, and the time lost is on average $B = T - C + \frac{C}{2} = T - \frac{C}{2}$.

The time lost for each failure is

$$D + R + \omega C + \frac{T - C}{T}A + \frac{C}{T}B = D + R + \omega C + \frac{T}{2}.$$

Finally,

$$\mathscr{T}_{\text{fails}} = \frac{\mathscr{T}_{\text{final}}}{\mu}\left(D + R + \omega C + \frac{T}{2}\right).$$

We are now ready to express the total execution time:

$$\mathscr{T}_{\text{final}} = \mathscr{T}_{\text{ff}} + \mathscr{T}_{\text{fails}}$$

$$= \mathscr{T}_{\text{base}}\frac{T}{T - (1 - \omega)C} + \frac{\mathscr{T}_{\text{final}}}{\mu}\left(D + R + \omega C + \frac{T}{2}\right)$$

$$= \frac{T}{(T - (1 - \omega)C)\left(1 - \frac{D+R+\omega C+T/2}{\mu}\right)} \mathcal{T}_{\text{base}}$$

$$= \frac{T}{(T - a)\left(b - \frac{T}{2\mu}\right)} \mathcal{T}_{\text{base}},$$

where $a = (1 - \omega)C$ and $b = 1 - \frac{D+R+\omega C}{\mu}$.

This equation is minimized for

$$\mathcal{T}_{\text{Time}}^{\text{opt}} = \sqrt{2(1 - \omega)C(\mu - (D + R + \omega C))}. \tag{5.1}$$

Note that we retrieve the value given by Eq. (1.24) in Sect. 1.3.3.2. In the following, we let ALGOT be the checkpointing strategy that checkpoints with period $\mathcal{T}_{\text{Time}}^{\text{opt}}$.

5.2.2.2 Energy Consumption

In order to compute the total energy consumption of the execution, we consider the different phases during which the different powers introduced in Sect. 5.2.1.2 are used:

- First, we consume $\mathcal{P}_{\text{Static}}$ during each time-step of the execution. Indeed, even when a node fails and is shutdown, we still pay for the power of all the other nodes, for the cooling system, etc. The corresponding energy cost is $\mathcal{T}_{\text{final}} \mathcal{P}_{\text{Static}}$.
- Next, let $\mathcal{T}_{\text{Cal}}$ be the time during which the CPU is used, inducing a power overhead $\mathcal{P}_{\text{Cal}}$. $\mathcal{T}_{\text{Cal}}$ includes the base work $\mathcal{T}_{\text{base}}$, and $\mathcal{T}_{\text{re-exec}}$, the work that must be re-executed after each failure (which we multiply by the number of failures $\mathcal{T}_{\text{final}}/\mu$):
  - with probability $\frac{T-C}{T}$, the failure does not happen during a checkpoint, and the work to re-execute is $A = \omega C + \frac{T-C}{2}$;
  - with probability $\frac{C}{T}$, the failure happens during the execution of a checkpoint, and the work to re-execute is $B = \omega C + T - C + \frac{\omega C}{2}$.

We derive $\mathcal{T}_{\text{re-exec}} = \frac{T-C}{T}A + \frac{C}{T}B$, hence

$$\mathcal{T}_{\text{re-exec}} = \omega C + \frac{T^2 - C^2}{2T} + \frac{\omega C^2}{2T}.$$

Finally, we have:

$$\mathcal{T}_{\text{Cal}} = \mathcal{T}_{\text{base}} + \frac{\mathcal{T}_{\text{final}}}{\mu}\left(\omega C + \frac{T^2 - C^2}{2T} + \frac{\omega C^2}{2T}\right).$$

The corresponding energy consumption is $\mathcal{T}_{\text{Cal}} \mathcal{P}_{\text{Cal}}$.

- Let $\mathcal{T}_{I/O}$ be the time during which the I/O system is used, inducing a power overhead $\mathcal{P}_{I/O}$. This time corresponds to checkpointing and recovery from failures.

    - The total number of checkpoints that are taken in a fault-free execution is equal to the number of periods, $\frac{\mathcal{T}_{base}}{T-(1-\omega)C}$, and the time taken by checkpoints is therefore $\frac{\mathcal{T}_{base}C}{T-(1-\omega)C}$.
    - For each failure, there is an additional overhead:

1. the system needs to recover, which lasts R time-steps;
2. with probability $\frac{T-C}{T}$, the failure does not happen during a checkpoint, and there is no additional I/O overhead;
3. however, with probability $\frac{C}{T}$, the failure happens during a checkpoint, and the I/O time wasted is (in average) $\frac{C}{2}$.

Altogether, we obtain

$$\mathcal{T}_{I/O} = \frac{\mathcal{T}_{base}C}{T-(1-\omega)C} + \frac{\mathcal{T}_{final}}{\mu}\left(R + \frac{C^2}{2T}\right).$$

The corresponding energy consumption is $\mathcal{T}_{I/O}\mathcal{P}_{I/O}$.

- Finally, let $\mathcal{T}_{Down}$ be the total down time, incurring a power overhead $\mathcal{P}_{Down}$. We have

$$\mathcal{T}_{Down} = \frac{\mathcal{T}_{final}}{\mu}D,$$

and the corresponding energy cost is $\mathcal{T}_{Down}\mathcal{P}_{Down}$. This term is only included for full generality, as we expect to have $\mathcal{P}_{Down} = 0$ in most scenarios.

The final expression for the total energy consumed is

$$\begin{aligned}\mathcal{E}_{final} &= \mathcal{T}_{Cal}\mathcal{P}_{Cal} + \mathcal{T}_{I/O}\mathcal{P}_{I/O} + \mathcal{T}_{Down}\mathcal{P}_{Down} + \mathcal{T}_{final}\mathcal{P}_{Static}\\ &= \left(\mathcal{T}_{base} + \frac{\mathcal{T}_{final}}{\mu}\left(\omega C + \frac{T^2 - C^2}{2T} + \frac{\omega C^2}{2T}\right)\right)\mathcal{P}_{Cal}\\ &\quad + \left(\frac{\mathcal{T}_{final}}{\mu}\left(R + \frac{C^2}{2T}\right) + C\frac{\mathcal{T}_{base}}{T-(1-\omega)C}\right)\mathcal{P}_{I/O}\\ &\quad + \frac{\mathcal{T}_{final}}{\mu}D\mathcal{P}_{Down} + \mathcal{T}_{final}\mathcal{P}_{Static}.\end{aligned}$$

It is important to understand that $\mathcal{T}_{final} \neq \mathcal{T}_{Cal} + \mathcal{T}_{I/O} + \mathcal{T}_{Down}$, unless $\omega = 0$. Indeed, CPU and I/O activities are overlapped (and both consumed) when checkpointing. To ease the derivation of the optimal period that minimizes $\mathcal{E}_{final}$, we introduce some notations and let $\mathcal{P}_{Cal} = \alpha\mathcal{P}_{Static}$, $\mathcal{P}_{I/O} = \beta\mathcal{P}_{Static}$, and $\mathcal{P}_{Down} = \gamma\mathcal{P}_{Static}$. Reusing parameters $a = (1-\omega)C$ and $b = 1 - \frac{D+R+\omega C}{\mu}$ from Sect. 5.2.2.1, we obtain:

$$\frac{\mathcal{T}'_{\text{final}}}{\mathcal{T}_{\text{base}}} = \frac{-ab + \frac{T^2}{2\mu}}{(T-a)^2 \left(b - \frac{T}{2\mu}\right)^2}, \text{ and}$$

$$\frac{\mathcal{E}'_{\text{final}}}{\mathcal{P}_{\text{Static}}} = \frac{\mathcal{T}'_{\text{final}}}{\mu}\left(\alpha\omega C + \beta R + \gamma D + \frac{\alpha T}{2} - \frac{\alpha(1-\omega)C^2}{2T} + \frac{\beta C^2}{2T} + \mu\right)$$
$$+ \frac{\mathcal{T}_{\text{final}}}{2\mu}\left(\alpha + \frac{\alpha(1-\omega)C^2}{T^2} - \frac{\beta C^2}{T^2}\right) - \frac{\beta C \mathcal{T}_{\text{base}}}{(T-(1-\omega)C)^2}.$$

Then, letting $K = \dfrac{(T-a)^2\left(b - \frac{T}{2\mu}\right)^2}{\mathcal{P}_{\text{Static}}\mathcal{T}_{\text{base}}}$, we have:

$$K\mathcal{E}'_{\text{final}} = \frac{-ab+\frac{T^2}{2\mu}}{\mu}\left((\alpha\omega C + \beta R + \gamma D + \mu) + \frac{\alpha T}{2} + \frac{\alpha(1-\omega)C^2}{2T} + \frac{\beta C^2}{2T}\right)$$
$$+ \frac{(T-a)\left(b-\frac{T}{2\mu}\right)}{2\mu}\left(\alpha + \frac{\alpha(1-\omega)C^2 - \beta C^2}{T}\right) - \beta C\left(b - \frac{T}{2\mu}\right)^2$$
$$= T^3\left(\frac{1}{4\mu} - \frac{1}{4\mu}\right) + T^2\left(\frac{\alpha\omega C + \beta R + \gamma D}{2\mu^2} + \frac{b+\frac{a}{2\mu}}{2\mu} - \frac{\beta C}{4\mu^2} + \frac{1}{2\mu}\right)$$
$$+ T\left(-\frac{ab}{2\mu} - \frac{ab}{2\mu} + \frac{\beta Cb}{\mu} - 2\frac{(\alpha(1-\omega)-\beta)C^2}{4\mu^2}\right) - \beta Cb^2$$
$$- \frac{ab(\alpha\omega C + \beta R + \gamma D + \mu)}{\mu} - \left(\frac{b}{2\mu} - \frac{a}{4\mu^2}\right)(\alpha(1-\omega) - \beta)C^2$$
$$+ \frac{1}{T}\left((\alpha(1-\omega) - \beta)\frac{C}{2\mu} - (\alpha(1-\omega) - \beta)\frac{C}{2\mu}\right)$$
$$= T^2\left(\frac{\alpha\omega C + \beta R + \gamma D}{2\mu^2} + \frac{b}{2\mu} + \frac{a-\beta C}{4\mu^2} + \frac{1}{2\mu}\right)$$
$$+ T\left(\frac{(\beta C - a)b}{\mu} - 2\frac{(\alpha(1-\omega)-\beta)C^2}{4\mu^2}\right)$$
$$- \frac{ab(\alpha\omega C + \beta R + \gamma D + \mu)}{\mu} - \beta Cb^2$$
$$+ \left(\frac{b}{2\mu} + \frac{a}{4\mu^2}\right)(\alpha(1-\omega) - \beta)C^2.$$

Let $\mathcal{T}^{\text{opt}}_{\text{Energy}}$ be the only positive root of this quadratic polynomial in T: $\mathcal{T}^{\text{opt}}_{\text{Energy}}$ is the value that minimizes $\mathcal{E}_{\text{final}}$. In the following, we let ALGOE be the checkpointing strategy that checkpoints with period $\mathcal{T}^{\text{opt}}_{\text{Energy}}$.

As a side note, let us emphasize the differences with the approach of Meneses et al. [35] when restricting to the case $\omega = 0$ (because they only consider the blocking variant). For each failure, they consider that:

- energy lost due to re-execution is $\frac{T-2C}{2}\mathcal{P}_{\text{Cal}}$, while we have $\left(\frac{T-C}{T}\left(\frac{T-C}{2}\right) + \frac{C}{T}(T-C)\right)\mathcal{P}_{\text{Cal}} = \frac{T^2-C^2}{2T}\mathcal{P}_{\text{Cal}}$;
- energy lost due to I/O is $C\mathcal{P}_{\text{I/O}}$, while we have $\frac{C^2}{2T}\mathcal{P}_{\text{I/O}}$.

Theses differences come from our more detailed analysis of the impact of the failure location, which can strike either during the computation phase, or during the checkpointing phase, of the whole period.

5.2.3 Experiments

In this section, we instantiate the previous model with scenarios taken from current projections for Exascale platforms [21, 26, 42, 43]. We choose realistic values for all model parameters: this includes all types of power consumption ($\mathcal{P}_{\text{Static}}$, $\mathcal{P}_{\text{Cal}}$, $\mathcal{P}_{\text{I/O}}$ and $\mathcal{P}_{\text{Down}}$), all checkpoint parameters (C, R, D and ω), and the platform MTBF μ. We start with a word of caution: our choices for these parameters may be somewhat arbitrary, and do not cover the whole range of scenarios that can be investigated. However, a key feature of our model is its robustness: as long as μ is reasonably large in front of checkpoint times, the model is able to accurately predict the best period for execution time and for energy consumption.

The power consumption of an Exascale machine is capped to 20 MW. With 10^6 nodes, this represents a nominal power of 20 w per node. Let us express all power values in watts. A reasonable scenario is to assume that half this power is used for operating the platform, hence to let $\mathcal{P}_{\text{Static}} = 10$. The overhead due to computing would represent the other half, hence $\mathcal{P}_{\text{Cal}} = 10$. As for communications and I/Os, which are expected to cost an order of magnitude more than computing [43], we take an overhead of 100, hence $\mathcal{P}_{\text{I/O}} = 100$. A key parameter for the experimental study is the ratio

$$\rho = \frac{\mathcal{P}_{\text{Static}} + \mathcal{P}_{\text{I/O}}}{\mathcal{P}_{\text{Static}} + \mathcal{P}_{\text{Cal}}} = \frac{1 + \beta}{1 + \alpha}. \tag{5.2}$$

With our values, we get $\rho = 5.5$. Note that if we used $\mathcal{P}_{\text{Static}} = 5$ and kept the same overheads 10 and 100 for computing and I/O respectively, we would get $\mathcal{P}_{\text{Cal}} = 10$, $\mathcal{P}_{\text{I/O}} = 100$, and $\rho = 7$. These two representative values of ρ ($\rho = 5.5$ and $\rho = 7$) are emphasized by vertical arrows in the plots below on Fig. 5.1. As for $\mathcal{P}_{\text{Down}}$, the power during downtime, we use $\mathcal{P}_{\text{Down}} = 0$, meaning that during downtime we only account for the static power $\mathcal{P}_{\text{Static}}$ of the processors that are idle.

The Jaguar platform, with $N = 45,208$ processors, is reported to have experienced about one fault per day [46], which leads to an individual (processor) MTBF μ_{ind} equal to $\frac{45,208}{365} \approx 125$ years. Therefore, we set the individual (processor) MTBF to $\mu_{\text{ind}} = 25$ years. Letting the total number of processors N vary from $N = 219,150$ to $N = 2,191,500$ (future Exascale platforms), the platform MTBF μ varies from $\mu = 300$ min (5 h) down to $\mu = 30$ min. The experiments use resilience parameters that are representative of current and forthcoming large-scale platforms [7, 26]. We take $C = R = 10$ min, $D = 1$ min, and $\omega = 1/2$.

On Figs. 5.1, 5.2, and 5.3, we evaluate the impact of the ratio ρ (see Eq. (5.2)) on the gain in energy and loss in time of ALGoE with respect to ALGoT. The general trend is that using ALGoE can lead to significant gains in energy at the price of a small increase in execution time.

We then study in Figs. 5.4 and 5.5 the scalability of the approach on forthcoming platforms. We set the duration of the complete checkpoint and rollback (C and R, respectively) to 1 min, independently of the number of processors, and we let the downtime D equal to 0.1 min. It is reasonable to consider that checkpoint storage

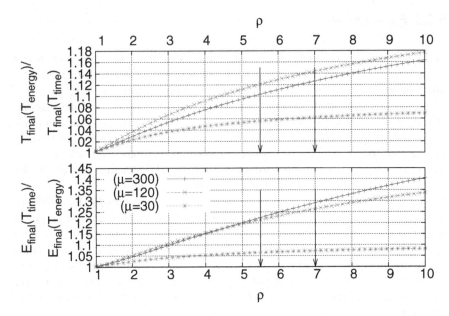

Fig. 5.1 Time and energy ratios as a function of ρ, with $C = R = 10$ min, $D = 1$ min, $\gamma = 0$, $\omega = 1/2$, and various values for μ

Fig. 5.2 Ratios of the different strategies with $C = R = 10$ min, $D = 1$ min, $\gamma = 0$, $\omega = 1/2$ as a function of μ and ρ: Energy ratio of ALGOT over ALGOE

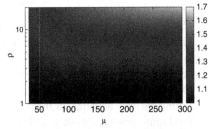

Fig. 5.3 Ratios of the different strategies with $C = R = 10$ min, $D = 1$ min, $\gamma = 0$, $\omega = 1/2$ as a function of μ and ρ: Execution time ratio of ALGOE over ALGOT

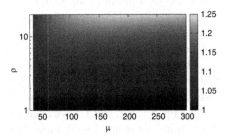

time will not increase with the number of nodes in the future, but on the contrary will remain constant. Indeed, system designers are studying a couple of alternative approaches. One consists of providing each computing node with local storage capability, ensuring through hardware mechanisms that this storage will remain available

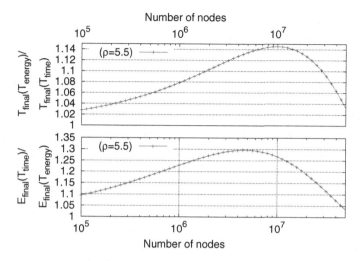

Fig. 5.4 Ratios of total energy and time for the two period strategies, as a function of the number of nodes, with $\mu = 120$ min for 10^6 nodes, $C = R = 1$ min, $D = 0.1$ min, $\gamma = 0$, $\omega = 1/2$: Time and energy ratios, as a function of the number of nodes, when $\rho = 5.5$

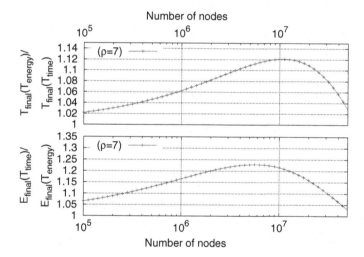

Fig. 5.5 Ratios of total energy and time for the two period strategies, as a function of the number of nodes, with $\mu = 120$ min for 10^6 nodes, $C = R = 1$ min, $D = 0.1$ min, $\gamma = 0$, $\omega = 1/2$: Time and energy ratios, as a function of the number of nodes, when $\rho = 7$

during a failure of the node. Another approach consists of using the memory of the other processors to store the checkpoint, pairing nodes as "buddies," thus allowing to take advantage of the high bandwidth capability of the high speed network to design a scalable checkpoint storage mechanism [22, 37, 40, 47].

The MTBF for 10^6 nodes is set to 2 hours, and this value scales linearly with the number of components. Given these parameters, Figs. 5.4 and 5.5 show (i) the execution time ratio of ALGOE over ALGOT, and (ii) the energy consumption ratio of ALGOT over ALGOE, both as a function of the number of nodes. Figures 5.4 and 5.5 confirm the important gain in energy that can be achieved, namely up to 30 % for a time overhead of only 12 %. When the number of nodes gets very high (up to 10^8), then we observe that both energy and time ratios converge to 1. Indeed, when C becomes of the order of magnitude of the MTBF, then both periods $\mathscr{T}_{\text{Time}}^{\text{opt}}$ and $\mathscr{T}_{\text{Energy}}^{\text{opt}}$ become close to C to account for the higher failure rate.

5.2.4 Summary

In this section, we have provided a detailed analysis to compute the optimal check-pointing period, when the checkpointing activity can be partially overlapped with computations. We have considered two distinct objectives: either the goal is to minimize the total execution time, or it is to minimize the total energy consumption. Because of the different power consumption overheads due to computations and I/Os, we obtain different optimal periods.

We have instantiated the formulas with values derived from current and future Exascale platforms, and we have studied the impact of the power overhead due to I/O activity on the gains in time and energy. With current values, we can save more than 20 % of energy with an MTBF of 300 min, at the price of an increase of 10 % in the execution time. The maximum gains are expected for a platform with between 10^6 and 10^7 processors (up to 30 % energy savings).

Our analytical model is quite flexible and can easily be instantiated to investigate scenarios that involve a variety of resilience and power consumption parameters.

5.3 Energy-Aware Fault-Tolerant Protocols for HPC Applications: A Methodology Based on Energy Estimation

Although some devices allow to measure the power and energy consumption of a protocol [14], measuring the energy consumption requires always to run the procotol at a large scale and this in all execution contexts. To reduce the number of measurements, we must be able to estimate accurately the energy consumption of a protocol, for any execution context and for any experimental platform. The advantage of such energy estimation is to evaluate the energy consumption of a protocol without pre-executing in each execution context, and in this order to be able to choose the fault-tolerant protocol that consumes less energy.

In order to adapt the energy estimations to the execution platform, we need to collect a set of power measurements of the nodes of the platform during the various operations that compose the fault-tolerant protocols. However, we learned from [15], that the nodes of a same cluster can have an heterogeneous idle power consumption, while they have the same extra power consumption due to the execution of an giving operation. We deduce that we need to measure the idle power consumption of each node of a same cluster but we need to measure the extra power consumption due to an operation only for each type of node. Moreover, in order to estimate the energy consumption according to the execution platform, we also need to measure the execution time of each operation on this platform. However, we have shown in [15] that the nodes of a cluster are homogeneous in terms of performance. We deduce that we do not need to measure the execution time due to an operation for each type of node. In order to adapt the energy estimations to the execution context, our estimation approach is also based on a description of the parameters execution provided by the user.

In this section, we explain our estimation methodology from the identification of the operations found in a fault-tolerant protocol to the energy estimation models of these operation, through a description of the calibration and the execution parameters that we need. We apply each step of our methodology to fault-tolerant protocols [17, 19]. In Sect. 5.3.1, we identify the various operations in the considered fault-tolerant protocols. Section 5.3.2 presents our methodology for calibrating the power consumption and the execution time of the identified operations. Section 5.3.3 shows how we estimate the energy consumption of the different operations by relying on the energy calibration and the different execution parameters. In Sect. 5.3.4, we evaluate the precision of the estimates for the considered fault-tolerant protocols by comparing then to the real energy measurements. In Sect. 5.3.5, we show how such energy estimations can be used in order to choose the energy-aware fault-tolerant protocol. Section 5.3.6 presents the conclusions of this section.

5.3.1 Identifying Operations in Fault-Tolerant Protocols

The first step of our methodology consists of identifying the various operations that we find in the different fault-tolerant protocols. An operation is a task that the fault-tolerant protocol may need to perform several times during the execution of an application.

As described in Sect. 5.1, we study the two families of fault-tolerant protocols: coordinated and uncoordinated protocols. For each of these two families, we distinguish two major phases: on the one hand, the checkpointing that occurs during a fault-free execution (i.e., without failure) of an application, and on the other hand, the recovery which occurs whenever a failure occurs. In our study, we focus on the checkpointing phase.

We consider an application using the fault-tolerant protocol that is running on N nodes with p processes per node and where p is identical in the N nodes. In fault-tolerant protocols, we identify the following operations:

- Checkpointing: performed in both coordinated and uncoordinated protocols, it consists in storing a snapshot image of the current application state that can be later on used for restarting the execution in case of failure. In our study, we consider the system level checkpointing at the system level and not checkpointing at the application level. Such a choice is motivated by the fact that not all the applications embed global checkpointing and that we cannot select the optimal checkpointing interval with the applicative checkpointing. We consider the checkpointing provided in the Berkeley Lab Checkpoint/Restart library (BLCR), and available in the MPICH2 implementation. In checkpointing, the basic operation is to write a checkpoint of V_{data} size on a reliable media storage. For our study, we consider only the HDD since RAM is not reliable.
- Message logging: performed in uncoordinated protocols, it consists in saving on each sender process the messages sent on a specific storage medium (RAM, HDD, NFS, ...). In case of failure, thanks to message logging, only the crashed processes need to restart. In message logging, the basic operation is to write the message of V_{data} size on a given media storage. For our study, we consider the RAM and the HDD.
- Coordination: performed in coordinated protocols, it consists in synchronizing the processes before taking the checkpoints. If some processes have inflight messages at the coordination time, all the other ones are actively polling until these messages are sent. This ensures that there will be no orphan messages: messages sent before taking the checkpoints but received after checkpointing. When there is no more inflight message, all the processes exchange a synchronization marker. In coordination, the basic operations are the active polling during the transmission of inflight messages of V_{data} and the synchronization of $N \times p$ processes that occurs when there is no more inflight message.

In order to estimate the energy consumption of these operations, we need to take into account a large set of parameters. These operations are associated to parameters that depend not only on the protocols but also on the application features, and on the hardware used. Thus, in order to estimate accurately the energy consumption due to a specific implementation of a fault-tolerant protocol, the estimator needs to take into consideration all the protocol parameters (checkpointing interval, checkpointing storage destination, etc.), all the application specifications (number of processes, number, and size of messages exchanged, volume of data written/read by each process, etc.) and all the hardware parameters (number of cores per node, memory architecture, type of hard disk drives, etc.).

- fault tolerance and application parameters: checkpointing interval, checkpointing storage destination, number of processes, number and size of messages exchanged between processes, type of storage media used (RAM, HDD, NFS, etc.), volume of data written/read by each process, etc.

- hardware parameters: number of nodes, number of sockets per node, number of cores per socket, network topology, memory architecture, network technologies (Infiniband, Gigabit Ethernet, proprietary solutions, etc.), type of hard disk drives (SSD, SATA, SCSI, etc.), etc.

We consider that a parameter is a variable of our estimator only if a variation of this parameter generates a significant variation of the energy consumption while all the other parameters are fixed. It is necessary to calibrate the execution platform by taking into account all the parameters to estimate the energy consumption.

5.3.2 Energy Calibration Methodology

Energy consumption depends strongly on the hardware used in the execution platform. For instance, the energy consumption of checkpointing depends on the checkpointing storage destination (SSD, SATA, SCSI, etc.), on the read and write speeds and on the access times to the resource. The goal of the calibration process is to gather energy knowledge of all the identified operations according to the hardware used in the supercomputer. To this end, we gather the information about the energy consumption of the operations by running a set of benchmarks allowing to collect at set of power measurements and execution times of the various operations. The goal of such calibration approach is to adapt to the supercomputer used, the energy evaluations computed from the theoretical estimation models, and this in order to make our energy estimations accurate on any supercomputer, regardless of specifications. Although this knowledge base has a significant size, it needs to be done only occasionally, for example, when there is a change in the hardware (like a new hard disk drive).

To estimate the energy consumption of a node performing an operation op, we need to obtain the power consumption of the node during the execution of op and the execution time of this operation. We know from [15] that the nodes from a same cluster are homogeneous in terms of performance. Therefore, we do need to measure and estimate the execution time due to an operation only for each type of nodes. Thus, the energy $\xi_{op}^{Node_i}$ consumed by a node i performing an operation op is:

$$\xi_{op}^{Node_i} = \rho_{op}^{Node_i} \cdot t_{op}$$

Analogously, the energy consumption $\xi_{op}^{Switch_j}$ of a $(switch)\, j$ during the operation op is:

$$\xi_{op}^{Switch_j} = \rho_{op}^{Switch_j} \cdot t_{op}$$

t_{op} is the time required to perform op by un type of nodes.
$\rho_{op}^{Node_i}$ is the power consumed by the node i during t_{op}.

$\rho_{op}^{Switch_j}$ is the power consumed by the switch j during t_{op}.

As a consequence, in order to calibrate the energy consumption, we need a calibrator for the power consumption described in Sect. 5.3.2.1 and a calibrator for the execution time described in Sect. 5.3.2.2.

5.3.2.1 Calibration of the Power Consumption ρ_{op}

We showed in [15] that the power consumption of a node i performing an operation op is composed of a static part, $\rho_{idle}^{Node_i}$, which is the power consumption of the node i when it is idle and a dynamic part $\Delta\rho_{op}^{Node_i}$, which is the extra power cost related to the operation op. We have shown that $\rho_{idle}^{Node_i}$ can be different even for identical nodes from homogeneous clusters. Therefore, we measure $\rho_{idle}^{Node_i}$ for each node i. We also have shown in [15] that $\Delta\rho_{op}^{Node_i}$ is the same for identical nodes running the same operation op. Consequently, we measure $\Delta\rho_{op}^{Node_i}$, for each operation op, once for each type of nodes,

In [15], we have also highlighted that the number p of processes used per node may influence the power consumed by the node. Therefore, we need to measure $\rho_{op}^{Node_i}(p)$ for every operation op and for different values of p. Thus, the power consumption $\rho_{op}^{Node_i}(p)$ of a node i during an operation op using p processes of this node is:

$$\rho_{op}^{Node_i}(p) = \rho_{idle}^{Node_i} + \Delta\rho_{op}^{Node}(p)$$

Analogously, the power consumption $\rho_{op}^{Switch_j}$ of a switch j during the operation op is:

$$\rho_{op}^{Switch_j} = \rho_{idle}^{Switch_j} + \Delta\rho_{op}^{Switch}$$

$\rho_{idle}^{Node_i}$ (or $\rho_{idle}^{Switch_j}$) is the power consumption of a node i (or of a switch j) when it is idle (i.e., switched on but executed nothing except the operating system) and $\Delta\rho_{op}^{Node_i}$ (or $\Delta\rho_{op}^{Switch_j}$) is the extra power consumption due to the execution of the operation op.

In order to compute $\Delta\rho_{op}^{Node}(p)$ (or $\Delta\rho_{op}^{Switch}$), we measure $\rho_{op}^{Node_i}(p)$ (or $\rho_{op}^{Switch_j}$) for a given node i (or a switch j) and subtract the static part of the power consumption which corresponds to the idle power consumption of the node i (or switch j). We measure $\rho_{op}^{Node_i}(p)$ (or $\rho_{op}^{Switch_j}$) by making the operation last a few seconds. Therefore, it is an *mean* extra power consumption because it is computed from the average of several power measurements (one every second).

Moreover, $\rho_{op}^{Node_i}(p)$ and so $\Delta\rho_{op}^{Node}(p)$ may vary depending on the number p of processes used by the node i. Therefore, we need to calculate $\Delta\rho_{op}^{Node}(p)$ and thus to measure $\rho_{op}^{Node_i}(p)$ for different values of p in order to be able to estimate $\Delta\rho_{op}^{Node}(p)$

for a number p of processes executing the operation op. To do this, we should be able to know how $\Delta\rho_{op}^{Node}(p)$ evolves according to p (and this for each type of nodes). We do not know such information a priori. To this end, we rely on four possible models presented in the table below:

Linear	$\Delta\rho_{op}^{Node}(p) = \alpha p + \beta$
Logarithmic	$\Delta\rho_{op}^{Node}(p) = \alpha ln(p) + \beta$
Power	$\Delta\rho_{op}^{Node}(p) = \beta p^{\alpha}$
Exponential	$\Delta\rho_{op}^{Node}(p) = \alpha^{p} + \beta$

For each type of nodes, we measure $\Delta\rho_{op}^{Node}(p)$ for five different numbers of processes:

- the smallest possible value p denoted p_{min}, which is equal to 1;
- the highest possible value p denoted p_{max}, which corresponds to the number of cores available in the node;
- the median value denoted p_2 which corresponds to half of the number of cores available in the node;
- the number p_1 which is located in the middle of the interval $[p_{min}; p_2]$;
- the number p_3 which is located in the middle of the interval $[p_2; p_{max}]$

Then, we determine thanks to the least squares method [41] the coefficients (α and β) of each of the four models according to the five measured values for $\Delta\rho_{op}^{Node}(p)$. We compute the coefficient of determination R^2 corresponding to each of the four adjusted models obtained with the least squares method. We consider $\Delta\rho_{op}^{Node}(p)$ evolves according to the adjusted model for which the coefficient of determination is the highest one (i.e., that is to say, the closest to 1).

For our measurements of $\Delta\rho_{op}^{Node}(p)$ (deduced from measurements of $\rho_{op}^{Node_i}(p)$), we use an external wattmeter capable to provide us the mean power measurements with a sufficiently high frequency (1 Hz). We have shown in [14] that the OMEGAWATT wattmeter is a good candidate to collect such power measurements.

5.3.2.2 Calibration of the Execution Time t_{op}

The execution time t_{op} depends on one or many parameters according to the operation op. To take into account the possible effects of congestion, we consider that the number p of the same process node performing the same operation simultaneously op is a parameter to consider in our calibration of t_{op}. For example, this may occur if multiple processes on the same node try to write data simultaneously on the local hard drive.

To calibrate t_{op}, we need to measure the execution time by varying different parameters. To do this, we measure t_{op} for five values uniformly distributed between the minimum and maximum for each parameter (while fixing all the other parameters). The five values of each parameter are chosen similarly to what we have previously reported with the parameter p for the calibration of $\Delta \rho_{op}^{Node}(p)$.

We consider two cases:

1. "Known model" case: we know a model a model where t_{op} evolves with respect to the parameters. We know it from the literature, with the knowledge of the algorithm used in the operation op or resource requested by the operation op. In this case, we determine the coefficients of the theoretical model using the least squares method [41] based on the values of the five parameters.
2. "No model known" case: we do not know how t_{op} evolves with respect to the parameters. In this case, for each parameter, we proceed to the determination by the adjusted least squares method as presented for the calibration of $\Delta \rho_{op}^{Node}(p)$ relying on the four models (linear, logarithmic, exponential, and power).

To measure t_{op}, we instrument the code of the algorithm or the protocol of the operation op, in order to obtain the corresponding execution time. To ensure that the calibration of the execution time is accurate, we realize each measurement 30 times and we compute the mean value of the 30 measurements.

5.3.2.3 Models Used for the Execution Times of the Identified Fault-Tolerant Operations

In this section, we describe the models used for the execution times of each operation of the fault-tolerant protocols. For each operation op, t_{op} depends on different parameters.

We remind that the calibration of t_{op} is required for each type of nodes. In other words, we do not need to calibrate t_{op} on all nodes when they are all identical.

For each type of nodes, the time $t_{checkpointing}$ required for checkpointing a volume of data V_{data} is:

$$t_{checkpointing}(p, V_{data}) = t_{access}(p) + t_{transfer}(p, V_{data}) = t_{access}(p) + \frac{V_{data}}{r_{transfer}(p)}$$

Similarly, the time $t_{logging}$ required to log a message with a size equal to V_{data} is:

$$t_{logging}(V_{data}) = t_{access} + t_{transfer}(V_{data}) = t_{access} + \frac{V_{data}}{r_{transfer}}$$

p is the number of processes within the same node simultaneously trying to perform the checkpointing operation. t_{access} is the time required to access the storage media where the checkpoint will be saved or the message logged. $t_{transfer}$ is the time required to write data size V_{data} on the storage medium. $r_{transfer}$ is the transmission rate when writing on storage medium.

In the case of checkpointing, t_{access} and $r_{transfer}$ (and $t_{transfer}$) depend on the number p of processes per node since the p processes save their checkpoints simultaneously on the same storage media as the frequency of checkpoints writing is the same for all processes of an application.

A message is logged on a storage medium once it has been sent by a process of the node through the network interface used by the node. Thus, if several processes of the node try to send messages, there will be a traffic congestion at the network interface and the time for the current message will overlap the time of writing the message previously sent. In other words, this means that we consider that we can not find themselves in a situation where multiple messages are logged simultaneously by p processes of the node. Therefore, in the case of message logging, t_{access} and $r_{transfer}$ (and $t_{transfer}$) do not depend on the number p process per node.

As explained in Sect. 5.3.2.2, we measure $t_{checkpointing}$ considering both p and V_{data} parameters.

We know the theoretical model of $t_{checkpointing}$ based on V_{data} so for this parameter, we proceed to the determination of the coefficients of the theoretical model as explained in the case "with known model" (Sect. 5.3.2.2).

For p parameter, we do not have theoretical model giving $t_{checkpointing}$ based on p and therefore proceed as explained in the case of "no known model" (Sect. 5.3.2.2).

Regarding $t_{logging}$, it depends only on V_{data} and we have the theoretical model giving $t_{logging}$ depending on this parameter. So we proceed as explained in the "with known model" case.

We calibrate $t_{checkpointing}$ and $t_{logging}$ with respect to various storage media available on each node of the platform (RAM, local hard disk, flash SSD, etc.).

As we consider checkpointing at system-level, coordinated protocol requires a coordination between all processes.

The execution time for coordination between all processes is:

$$t_{coordination}(N, p, V_{data}) = t_{polling}(V_{data}) + t_{synchro}(N, p)$$
$$= \frac{V_{data}}{R_{transfer}} + t_{synchro}(N, p)$$

p is the number of processes of the node i trying to perform coordination. $t_{synchro}(N, p)$ is the time required to exchange a marker synchronization between all processes. $t_{synchro}(N, p)$ depends on the number of nodes and the number of processes per node involved in the synchronization. We do not have a theoretical model for $t_{synchro}(N, p)$ neither in terms of N nor based on p. For the calibration, we proceed as explained in the "without known model" case (Sect. 5.3.2.2). $t_{polling}(V_{data})$ is the time required to finish transmitting the messages being transmitted at the time of coordination. In other words, $t_{polling}(V_{data})$ is equal to the time required to transfer the larger application message. $R_{transfer}$ is the transmission rate in the network infrastructure used for the platform.

Regarding the polling time, $t_{polling}(V_{data})$, we have a theoretical model giving $t_{polling}(V_{data})$. For the calibration, we proceed as in the "known model" case for V_{data} parameter.

5.3.3 Energy Estimation Methodology

We have previously described how we realize the energy calibration. Once the calibration is done, the estimator is able to provide estimates of the energy consumed by the various operations identified for fault-tolerant protocols. Figure 5.6 shows the framework components related to the estimation of the energy consumed.

We can now describe how to estimate the energy consumed by each of the identified operations. To this end, we rely on the parameters provided by the user and the data measured by our calibrator.

Once the administrator has provided the hardware settings of the platform, the calibrator performs the various steps required to build the knowledge base on the power consumption and the execution time of the various identified operations. Then, based on the calibration results and a description of the application (the application memory size, etc.) and runtime parameters (number of nodes used, number of processes per node, etc.) provided by the user, the estimator calculates the energy consumption of different fault-tolerant protocols.

The parameters that we get from the user for the estimation depend on each operation to estimate. In case these parameters correspond to the values that we have measured during calibration, estimation directly uses these values to calculate the energy consumed by the operation. If this is not the case, that is to say, if there is a lack of measurement points in the calibrator, the estimator uses the models created with the least squares method [41] during calibration.

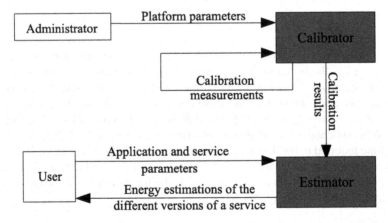

Fig. 5.6 Framework to estimate the energy consumption of fault-tolerant protocols

This section describes how we estimate the energy consumed by each operation identified in the considered fault-tolerant protocols. For this, we show the necessary information: the parameters provided by the user and the data measured by our calibrator.

5.3.3.1   Checkpointing

To estimate the energy consumption of checkpointing, the estimator gets from the user the total memory size required by the application to run, the number of nodes N and the number p of processes per node. From this information, the estimator calculates the average memory size V_{memory}^{mean} required by each process (total memory size divided by the number of processes). Then the estimator gets from the calibrator the extra power consumption $\Delta\rho_{checkpointing}(p)$ and the execution time $t_{checkpointing}(p, V_{memory}^{mean})$ depending on the models obtained by the least squares method in the step of the calibration. It also gets the measurement $\rho_{idle}^{Node_i}$ for each node i. We denote respectively by $\xi_{checkpointing}^{Node_i}(p)$ and $\rho_{checkpointing}^{Node_i}(p)$ the energy consumption and the average power consumption of each node i performing checkpointing. The estimation of the energy consumption of a single checkpointing is given by:

$$
\begin{aligned}
E_{checkpointing} &= \sum_{i=1}^{N} \xi_{checkpointing}^{Node_i}(p) \\
&= \sum_{i=1}^{N} \rho_{checkpointing}^{Node_i}(p) \cdot t_{checkpointing}(p, V_{memory}^{mean})) \\
&= t_{checkpointing}(p, V_{memory}^{mean}) \cdot \sum_{i=1}^{N} (\rho_{idle}^{Node_i} + \Delta\rho_{checkpointing}(p)) \\
&= t_{checkpointing}(p, V_{memory}^{mean}) \cdot \left(N \cdot \Delta\rho_{checkpointing}(p) \right. \\
&\quad \left. + \left(\sum_{i=1}^{N} \rho_{idle}^{Node_i} \right) \right)
\end{aligned}
$$

5.3.3.2   Message Logging

To estimate the energy consumption of message logging, the estimator gets from the user the number of nodes N, the number p of processes per node, the number and total size of all messages sent during the application that he wants to run.

With this information, the estimator calculates the average volume V_{data}^{mean} of data sent and therefore logged on each node (total size of all messages sent divided by the number of nodes N). Then, the estimator gets from the calibrator the extra power consumption $\Delta\rho_{logging}$ and the execution time $t_{logging}(p, V_{data}^{mean})$ depending on the models obtained with least squares method in the step of the calibration. It also receives the measurement of $\rho_{idle}^{Node_i}$ for each node i.

The estimation of the energy consumption of messages logging is given by:

$$
\begin{aligned}
E_{logging} &= \sum_{i=1}^{N} \xi_{logging}^{Node_i}(p) \\
&= \sum_{i=1}^{N} \rho_{logging}^{Node_i}(p) \cdot t_{logging}(V_{data}^{mean}) \\
&= t_{logging}(V_{data}^{mean}) \cdot \sum_{i=1}^{N}(\rho_{idle}^{Node_i} + \Delta\rho_{logging}(p)) \\
&= t_{logging}(V_{data}^{mean}) \cdot \left(N \cdot \Delta\rho_{logging}(p) + \left(\sum_{i=1}^{N} \rho_{idle}^{Node_i}\right)\right)
\end{aligned}
$$

5.3.3.3 Coordination

We remind that the coordination is divided into two phases: the active polling during the transmission of the inflight messages, followed by the synchronization of all processes. To estimate the energy consumption of the coordination, the estimator calculates the average message size $V_{message}^{mean}$ as the total size of messages divided by the total number of messages exchanged. The estimator also uses the total number of nodes N and the number of processes per node p. Then the estimator gets from the calibrator the extra power consumption $\Delta\rho_{sync}(p)$ and the execution time $t_{sync}(N, p)$ depending on the models obtained with the least squares method in calibration step. It also receives the measurement $\rho_{idle}^{Node_i}$ for each node i. The estimation of the energy consumption $E_{synchro}$ of synchronization is given by:

$$
\begin{aligned}
E_{synchro} &= \sum_{i=1}^{N} \xi_{synchro}^{Node_i}(N, p) \\
&= \sum_{i=1}^{N} \rho_{synchro}^{Node_i}(p) \cdot t_{synchro}(N, p) \\
&= t_{synchro}(N, p) \cdot \sum_{i=1}^{N}(\rho_{idle}^{Node_i} + \Delta\rho_{synchro}(p)) \\
&= t_{synchro}(N, p) \cdot \left(N \cdot \Delta\rho_{synchro}(p) + \left(\sum_{i=1}^{N} \rho_{idle}^{Node_i}\right)\right)
\end{aligned}
$$

Regarding active polling, the estimator gets from the calibrator the extra power consumption $\Delta\rho_{polling}(p)$ and the execution time $t_{polling}(N, p, V^{mean}_{message})$ depending on the models obtained with least squares method in the calibration step. The estimation of the energy consumption of active polling is given by:

$$
\begin{aligned}
E_{polling} &= \sum_{i=1}^{N} \xi^{Node_i}_{polling}(N, p) \\
&= \sum_{i=1}^{N} \rho^{Node_i}_{polling}(p) \cdot t_{polling}(V^{mean}_{message}) \\
&= t_{polling}(V^{mean}_{message}) \cdot \sum_{i=1}^{N} (\rho^{Node_i}_{idle} + \Delta\rho_{polling}(p)) \\
&= t_{polling}(V^{mean}_{message}) \cdot \left(N \cdot \Delta\rho_{polling}(p) + \left(\sum_{i=1}^{N} \rho^{Node_i}_{idle} \right) \right)
\end{aligned}
$$

The estimator computes the energy consumption of coordination as follows:

$$
E_{coordination} = E_{polling} + E_{synchro}
$$

5.3.4 Validation of the Estimations

To validate our estimations, we perform various real applications of high-performance computing with different fault-tolerant protocols on a homogeneous cluster of the experimental distributed platform for large-scale computing, Grid'5000[6], then we compare the energy consumption actually measured to the energy consumption evaluated by our estimator. For the experiments to validate our estimations, we used a cluster of the Grid'5000 distributed platform. The cluster we used for our experiments offers 16 identical nodes Dell R720. Each node contains 2 Intel Xeon CPU 2.3 GHz, with 6 cores each; 32 GB of memory; a 10 Gigabit Ethernet network; a SCSI hard disk with a storage capacity of 598 GB. We monitor this cluster with an energy-sensing infrastructure of external wattmeters from the SME Omegawatt. This energy-sensing infrastructure, which was also used in [13], enables to get the instantaneous consumption in Watts, at each second for each monitored node[12]. Logs provided by the energy-sensing infrastructure are displayed lively and stored into a database, in order to enable users to get the power and the energy consumption of one or more nodes between a start date and an end date. We ran each experiment 30 times and computed the mean value over the 30 values. We use the same notations as in previous sections: N is the number of nodes, p is the number of processes, and op denotes one of the identified operations (checkpointing, message logging, etc.).

5.3.4.1 Calibration Results of the Platform

In this section, we present some of the calibration results on the considered platform according to the methodology described in Sect. 5.3.2. The considered platform is composed only of identical nodes: thus there is only one type of node. They are interconnected using a single network switch.

Calibrating the power consumption

First, we measure the idle power consumption ρ_{idle}^i for each node i of the experimental cluster. Figure 5.7 shows the idle power consumption of the 16 nodes belonging to the considered cluster. From this figure, even if the cluster is composed of homogeneous nodes, we notice the need to calibrate the electrical power when idle of each node.

For each identified operation op and for each of the cluster nodes, we calibrate with OMEGAWATT the average additional cost of electrical power due to the op operation, $\Delta\rho_{op}(p)$, as explained in Sect. 5.3.2.1. Since each node of the *Taurus* cluster has 12 processing cores, the five values of p we choose to calibrate $\Delta\rho_{op}(p)$ are 1, 4, 6, 9, and 12 processes per node. Figure 5.8 shows the measurements $\Delta\rho_{op}(p)$ for the five values of p and for each operation op identified in fault-tolerant protocols.

We note in Fig. 5.8 that for some operations, $\Delta\rho_{op}(p)$ does vary depending on the number of cores per node that perform the same operation. For some operations, such as checkpointing $\Delta\rho_{op}(p)$ is almost a constant function of p. For $\Delta\rho op(p)$ of these operations, we obtain one of the four models of the calibrator (see Sect. 5.3.2.1) with a coefficient α very close to 0 and a value of β very close to the constant value of $\Delta\rho_{op}(p)$ (i.e., that is to say, quasi-stationary model). For example, the model of $\Delta\rho_{checkpointing}(p)$ adjusted by the least squares method for the five values of p is:

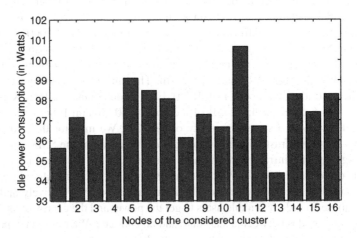

Fig. 5.7 Idle power consumption of the nodes of the cluster *Taurus*

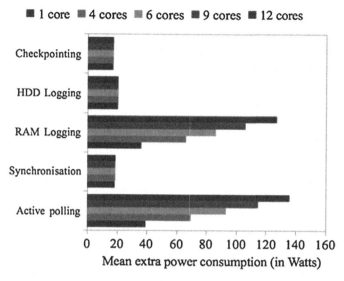

Fig. 5.8 Extra power consumption of fault-tolerant operations

$$\Delta \rho_{checkpointing}(p) = 17.22 \cdot p^{0.0084271}$$

Although the fitted model is a power model, the very low coefficient α implies that $\Delta \rho_{checkpointing}(p)$ is a quasi-stationary function p. The coefficient of determination R^2 corresponding to this model is 0.976, which is very close to 1.

For other operations such as *RAM logging*, $\Delta \rho_{op}(p)$ increases with p. For example, the model of $\Delta \rho_{RAM\_logging}(p)$ obtained in the calibration is:

$$\Delta \rho_{RAM\_logging}(p) = 35.237 \cdot p^{0.50158}$$

The fact that α is very close to 0.5 means that $\Delta \rho_{RAM\_logging}(p)$ is almost expressed in terms of $\sqrt{p}$. The coefficient of determination R^2 corresponding to this model is 0.992, which is also very close to 1.

In addition, we measure the energy consumption when idle of 10 Gigabit Ethernet switch for 300 s followed by the electrical power during heavy network traffic for 300 s. To measure its electrical power when idle, we ensure that there is no network traffic by turning off all nodes that are interconnected by the network switch. To measure its electrical power during heavy network traffic, we run `iperf` in server mode on one of the nodes and `iperf` in client mode on all other interconnected nodes. Figure 5.9 shows the electrical measurements.

From Fig. 5.9, we note that the electric power switch remains almost constant throughout the duration of the experiment. In other words, the electrical power network switch does not vary depending on the network traffic. This means that $\Delta \rho_{op}^{Switch}$ is (almost) equal to 0 for all operations ($\forall op, \rho_{op}^{Switch_j} = \rho_{idle}^{Switch_j}$). A recent

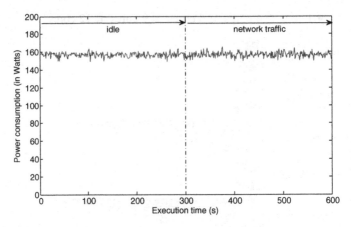

Fig. 5.9 Power consumption of the switch when idle for 300 s and with an intense network load for 300 s

study [29, 34] confirms this fact in evaluating and demonstrating that the electrical power of multiple network devices is not affected by network traffic. That said, even if the electrical power of a network switch would depend on network traffic, our approach to calibration would allow to take into account in measuring $\Delta \rho_{op}^{Switch}$ for each operation op.

Calibration of the execution time

Based on the methodology presented in Sect. 5.3.2.2, we calibrate the execution time for each operation on each type of node of the experimental platform.

To calibrate the execution time of checkpointing on local hard drive, we consider a variable number of cores per node simultaneously checkpointing and we measure the time for different sizes of checkpoints V_{data} for one node of the experimental platform. Each node process saves a checkpoint with a size equal to V_{data}. In other words, when there are p processes that save checkpoints simultaneously a volume of $p \cdot V_{data}$ is saved on the local hard drive. Figure 5.10 shows the measured time for checkpointing on a node of the experimental platform. As explained in Top, we choose 1, 4, 6, 9, and 12 processes per node for the five values of p and 0 MB, 500 MB, 1000 MB, 1500 MB, and 2000 MB for the five values of V_{data}. The choice of 2000 MB as the maximum size of checkpoint is motivated by the fact that each node has only 32 GB of memory that can be shared by 12 processing cores. For different values of p, figure shows how evolves $t_{checkpointing}$ with respect to V_{data}.

First, we observe that the curves have a linear trend according to V_{data} for p fixed. For example, for $p = 4$, the model for $t_{checkpointing}$ adjusted by the least squares method from the five values of V_{data}:

$$t_{checkpointing}(4, V_{data}) = \frac{1}{0.56569 \cdot 10^9} \cdot V_{data} + 0.09433 \cdot 10^{-3}$$

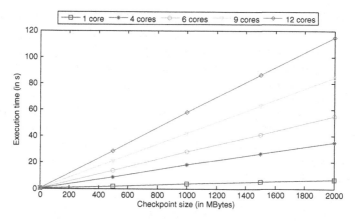

Fig. 5.10 Calibration of checkpointing on local hard drive

We also note that for V_{data} fixed $t_{checkpointing}$ increases when p grows this is because of the congestion of the input–output generated by concurrent access by p process on local hard drive. For $V_{data} = 1000$ MB the model of $t_{checkpointing}$ adjusted by the least squares method from the five values of p:

$$t_{checkpointing}(p, 1000\text{MB}) = 4.91359 \cdot p - 1.5026$$

If, for example, we want to estimate the time $t_{checkpointing}(3, 800\text{MB})$, that is to say, for values of p and V_{data} which both are not belonging to the five measured values, we calculate:

- on one side: $t_{checkpointing}(1, 800\text{MB})$, $t_{checkpointing}(4, 800\text{MB})$, $t_{checkpointing}(6, 800\text{MB})$, $t_{checkpointing}(9, 800\text{MB})$ and $t_{checkpointing}(12, 800\text{MB})$, respectively from the equations $t_{checkpointing}(1, V_{data})$, $t_{checkpointing}(4, V_{data})$, $t_{checkpointing}(6, V_{data})$, $t_{checkpointing}(9, V_{data})$ and $t_{checkpointing}(12, V_{data})$;
- on the other side: $t_{checkpointing}(3, 0\text{MB})$, $t_{checkpointing}(3, 500\text{MB})$, $t_{checkpointing}(3, 1000\text{MB})$, $t_{checkpointing}(3, 1500\text{MB})$, $t_{checkpointing}(3, 2000\text{MB})$, respectively from the equations $t_{checkpointing}(p, 0\text{MB})$, $t_{checkpointing}(p, 500\text{MB})$, $t_{checkpointing}(p, 1000\text{MB})$, $t_{checkpointing}(p, 1500\text{MB})$, $t_{checkpointing}(p, 2000\text{MB})$.

From the calculated values $t_{checkpointing}(1, 800\text{MB})$, $t_{checkpointing}(4, 800\text{MB})$, $t_{checkpointing}(6, 800\text{MB})$, $t_{checkpointing}(9, 800\text{MB})$ and $t_{checkpointing}(12, 800\text{MB})$, we determine by the least squares method, the model giving $t_{checkpointing}(p, 800\text{MB})$ as a function of p (as explained in Sect. 5.3.2.2) and calculate the determination coefficient R^2 corresponding to the adjusted model.

Similarly, from the values $t_{checkpointing}(3, 0\text{MB})$, $t_{checkpointing}(3, 500\text{MB})$, $t_{checkpointing}(3, 1000\text{MB})$, $t_{checkpointing}(3, 1500 \text{ MB})$, $t_{checkpointing}(3, 2000 \text{ MB})$, we determine the model giving $t_{checkpointing}(3, V_{data})$ as a function of V_{data} and calculate the determination coefficient R^2 corresponding to the thereby adjusted model.

Then between $t_{checkpointing}(p, 800MB)$ and $t_{checkpointing}(3, V_{data})$, we choose the model for which the determination coefficient is the closest to 1. Then we calculate $t_{checkpointing}(3, 800MB)$ with the choosen model.

Figure 5.11 presents the execution time for message logging in RAM and on a HDD. To calibrate the execution time of message logging on memory or on disk, we measure the time for different message sizes V_{data} for one node of the experimental platform. The values choosen for V_{data} are 0 KB, 500 KB, 1000 KB, 1500 KB et 2000 KB. As explained in Sect. 5.3.2.3, we do not need to calibrate $t_{logging}$ as a function of p because the processes do not write simultaneously the messages on the medium storage due to the contention during message sending. We measure the execution time when a single process ($p = 1$) of the node executes the message logging operation.

We observe that the curves have a linear trend and this as well for message logging on RAM on local hard drive. The message logging time on local hard drive is higher than the RAM one and this regardless of the size of the logged message. Similarly to checkpointing, we get the following adjusted models for $t_{logging}$:

$$\text{In RAM} \qquad : t_{logging}(V_{data}) = \frac{1}{4.4342 \cdot 10^9} \cdot V_{data} + 0.0426 \cdot 10^{-3}$$
$$\text{On the local HDD} : t_{logging}(V_{data}) = \frac{1}{1.0552 \cdot 10^9} \cdot V_{data} + 0.0858 \cdot 10^{-3}$$

Regarding coordination, we need to calibrate the time of the synchronization as well as the transfer time of a message.

To calibrate the synchronization time $t_{synchro}(N, p)$ of Np process, we measure this time for different values of N and for different values of p. The measured values

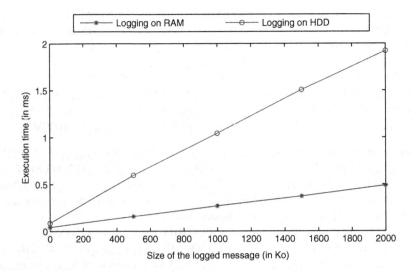

Fig. 5.11 Calibration of message logging on RAM and local disk

of p and N are chosen as explained in Sect. 5.3.2.2. In our case, the measured values of p are 1, 4, 6, 9 and 12, while the measured values for N are 1, 4, 8, 12, and 16. Figure 5.12 presents the synchronization time measured by the calibrator. For example, point 4 cores/8 nodes is the time required to synchronize 32 processes, 32 uniformly distributed over 8 nodes. First, we find that the time to synchronize processes located on the same node is lower than for processes located on different nodes. Indeed, it requires much less time to synchronize processes located on the same node than for processes located on different nodes. The transmission rate of the network is much lower than the transmission rate within a single node.

For example, for $p = 4$, the model for $t_{synchro}$ adjusted by the least squares method from the five values of N is:

$$t_{synchro}(N, 4) = 0.0103757 \cdot ln(N) + 0.00445945$$

For $N = 8$, the model for $t_{synchro}$ adjusted by the least squares method from the five values of N is:

$$t_{synchro}(8, p) = 0.00443799 \cdot ln(p) + 0.02225942$$

If, for example, we want to estimate the time $t_{synchro}(N, p)$, that is to say, for values of N and p which booth are not belonging to the five measured values, then we proceed in a manner similar to that explained for $t_{checkpointing}(p, V_{data})$.

We calibrate the time needed to transfer a message (i.e., the active polling occurring at the time of coordination) on the experimental platform by varying the of size V_{data} of the message to transfer. To do this, we measure the execution time

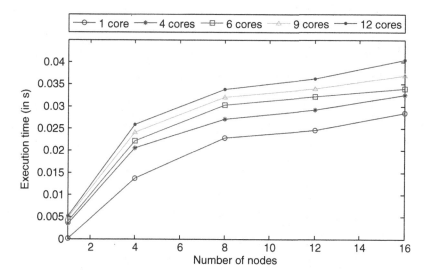

Fig. 5.12 Calibration of the synchronization time of the experimental platform

$t_{polling}(V_{data})$ to transfer a message sent using *MPI_Send* by a process located on a given node to a process on a different node. In the general case, we must make this measurement for each pair of processes at different levels of the network hierarchy, i.e., for two processes that need to cross a single network switch, then for two processes that need to cross two network switches, etc. In our experimental platform, a single network switch interconnects all nodes so we only need to measure the time for a couple of processes on different nodes.

To calibrate the execution time to transfer a message over the network, we measure the time for different message sizes V_{data} for a couple of processes located on two separate nodes. The values chosen for V_{data} are 0 KB, 500 KB, 1000 KB, 1500 KB, and 2000 KB. On Fig. 5.13, we present the calibration of the transfer time of a message.

The measured transfer time depends linearly on the size of the message transferred. Similarly to checkpointing, we get the following adjusted model for $t_{polling}$:

$$t_{polling}(V_{data}) = \frac{1}{0.60148 \cdot 10^9} \cdot V_{data} + 3.6222 \cdot 10^{-3}$$

5.3.4.2 Accuracy of the Estimations

In this section, we seek to compare the energy consumption achieved by our estimator once the calibration is made (but before executing the application) to the energy actually measured by the meters OMEGAWATT during the execution of the application.

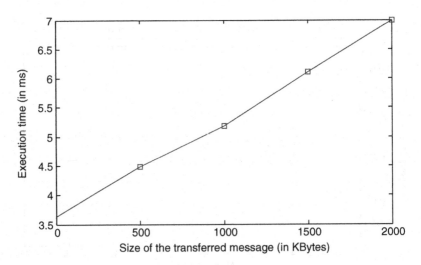

Fig. 5.13 Calibration of the active polling for the experimental platform

Fig. 5.14 Energy
estimations (in kJ) of
operations related to fault
tolerance

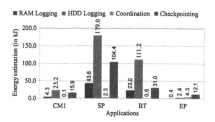

Fig. 5.15 Relative
differences (in %) between
the estimated and measured
energy consumption of the
operations related to fault
tolerance

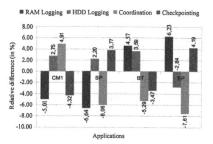

We consider four HPC applications: CM1[1] with a 2400 × 2400 × 40 resolution
and 3 NAS[2] of class D (SP, BT, et EP) executed on 144 processes (i.e., 12 nodes with
12 cores per node) of the considered cluster.

With the infrastructure of external wattmeter OMEGAWATT, we measure for each
application the energy consumption during the execution of the application with and
without activation of the fault-tolerant protocols. Specifically, we instrumented the
source code implementations of the different protocols of fault tolerance in order
to enable/disable each of the operations described above: checkpointing, message
logging (on local disk or RAM disk), and coordination. Thus, we obtain the actual
energy consumption for each operation. Each energy measurement is performed 30
times, and we consider the average values.

As concerns the uncoordinated protocol, we estimated and measured the energy
consumption of all message logging. As concerns the coordinated protocol, we estimated and measured the energy consumption of a single checkpointing and therefore
for one single coordination. To measure the energy consumption of a single checkpointing, we used a checkpoint interval greater than half of the application duration.
Thus, the first (and only) checkpoint will occur in the second half of the application.

In Fig. 5.14, we show the energy estimations for different operations identified
in the protocols of fault tolerance. In Fig. 5.15, we show the relative differences
(in percent) between the estimated and the actual energy consumption. Figure 5.15
shows that the energy estimations provided in Fig. 5.14 are accurate. Indeed, the
relative differences between the estimated and measured energy consumption is low.
The worst estimate shows a gap of 7.6 % compared to the measured coordination
with EP value. The average deviation of all tests is 4.9 %.

---

[1] Cloud Model 1: http://www.mmm.ucar.edu/people/bryan/cm1/.

[2] NAS: http://www.nas.nasa.gov/publications/npb.html.

In comparison with message logging and checkpointing, we find that we estimate a little less coordination. This is due to the fact that this process takes much less time than message logging. This is also due to the fact that this operation is evaluated from the estimated two sub-operations ($t_{polling}$ et $t_{synchro}$) which generates more inaccuracies in our estimation We will show in Sect. 5.3.5 how such energy estimations can reduce energy consumption related to protocols of fault tolerance when they are known before pre-executing the application.

5.3.5 Energy-Aware Choice of Checkpointing Protocols

In this section, we show how we can rely on energy estimations in order to reduce the energy consumption of the different fault-tolerant protocols executed with high-performance computing applications. The fault-tolerant protocol that consumes less energy may change depending on the considered application. The energy estimation that we are able to provide allows the users to choose the best fault-tolerant protocol in terms of energy consumption according to the execution context. By making such choice, the user is able to reduce the energy consumption of the executed fault-tolerant protocols.

The two families of fault-tolerant protocols that we considered are the coordinated and the uncoordinated protocols. We consider the 4 HPC applications that we studied in Sect. 5.3.4.2: CM1 with a resolution of $2400 \times 2400 \times 40$ and 3 NAS in Class D (SP, BT, and EP) running over 144 processes (i.e., 12 nodes with 12 cores per node). For each application and for each fault-tolerant protocol, we estimate the energy consumption by considering the different operations that we have identified in Sect. 5.3.1. First, we highlight that the energy consumption of a fault-tolerant operation depends highly on the application. Then, we show how the energy estimations of the different operations identified in Sect. 5.3.1 help the user in the choice of the fault-tolerant protocol that consumes the less energy.

Figure 5.14 shows that energy consumption of the operations are not the same from one application to another. For instance, the energy consumption of RAM logging in SP is more than 10 times the one in CM1. This is because CM1 exchanges much less messages compared to SP. Another example is that checkpointing in CM1 is more than 20 times the one in EP. Indeed, the execution time of CM1 is much higher than EP so the number of checkpoints is more important in CM1. Moreover, the volume of data to checkpoint is more important in CM1 as it involves a more important volume of data in memory.

We can obtain the overall energy estimation of the entire fault-tolerant protocols by summing the energy consumptions of the operations considered in each protocol. For fault-free uncoordinated checkpointing, we add the energy consumed by checkpointing to the energy consumption of message logging. For fault-free coordinated checkpointing, we add the energy consumed by checkpointing to the energy consumption of coordinations.

Both of uncoordinated and coordinated protocols rely on checkpointing. To obtain a coherent global state, checkpointing is combined with message logging in unco-ordinated protocols and with coordination in coordinated protocols. Therefore, to compare coordinated and uncoordinated protocols from an energy consumption point of view, we compare the energy cost of coordinations to message logging. In our experiments we considered message logging either in RAM or in HDD. Coordination will consume as much as there are still bulked messages that are being transferred at the moments of the processes synchronization. Message logging will consume as much as the number and the size of exchanged messages during the application are important.

Figure 5.14 shows that from one application to another the less energy consuming protocol is not always the same. In general, determining the less consuming protocol depends on the trade-off between the volume of logged data and the coordination cost. For BT, SP and CM1, the less energy consuming protocol is the coordinated protocol (Coordination values in Fig. 5.14 lower to the RAM and HDD logging values) since the volume of data to log for these applications is relatively important and leads to a higher energy consumption. Oppositely, the less energy consuming fault-tolerant protocol for EP is the uncoordinated one.

These conclusions are specific to the case where there is only one checkpointing and so one coordination during the execution of these applications. If the user is interested in more reliability, and this specifically for the applications that last long (several hours), he should choose a smaller checkpoint interval and so a higher number of checkpointing and coordinations. This checkpoint interval can influence the choice of the fault-tolerant protocol that consumes the less energy. Indeed, if for instance during the execution of SP, there are more than 19 checkpointing and therefore more than 19 coordinations, the energy consumption of coordinations will be higher than the one of RAM logging. As a consequence, as opposed to what we have seen previously, it would be better to use the uncoordinated protocol to reduce the energy consumption of fault tolerance.

This checkpoint interval can be selected by considering the models that define the optimal interval: the one that enables to maximize the reliability by minimizing the performance degradation [11, 45].

In case we use a higher number of processes for the execution of a same appli-cation, the energy consumption of coordination will be more important. However, the energy consumption of message logging may also increase since there may be more communications with an increased number of processes. Therefore, there will be more message to send and so more message to log.

Thus, by providing such energy estimations before executing the HPC applica-tion, we help the user to select the best fault-tolerant protocol in terms of energy consumption depending on the number of checkpoints that he would like to perform during the execution of his application.

5.3.6 Summary

In this section, we have presented an approach to accurately estimate the energy consumption of fault-tolerant protocols. We focused on the phase without failure. We considered the case of coordinated protocols and uncoordinated protocols.

This approach is to first identify the operations that we find in the different fault-tolerant protocols. Then, in order to adapt our theoretical models to the specificities of the considered platform, we perform an energy calibration that consists in gathering a set of measurements of the electrical power and execution times of each of the identified operations. To calibrate the considered platform, the calibrator collects parameters describing the execution platform, such as the number of nodes or the number of cores per node. With this calibration, energy estimations that we provide can adapt to any platform. Once the calibration is complete, the estimator is based on the calibration results as well as a description of the execution context to provide an estimation of the energy consumption of the fault-tolerant protocols.

We have shown that energy estimations are accurate for each fault-tolerant operation. Indeed, comparing the energy measurements for each operation to energy estimations that we are able to provide, we have shown that the relative differences were small. The relative differences between the estimates and energy measures are equal to 4.9 % on average and do not exceed 7.6 %.

Furthermore, we described the way to use our estimations in order to consume less energy. By providing energy consumption estimations before the execution of the application, we showed that it is possible to choose the fault-tolerant protocol which is consuming the less energy for a particular application in a given execution context.

5.4 Conclusion

In this chapter, we have focused on the combination of two of the main challenges faced by Exascale systems: resilience and energy consumption. Even though these challenges have mainly been tackled independently, they are strongly interrelated. We have reviewed the literature on energy-aware checkpointing strategies and detailed two main topics.

First, we have provided a detailed analysis to compute the optimal checkpointing period for a coordinated checkpointing protocol where the checkpointing activity can be partially overlapped with computations. We have considered different power consumption overheads for computations and I/Os in order to represent real-life systems in an accurate way. Experiments have shown that we can save more than 20 % in energy, at the price of an increase of 10 % in the execution time.

Then, we have considered both coordinated and uncoordinated protocols and explained how to estimate the energy consumption of these protocols. Energy estimations were shown to be accurate, and they can be used to reduce the energy consumption by allowing the user to use the protocol best suited for a particular application.

Besides, thanks to the energy estimations, understanding the energy behavior of the different fault-tolerant protocols allows us to consider other solutions in order to reduce the energy consumption of a fault-tolerant protocol. Indeed, by predicting the idle periods and the active polling periods, we would be able to apply some power saving capabilities such as slowing down resources (like DVFS [24, 27, 30, 32]) or even shutting down [28, 39] some components if these idle or active polling periods are long enough [38].

Acknowledgments This work was supported in part by the ANR *RESCUE* project and the Joint Laboratory for Petascale Computing between Inria and the University of Illinois at Urbana Champaign. Some experiments presented in this chapter were carried out using the Grid'5000 experimental testbed, being developed under the INRIA ALADDIN development action with support from CNRS, RENATER, and several Universities as well as other funding bodies (see https://www.grid5000.fr).

References

1. Alonso P, Dolz MF, Mayo R, Quintana-Ortí ES (2012) Energy-efficient execution of dense linear algebra algorithms on multi-core processors. Cluster Comput, May 2012
2. Alonso P, Dolz MF, Igual FD, Mayo R, Quintana-Ortí ES (2012) DVFS-control techniques for dense linear algebra operations on multi-core processors. Comput Sci—R&D 27(4):289–298
3. Aupy G, Benoit A, Melhem R, Renaud-Goud P, Robert Y (2013) Energy-aware checkpointing of divisible tasks with soft or hard deadlines. In: International green computing conference (IGCC), IEEE, p 1–8
4. Bosilca G, Bouteiller A, Brunet E, Cappello F, Dongarra J, Guermouche A, Herault T, Robert Y, Vivien F, Zaidouni D (2013) Unified model for assessing checkpointing protocols at extreme-scale. Concurr Comput: Pract Exp, Oct 2013, to be published. Also available as INRIA research report 7950 at http://graal.ens-lyon.fr/yrobert
5. Bouteiller A, Bosilca G, Dongarra J (2010) Redesigning the message logging model for high performance. Concurr Comput: Pract Exp 22(16):2196–2211
6. Cappello F, Caron E, Daydé MJ, Desprez F, Jégou Y, Primet PV-B, Jeannot E, Lanteri S, Leduc J, Melab N, Mornet G, Namyst R, Quétier B, Richard O (2005) Grid'5000: a large scale, Reconfigurable, controlable and monitorable grid platform. In IEEE/ACM Grid 2005, Seattle, Washington, Nov 2005
7. Cappello F, Casanova H, Robert Y (2011) Preventive migration vs. preventive checkpointing for extreme scale supercomputers. Parallel Process Lett 21(2):111–132
8. Cappello F, Geist A, Gropp B, Kale S, Kramer B, Snir M (2009) Toward exascale resilience. Int J High Perform Comput Appl 23:374–388
9. Chandy KM, Lamport L (1985) Distributed snapshots: determining global states of distributed systems. Trans Comput Syst 3(1)63–75. ACM, Feb 1985
10. Daly JT (2004) A higher order estimate of the optimum checkpoint interval for restart dumps. FGCS 22(3):303–312
11. Daly JT (2006) A higher order estimate of the optimum checkpoint interval for restart dumps. Future Generat Comp Syst 22(3):303–312
12. Dias de Assuncao M, Gelas J-P, Lefèvre L, Orgerie A-C (2010) The green grid5000: Instrumenting a grid with energy sensors. In: 5th international workshop on distributed cooperative laboratories: instrumenting the grid (IN-GRID 2010). Poznan, Poland, May 2010
13. Dias de Assuncao M, Orgerie A-C, Lefèvre L (2010) An analysis of power consumption logs from a monitored grid site. In: IEEE/ACM international conference on green computing and communications (GreenCom-2010). Hangzhou, China, p 61–68, Dec 2010

14. Diouri MEM, Dolz MF, Glück O, Lefèvre L, Alonso P, Catalán S, Mayo R, Quintana-Ortí ES (2013) Solving some mysteries in power monitoring of servers: take care of your wattmeters! In: Energy efficiency in large scale distributed systems (EE-LSDS). Vienna, Austria, April 2013, 22–24
15. Diouri MEM, Glück O, Lefèvre L (2013) Your cluster is not power homogeneous: take care when designing green schedulers. In: 4th IEEE international green computing conference (IGCC). Arlington, Virginia, June 2013
16. Diouri MEM, Gluck O, Lefèvre L, Cappello F (2012) Energy considerations in checkpointing and fault tolerance protocols. In: IEEE DSNW, pp 1–6
17. Diouri MEM, Glück O, Lefèvre L, Cappello F (2013) ECOFIT: a framework to estimate energy consumption of fault tolerance protocols during HPC executions. In: 13th IEEE/ACM international symposium on cluster, cloud and grid computing (CCGrid). Delft, Netherlands, May 2013
18. Diouri MEM, Gluck O, Lefevre L Cappello F (2013) Ecofit: a framework to estimate energy consumption of fault tolerance protocols for HPC applications. In: IEEE CCGRID, pp 522–529
19. Diouri MEM, Tsafack Chetsa GL, Glück O, Lefèvre L, Pierson J-M, Stolf P, Da Costa G (2013) Energy efficiency in high-performance computing with and without knowledge of applications and services. Intern J High Perform Comput Appl (IJHPCA) (to appear)
20. Dongarra J et al (2011) The international exascale software project roadmap. Int J High Perform Comput Appl 25(1)
21. Dongarra J, Beckman P, Aerts P, Cappello F, Lippert T, Matsuoka S, Messina P, Moore T, Stevens R, Trefethen A, Valero M (2009) The international exascale software project: a call to cooperative action by the global high-performance community. Int J High Perform Comput Appl 23(4):309–322
22. Dongarra J, Herault T, Robert Y (2013) Revisiting the double checkpointing algorithm. In: 15th workshop on advances in parallel and distributed computational models APDCM. IEEE Computer Society Press
23. Elnozahy ENM, Alvisi L, Wang Y-M, Johnson DB (2002) A survey of rollback-recovery protocols in message-passing systems. ACM Comput Surv 34(3):375–408
24. Etinski M, Corbalan J, Labarta J, Valero M (2010) Utilization driven power-aware parallel job scheduling. Comput Sci—Res Dev 25(3–4):207–216
25. Feng W-C, Feng X, Ge R (2008) Green supercomputing comes of age. IT Prof 10(1):17–23
26. Ferreira K, Stearley J, Laros JHI, Oldfield R, Pedretti K, Brightwell R, Riesen R, Bridges PG, Arnold D (2011) Evaluating the viability of process replication reliability for exascale systems. In: Proceedings of the ACM/IEEE conference on supercomputing
27. Freeh VW, Lowenthal DK, Pan F, Kappiah N, Springer R, Rountree B, Femal ME (2007) Analyzing the energy-time trade-off in high-performance computing applications. IEEE Trans Parallel Distrib Syst 18(6):835–848
28. Hermenier F, Loriant N, Menaud J-M (2006) Power management in grid computing with xen. In: frontiers of high performance computing and networking—ISPA 2006 international workshops, FHPCN, XHPC, S-GRACE, GridGIS, HPC-GTP, PDCE, ParDMCom, WOMP, ISDF and UPW. Lecture Notes in Computer Science, vol 4331, pp 407–416, Sorrento, Italy, 4–7 Dec 2006
29. Hlavacs H, da Costa G, Pierson J-M (2009) Energy consumption of residential and professional switches. In IEEE CSE
30. Hotta Y, Sato M, Kimura H, Matsuoka S, Boku T, Takahashi D (2006) Profile-based optimization of power performance by using dynamic voltage scaling on a PC cluster. In: Proceedings of the 20th international in parallel and distributed processing symposium (IPDPS)
31. Hsing Hsu C, chun Feng W, Archuleta JS (2005) Towards efficient supercomputing: a quest for the right metric. In: Proceedings of the high performance power-aware computing workshop
32. Hsu C-Hchun Feng W (2005) A power-aware run-time system for high-performance computing. In Supercomputing, 2005. Proceedings of the ACM/IEEE SC 2005 Conference, 2005
33. Laros III JH, Pedretti KT, Kelly SM, Shu W, Vaughan CT (2012) Energy based performance tuning for large scale high performance computing systems. In: Proceedings of the 2012 symposium on high performance computing, HPC'12. San Diego, CA, pp 6:1–6:10

34. Mahadevan P, Sharma P, Banerjee S, Ranganathan P (2009) A power benchmarking framework for network devices. In: Networking 2009 conference, Aachen, Germany, 11–15 May 2009, p 795–808
35. Meneses E, Sarood O, Kalé LV (2012) Assessing energy efficiency of fault tolerance protocols for HPC systems. In: Proceedings of the 2012 IEEE 24th international symposium on computer architecture and high performance computing (SBAC-PAD 2012), New York, Oct 2012
36. Netzer RHB, Xu J (1995) Necessary and sufficient conditions for consistent global snapshots. IEEE Trans Parallel Distrib Syst 6(2):165–169
37. Ni X, Meneses E, Kalé LV (2012) Hiding checkpoint overhead in HPC applications with a semi-blocking algorithm. In: Proceedings of the IEEE International Conference on Cluster Computing. IEEE Computer Society
38. Orgerie A-C, Lefevre L, Gelas J-P (2008) Save watts in your grid: green strategies for energy-aware framework in large scale distributed systems. In ICPADS 2008: The 14th IEEE international conference on parallel and distributed systems, Melbourne, Australia, Dec 2008
39. Pinheiro E, Bianchini R, Carrera EV, Heath T (2001) Load balancing and unbalancing for power and performance in cluster-based systems. In: Workshop on compilers and operating systems for low power
40. Rajachandrasekar R, Moody A, Mohror K, Panda DKD (2013) A 1 PB/s file system to checkpoint three million MPI tasks. In: Proceedings of the 22nd international symposium on high-performance parallel and distributed computing, HPDC '13, ACM. New York, pp 143–154
41. Rao C, Toutenburg H, Fieger H, Heumann C, Nittner T, Scheid S (1999) Linear models: least squares and alternatives. Springer series in statistics
42. Sarkar V et al (2009) Exascale software study: software challenges in extreme scale systems. White paper. http://users.ece.gatech.edu/mrichard/ExascaleComputingStudyReports/ECSSreport101909.pdf
43. Shalf J, Dosanjh S, Morrison J (2011) Exascale computing technology challenges. In: VEC-PAR'10, the 9th International Conference on high performance computing for computational science, LNCS 6449, pp 1–25. Springer
44. Young JW (1974) A first order approximation to the optimum checkpoint interval. Commun ACM 17(9):530–531
45. Young JW (1974) A first order approximation to the optimum checkpoint interval. Commun ACM 17(9):530–531 Sept
46. Zheng G, Ni X, Kalé LV (2012) A scalable double in-memory checkpoint and restart scheme towards exascale. In: Dependable systems and networks workshops (DSN-W)
47. Zheng G, Shi L, Kalé LV (2004) FTC-Charm++: an in-memory checkpoint-based fault tolerant runtime for Charm++ and MPI. In: Proceedings of the IEEE international conference on cluster computing. IEEE Computer Society

Index

© Springer International Publishing Switzerland 2015
T. Herault and Y. Robert (eds.), *Fault-Tolerance Techniques*
for High-Performance Computing, Computer Communications and Networks,
DOI 10.1007/978-3-319-20943-2

Printed in the United States
By Bookmasters